Design of Electrical Transmission Lines

Dedicated to Shirdi Saibaba

Design of Electrical Transmission Lines

Structures and Foundations

Sriram Kalaga
Ulteig Engineers Inc.

Prasad Yenumula
Duke Energy

Volume I

CRC Press is an imprint of the
Taylor & Francis Group, an **informa** business
A BALKEMA BOOK

Cover illustration: Power lines, Copyright: Jason Lee/www.shutterstock.com

Published by:
CRC Press/Balkema
P.O. Box 447, 2300 AK Leiden, The Netherlands
e-mail: Pub.NL@taylorandfrancis.com
www.crcpress.com – www.taylorandfrancis.com

First issued in paperback 2021

Typeset by MPS Limited, Chennai, India

ISBN-13: 978-1-03-209729-9 (pbk)
ISBN-13: 978-1-138-00091-9 (hbk)

Visit the Taylor & Francis Web site at
http://www.taylorandfrancis.com

and the CRC Press Web site at
http://www.crcpress.com

Library of Congress Cataloging-in-Publication Data

Names: Kalaga, Sriram, author. | Yenumula, Prasad, author.
Title: Design of electrical transmission lines : structures and foundations /
Sriram Kalaga, Ulteig Engineers, Prasad Yenumula, Duke Energy.
Description: Leiden, The Netherlands : CRC Press/Balkema, [2017] | Includes
bibliographical references and index.
Identifiers: LCCN 2016026908 (print) | LCCN 2016031338 (ebook) | ISBN
9781138000919 (hbk : alk. paper) | ISBN 9781315755687 (eBook)
Subjects: LCSH: Overhead electric lines. | Electric power distribution–High
tension.
Classification: LCC TK3231 .K236 2016 (print) | LCC TK3231 (ebook) | DDC
621.319/22–dc23
LC record available at https://lccn.loc.gov/2016026908

Table of contents

Foreword

I am honored and thankful to be asked to write the foreword to this timely textbook, *Design of Electrical Transmission Lines – Structures and Foundations*. As an industry, we find ourselves at a crossroads. Many industry professionals are nearing retirement as part of the large 'baby boom' generation. They will take decades of knowledge and experience with them. Engineering curricula typically do not include electric utility design courses. We learn the basic engineering principles and then, over time, learn how to apply these facts to our industry.

As this transition occurs, many of us are concerned about the need for effective and timely knowledge transfer. How can we pass on this critical knowledge to the next generation?

Design of electrical transmission structures requires proper application of fundamental theories of strength of materials, engineering mechanics, structures, soil mechanics and electrical engineering. Knowledge of applicable industry codes and standards is also necessary as they govern the design process. Traditionally, engineers learn the design process on the job, from their mentors, colleagues, at seminars and workshops, and from utility proprietary manuals and other tools. In the absence of a specific reference book that contains this guidance, the learning process can take years.

I am encouraged that Sriram Kalaga and Prasad Yenumula have created this textbook in an attempt to bridge that gap. They have taken their many combined years of experience and put them into a single location for the benefit of the rest of us. *Design of Electrical Transmission Lines – Structures and Foundations* will provide industry professionals a valuable resource from which to learn. The detailed overview and design instruction, along with references to applicable standards, will help younger industry professionals more quickly understand the basic design principles. I also believe readers will benefit from the many detailed sample problems, design tables, hardware information and line design illustrations.

I trust that you will find value in spending time in this book. It will prove to be a valuable resource in your electric utility career!

Marlon W. Vogt, PE, F. SEI.
Account Executive, Ulteig Engineers Inc.
Cedar Rapids, Iowa

Foreword

Foreword

The authors of *"Design of Electrical Transmission Lines – Structures and Foundations"* have created a unique engineering book for utility engineers. The subject is presented from the viewpoint of civil engineers; however their presentation can greatly benefit anyone involved with engineering of transmission lines.

The main purpose of this book is to assist utility engineers in understanding basic design of transmission line structures and foundations. For young engineers it is a great resource for learning, understanding, and applying the engineering principles required to successfully design transmission line structures and foundations. While for the mature engineer the book becomes a quick reference which can be used to refresh their knowledge of a particular subject.

Many of the theories and methods in the book have sample problems to aid in their understanding. These sample problems also provide excellent "blueprints" for applying these theories and methods in real life applications. Illustrations, photos, charts, and graphs are also effectively used throughout the book to define the subject matter.

It has been an honor to write a foreword to this book written by Sriram Kalaga and Prasad Yenumula. The book's comprehensive engineering approach reflects their combined knowledge and experience of transmission lines.

May this book prove enjoyable and valuable to utility engineers for many years!

James A. Robinson, Jr, PE
Principal Engineer
Duke Energy
Charlotte, North Carolina

Preface

Electrical power is now an indispensable requirement for the comfort, safety and welfare of mankind in the 21st century. No matter what the source of power generation is, its final destination is the abode of the individual consumer – a person, industry, machine or organization. This book deals with the how, what and where the many engineering disciplines collaborate to make that journey happen.

The design of overhead electrical transmission lines is a unique activity which involves direct or indirect contributions of many other disciplines, both engineering and others. The word "electrical" just implies that the main focus is transmittal of electrical power or energy from one point to another. But that movement of power also requires conductors, insulators, supporting structures (or pylons), connecting hardware, good anchorage into ground while satisfying myriad technical rules, governmental regulations and guidelines aimed at safety and reliability. This calls for the involvement of civil engineers (structural and geotechnical), electrical engineers, surveyors (analog and digital), drafters (CAD) and finally construction contractors who build what we design. Since transmission lines often begin and end at substations, specialists in substation design and protection and control are also involved.

In most areas of the world, the term "transmission structures" usually means steel lattice towers. In the West, they however encompass a bewildering range of structural systems and configurations, materials, hardware and construction practices. The industry now employs steel (both tubular as well as lattice), prestressed concrete, wood (natural and laminated) and composites as primary materials. Polymer insulators often replace traditional porcelain and glass units; high temperature low sag (HTLS) and vibration-resistant conductors with superior sag-tension characteristics are available for longer spans. Fiber optic ground wires now serve a dual purpose: shielding against lightning strikes as well as communication. The advent of powerful digital computers enabled modeling and analysis of not only individual structures comprising a transmission line but also the *entire line* in one session.

However, the knowledge related to the activity is scattered mostly in design guides, standards and manuals and not available in a form amenable for larger public utilization. Though the basic principles of transmission line design are more or less the same all over the world, different regions impose different rules and regulations, mostly associated with safety and reliability. As of now, there is no single reference book which covered these topics. We hope to fill that gap with this book.

The present book is organized into 6 chapters.

Chapter 1 presents a brief introduction of the history of transmission line structures through the years and the current state of the art.

Chapter 2 provides an overview of the general design criteria – Electrical, Structural and Geotechnical – associated with transmission structures. Also discussed are computer programs, various codes and standards and specifications governing both material as well as construction of such structures.

Chapter 3 deals with modeling, structural analysis and design of various types of transmission structures. The importance of form, function and purpose of the structural configuration are discussed in detail as well as material type influencing such selection. Structures of wood, steel (lattice and polygonal poles), concrete and fiber-reinforced polymer (FRP) are covered.

Chapter 4 deals with geotechnical aspects of foundation analysis and design for various types of transmission structures. The importance of soil data, function and purpose of the foundation are discussed as well as popular computer programs used in foundation analysis and design. Various types of anchors used in guying are also reviewed.

Chapter 5 provides an overview of design deliverables – from the engineer to the utility – which form the core of the design documentation.

Chapter 6 presents a brief review of current research of direct relevance to transmission lines and structures.

Worked out design examples and problems are provided in each chapter, where necessary. Calculations for all problems cover both the English and SI units. Appendices containing various tables of data on transmission materials (poles, conductors, shield wires, insulators, guy wires etc.) are given. Although the focus of this book is U.S. design procedures and standards, relevant information on codes of other countries is given in Appendix 15.

A comprehensive design of a small transmission line is illustrated in Appendix 1.

Most analysis procedures discussed in this book are basically non-linear in nature; however, a discussion of these methods is beyond the scope of the book. It is expected that students and engineers perusing the book possess some basic knowledge of mechanics of materials, structural engineering (steel, concrete and wood design), basics of soil mechanics and foundation design.

This book is an undertaking to bring about the merger of the authors' individual association with the world of high-voltage transmission lines, structures and foundations in North America. It is also aimed at presenting the material in a form useful as a textbook for educators at universities. We hope the book will be a useful reference for everyone involved with transmission structures.

We are indebted to many of our colleagues, mentors and students, who, with their helpful suggestions and encouragement, have provided critical input for this work.

Although we have spared no effort to eliminate typos and errors, we recognize that any work of this magnitude cannot weed out all; the authors wish to thank in advance all readers and users who will be kind enough to draw attention to any inadvertent errors.

Sriram Kalaga
St. Paul, Minnesota

Prasad Yenumula
Raleigh, North Carolina

Acknowledgments

This book is an undertaking based on the authors' individual professional and academic association with the high-voltage transmission lines, structures and foundations for nearly two decades. As such, we have benefited from the wisdom and input of various other engineers, design professionals and contractors.

The authors wish to express their gratitude to the Engineering Division of Rural Utilities Service (RUS) of United States Department of Agriculture (USDA) for their generous permission to reproduce material from various RUS Bulletins. The reader will note that many figures used in the book are modified versions of RUS originals drawings, reformatted to fit the scope of this book. We also wish to acknowledge the usage of various drawings from Allgeier Martin and Associates as indicated.

We also wish to thank the T & D Engineering Department at Hughes Brothers Inc., Seward, Nebraska for their permission to reuse content from their design manuals and catalogs.

We also appreciate the kind permission given to us by ASCE, ACI, CEATI International, EPRI, IEEE, PLS (Power Line Systems), Hubbell Power Systems, Trinity Meyer Utility Structures, LLC and NRC Research Press to reproduce portions of their publications in various segments of this book.

Special appreciation is extended to Dr. Peter McKenny (Director) and Ms. Jilliene McKinstry (Assistant Director) and to the students of Transmission and Distribution Online Program, Gonzaga University, Spokane, Washington for their support during the process of writing the book.

We are indebted to many of our colleagues, mentors and students, who, with their helpful suggestions and encouragement, have provided critical input for this work. Special thanks to the following individuals for their review of the book material:

- Dr. Alfredo Cervantes, Consulting Structural Engineer, Ph.D., PE, Dallas, Texas
- Mr. Brad Fossum, PE, Technical Manager, Ulteig Engineers, St. Paul, Minnesota
- Mr. Parvez Rashid, PE, American Transmission Company, Milwaukee, Wisconsin
- Mr. Anil Ayalasomayajula, PE, Duke Energy, Raleigh, North Carolina
- Mr. James A. Robinson, Jr., PE, Duke Energy, Charlotte, North Carolina
- Mr. Marlon Vogt, PE, F.SEI., Account Executive, Ulteig Engineers, Cedar Rapids, Iowa and Member, ASCE Committee on Steel Transmission Poles

We are also grateful to Mr. K. Gopinath of Hyderabad, India for his help with the drawings of this book.

Sriram Kalaga
St. Paul, Minnesota

Prasad Yenumula
Raleigh, North Carolina

Acknowledgments

About the authors

Dr. Prasad Yenumula is currently a Principal Engineer from the Transmission Line Engineering System Standards of Duke Energy. He earned his Bachelor's, Master's and doctoral degrees in Civil engineering along with an MBA degree with a specialization in Global Management. With a post-doctoral fellowship in engineering, he published more than 50 research papers in various journals and conferences. He worked as a line design engineer, line standards engineer and lines asset manager in the US & Canada for over 20 years. He was responsible for managing and leading a number of line projects and special assignments, along with developing various technical standards & specifications.

He is also a professor (adjunct) with Gonzaga University, Washington and contributes to the development and teaching of the on-line transmission & distribution engineering Master's program. He also teaches business students at the University of Phoenix. He offered training classes in the areas of line design, standards and line design software. He was invited to make several presentations, was a reviewer of research papers, an advisor to Master's students and an examiner for doctoral students.

He is currently the Chair of line design task force of the Electric Power Research Institute (EPRI). He is also the current Chair of CEATI (Center for Energy Advancement through Technological Innovation) International's TODEM interest group. He is a member (*alt*) of NESC (National Electrical Safety Code) Subcommittee-5 (Strength and Loadings), EEI's (Electric Edison Institute) NESC/Electric Utilities Representative Coordinating Task Force and ASCE (American Society of Civil Engineers), ISSMGE, SEI and DFI. He is Duke's industry advisor to NATF, NEETRAC, EPRI and CEATI. He is also in various other national standard committees such as ASCE 10 on lattice towers, ANSI C29 on insulators and on the ASCE-FRP Blue Ribbon Panel review team. He received thirteen awards for his engineering, research and teaching efforts which include best Ph.D. thesis, best papers and four faculty of the year awards.

Dr. Sriram Kalaga is currently a Senior Engineer in the Transmission Line Division of Ulteig Engineers. He holds Bachelor's, Master's and doctoral degrees in Civil engineering with a specialization in Structural Engineering and Mechanics. He has published more than 35 research papers in various journals and conferences. His research background included finite element methods, buckling of beam-columns, nonlinearities, low-cost composites and reliability-based approaches.

He has been involved in transmission line design in the US for 16 years while his overall experience as a Civil Engineer spans over 36 years. As a consulting engineer, he developed various in-house design manuals and technical specifications related to transmission structures.

He conducts regular seminars on transmission line design and also participates in teaching workshops on transmission structural design. He is currently a guest editor for the International Journal of Civil Engineering and has advised graduate students on their theses. In addition to being a member of ASCE, AISC and ACI, he is currently serving on the ASCE-FRP Blue Ribbon Committee on composite transmission structures. He is a licensed professional engineer in several states in the USA.

Special thanks

To my wife *Shanti*

Sriram Kalaga
St. Paul, Minnesota

To my wonderful family: *Aruna, Rohith* and *Sarayu*

Prasad Yenumula
Raleigh, North Carolina

Notice to the reader

The information contained in this book has been reviewed and prepared in accordance with established engineering principles. Although it is believed to be accurate, this information should be used for specific applications only after competent professional examination of its suitability and applicability by experienced and licensed professional engineers or designer. The authors and publisher of this book make no warranty of any kind, express or implied, with regard to the material contained in this book nor shall they, or their respective employers, be liable for any special or consequential damages resulting, in whole or in part, from the reader's or user's reliance upon this material.

List of abbreviations

The following abbreviations are used in this book.

AAAC	All Aluminum Alloy Conductor
AAC	All Aluminum Conductor
AC	Alternating Current
AACSR	Aluminum Alloy Conductor Steel Reinforced
ACCR	Aluminum Conductor Composite Reinforced
ACAR	Aluminum Conductor Alloy Reinforced
ACCC/TW	Aluminum Conductor Composite Core/Trapezoidal Wire
ACI	American Concrete Institute
ACSR	Aluminum Conductor Steel Reinforced
ACSS	Aluminum Conductor Steel Supported
ACSR/AW	Aluminum Conductor Steel Reinforced/Aluminum Clad
ACSR/SD	Aluminum Conductor Steel Reinforced/Self Damping
ACSR/TW	Aluminum Conductor Steel Reinforced/Trapezoidal Wire
AISC	American Institute of Steel Construction
ALCOA	Aluminum Corporation of America
ANSI	American National Standards Institute
ASCE	American Society of Civil Engineers
ASTM	American Society for Testing and Materials
AWAC	Aluminum Clad Steel, Aluminum Conductor
AWG	American Wire Gauge
BIA	Bureau of Indian Affairs
BLM	Bureau of Land Management
BS	British Standard
CADD	Computer-Aided Design and Drafting
CCI	Corus Construction and Industrial
CEQ	Council on Environmental Quality
CFR	Council of Federal Regulations
CIGRE	International Council on Large Electric Systems
COE	Corps of Engineers
CSA	Canadian Standards Association
DF	D-Fir Douglas-Fir
DOE	Department of Energy
ECCS	European Convention for Constructional Steelwork

ECES	European Committee for Electrotechnical Standardization
EHV	Extra High Voltage
EIA	Electronic Industries Association
EIS	Environmental Impact Statement
ENA	Energy Networks Association
EPA	Environmental Protection Agency
EPRI	Electric Power Research Institute
ESCSA	Essential Services Commission of South Australia
FAA	Federal Aviation Agency
FEMA	Federal Emergency Management Agency
FERC	Federal Energy Regulatory Commission
FHA	Federal Highway Administration
FLPMA	Federal Land Policy Management Act
FO	Fiber Optic
FS	Forest Service
FWS	Fish and Wildlife Service
GIS	Geographic Information Systems
GPS	Global Positioning System
HB	Hughes Brothers
IEC	International Electrotechnical Commission
IEEE	Institute of Electrical and Electronics Engineers
IS	Indian Standard
INS	Indian National Standards
Kcmil	Thousand Circular mils
LF	Load Factor
LiDAR	Light Detection and Radar
M&E	Mechanical and Electrical
MCOV	Maximum Continuous Over Voltage
MOR	Modulus of Rupture
MVA	Mega Volt Amperes
NEPA	National Environmental Protection Act
NERC	North American Electric Reliability Corporation
NESC	National Electrical Safety Code
OLF	Over Load Factor
OHGW	Overhead Ground Wire
OPGW	Optical Ground Wire
PLS	Power Line Systems
REA	Rural Electrification Administration
RI	Radio Interference
ROW	Right of Way
RUS	Rural Utilities Service
SA	Standards Australia
SF	Strength Factor
SML	Specified Mechanical Load
SNZ	Standards New Zealand
SYP	Southern Yellow Pine

TIN	Triangular Irregular Network
TP	Transpower
TVI	Television Interference
UHV	Ultra High Voltage
USACE	U.S. Army Corps of Engineers
USDA	United States Department of Agriculture
USBR	United States Bureau of Reclamation
USDI	United States Department of Interior
USGS	United States Geological Survey
WRC	Western Red Cedar

List of figures

List of tables

Conversion factors

1 Pa = 1 N/mm^2
1 kPa = 1 kN/mm^2
1 MPa = 1000 kN/mm^2

Length
1 inch = 25.4 mm
1 foot = 30.48 cm or 0.3048 m
1 mile = 1.609 kilometers

Area
1 sq. inch = 6.452 sq. cm
1 sq. foot = 929.03 sq. cm
1 sq. foot = 0.0929 sq. m
1 sq. yard = 0.836 sq. m
1 sq. mile = 2.59 sq. km

Volume
1 cu. in. = 16.39 cu. cm
1 cu. ft. = 0.0283 cu. m
1 cu. yd. = 0.764 cu. m

Moment of Inertia
1 in^4 = 41.623 cm^4

Section Modulus
1 in^3 = 16.387 cm^3

Force
1 lb. = 4.45 N
1 lb. = 0.4536 kg
1 kip = 1000 lb. = 4.45 kN
1 lb./ft (plf) = 14.594 N/m
1 lb./in = 175.13 N/m

Pressure
1 psi = 6.895 kPa
1 psf. = 47.88 Pa
1 ksf. = 47.88 kPa
1 ksi. = 6.895 MPa

Moment
1 lb.-in = 0.113 N-m
1 lb.-ft. = 1.356 N-m
1 kip-ft. = 1.356 kN-m

Density
1 lb./in^3 = 271.45 kN/m^3
1 lb./ft^3 = 0.157 kN/m^3
1 lb./ft^3 = 16.02 kg/m^3

Speed
1 foot/sec = 0.375 meters/sec
1 mph = 1.609 kmph
1 mph = 0.447 meters/sec

Temperature
1 deg. C = (5/9)(F − 32)
1 deg. F = 1.8C + 32

Chapter 1

Introduction

The design of overhead electrical transmission lines is an activity which involves contributions of many disciplines, both engineering and others. Movement of electric power requires supporting structures, conductors to carry the current, insulators to provide safe distance of charged conductors from supporting structures and appropriate connecting hardware all meeting standards for safety and reliability. Although transmission lines are primarily conduits of electrical energy, the design of those supporting structures calls for the involvement of civil engineers – structural and geotechnical.

This chapter takes a brief look at the origins of electrical transmission, the early structural systems used and the 100-year journey from those humble beginnings to the current world defined by technological advances and computers.

1.1 HISTORY OF ELECTRICAL TRANSMISSION

Count Alessandro Volta (1745–1827), the Italian physicist and inventor of the battery, was the first person to suggest the idea of a transmission line by writing in 1777 "*... the igniting spark could be transported from Como to Milan with barbed wire supported by wooden poles planted here and there ...*" The structures in use those days for telegraph poles were wooden poles with zinc iron barbed wire supported by porcelain insulators fixed to the pole with screws and bolt hooks (TERNA, 2013).

The first industrial transmission line ran somewhere between Tivoli and Rome in 1882. The line carried a 5.1 kV single-phase circuit supported by metal fixtures made of double beams, concrete bases and insulators mounted on bolt hooks with wires made of copper. On September 16, 1882, Miesbach in Germany became the starting point for the first long distance transmission of electric power in the world. A 2.4-kilovolt direct current (DC) power transmission line transferred electricity from Miesbach over a distance of 31 miles (50 km) to Munich. However, the first long distance transmission of electrical energy occurred in 1884 during the Turin Expo. A 3 kV single phase current was sent over a 26-mile (42 km) line from Como to Lanzo, Italy. The supports were wooden poles and bell insulators with bronze wires were used.

The construction of the first three-phase 12 kV alternating current (AC) overhead transmission line took place in 1891 between Lauffen and Frankfurt, about 112 miles (180 km), coinciding with the International Electricity Exhibition in Frankfurt. Back in Italy, the Tivoli-Rome line was followed in 1898 with a 20-mile (32 km) line between

Paderno and Milan: the first 3-phase circuit with metal pylons and delta-type multiple bell insulators with copper wires.

In the United States, the first power transmission line operated at 4 kV. It went into operation in June 1889 between Willamette Falls and downtown Portland in Oregon, running about 13 miles (21 km). In 1912, the first 110 kV overhead transmission line was constructed between Croton and Grand Rapids, Michigan. The year 1913 saw the construction of the biggest and longest high-voltage line – the 150 kV Big Creek Line in California – which spanned 250 miles (402 km).

The following years witnessed technical advances and rapid developments everywhere. The first 220 kV lines were constructed in Germany and Italy in 1928; by 1936, a 287 kV line was built between Hoover Dam in Nevada and Los Angeles, California. Sweden built the world's first 380 kV line from Harsprannget to Stockholm, running 596 miles (959 km), in 1953. At the same time, American Electric Power (AEP) constructed the first 345 kV transmission line. In most cases, the average design spans between structures ranged from 1000 to 1500 ft (305 to 457 m); almost all lines used aluminum-steel conductors, bell insulators and latticed steel towers.

Hydro-Quebec in 1965 built Canada's first 735 kV overhead line; soon, Russia and USA built overhead lines at 765 kV – then the largest voltage in the world. A 1200 kV line was commissioned in the Soviet Union (now Russia) in 1982.

1.2 TRANSMISSION STRUCTURES

Historically, the term "transmission structures" usually implied iron or steel latticed towers. The early "pylons" dating back to 1829 were iron structures; the basic shape of later pylons was mostly inspired by the famous Eiffel Tower. Figure 1.1 shows some of the early shapes and forms of transmission structures.

Figure 1.1 Early Forms of Transmission Structures (Source: 130 years of History for Electricity Transmission, TERNA, Rome, Italy, 2013).

Single wood poles directly-embedded into the ground formed the bulk of the transmission structures family for a greater part of the 20th century. France in the late 1900s and later Belgium in 1924 began producing concrete poles. The first steel tubular transmission pole in USA was erected in 1958.

Wood H-frames and lattice steel towers became popular later on, dictated mostly by height, availability, strength and urban convenience. Prestressed concrete poles are also used in various places. The world record for the largest transmission structure is now held by China's 500 kV Yangtze River Crossing double-circuit tower, 1152 ft (351 m) tall, supporting a maximum span of 7667 ft (2337 m) and weighing 8.4 million lbs (3.81 million kgs).

Little historical information is available on how foundations were designed for transmission structures in the early days. It is conceivable that some rule of thumb and field tests were used while determining how much a pole needs to be embedded into the ground. One of the earliest discussions on soil behavior in wood H-frames can be traced to 1943 (Hughes Brothers, 1943). Figure 1.2 illustrates the earth pressures below the ground on the legs of an H-frame.

1.3 CURRENT STATE OF THE ART

The present state of technology encompasses a wide range of structural systems and configurations, materials, hardware and construction practices. The utility industry now uses wood poles (Figures 1.3a and b), tubular steel poles (Figures 1.4a, b and c) as well as steel lattice towers (Figure 1.5), spun prestressed concrete poles (Figure 1.6), laminated wood poles (Figure 1.7) and composite poles (Figures 1.8a and b) as primary structural elements. Fiberglass cross arms and braces are increasingly used on under-build distribution circuits on transmission poles. Helical screw anchors, easy to install in a variety of soils, are becoming very common in guying applications.

Figure 1.4d shows a typical reinforced concrete drilled shaft foundation for a steel pole. The main components of the foundation include a base plate welded to the pole bottom, anchor bolts connecting the base plate to the concrete pier and longitudinal and transverse (ties or spirals) reinforcement in the pier. (See Chapter 4 for more details on the analysis, design and construction of drilled shafts).

Polymer insulators have often replaced conventional porcelain and glass units. High-temperature, low sag, vibration-resistant conductors with superior sag-tension characteristics are available for longer spans. Fiber optic ground wires now serve a dual purpose: protecting against lightning strikes *and* providing communication channels. The advent of powerful digital computers and software such as PLS-CADD™, PLS-POLE™ and TOWER™ enabled accurate modeling and analysis of not only individual structures comprising a transmission line but also the *entire line* in a single modeling session. The programs also perform foundation capacity checks. Plan and Profile drawings can now be digitally processed, printed and saved in various formats and sizes.

Although direct embedment as well as concrete drilled shafts are commonly used as foundations, micropiles are also becoming popular in situations where conventional foundations are difficult to build. Powerful computer programs such

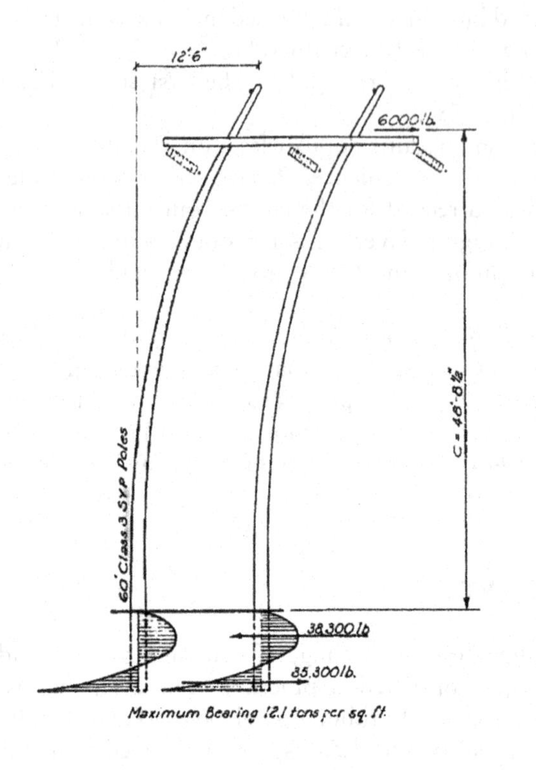

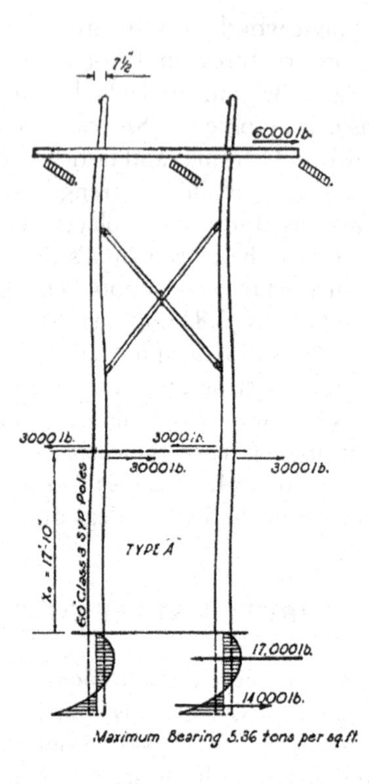

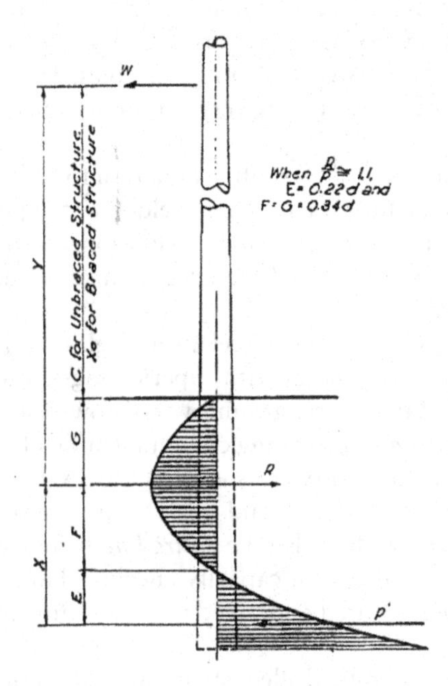

[SOURCE : HUGHES BROS.]

1 ft = 30.5 cm 1 in = 25.4 mm 1 lb = 4.45 N

Figure 1.2 Earth Pressures in H-Frames.

Figure 1.3a Wood H-Frame.

as CAISSON™, MFAD™ and LPILE™ are enabling accurate analysis of transmission foundations in complex soil profiles. Helical screw anchors can now be analyzed with HeliCAP™. LPILE also facilitates design of concrete piers with steel casings when required to handle large lateral loads under deflection and rotation constraints.

1.4 DESIGN PROCESSES

The process of designing an overhead transmission line involves various phases or sub-processes. From a Project Management (PM) perspective, there are five (5) main sequential phases as shown in Figure 1.9. This book is concerned with *Detailed Engineering Design,* which includes structures and foundations.

The basic contents of each of the sub-processes are shown in Figures 1.10, 1.11 and 1.12. The other sub-processes will be discussed in the companion volume. For

Figure 1.3b Wood Pole with Steel Davit Arms.

small projects, a PM-based process is not necessary; however, for large, time- and cost-constrained projects, where several entities and stakeholders interact, a formal PM-based approach is warranted. In such situations, project managers coordinate the design with engineers and serve to optimize costs, schedules and finally quality assurance and control.

A brief description of each component of *Detailed Engineering Design* block is given below.

Structure Locations: Before any design takes place, the engineer must finalize the alignment or route of the transmission line taking into consideration the various constraining criteria. These criteria include Right of Way (property boundary) and landowner issues, soil conditions, road and other clearances, cost and regulatory issues and construction access. Potential structure locations are identified and structures are "spotted" or placed at these points. Survey coordinates of the points when input to

Figure 1.4a Steel Pole with V-String Insulators.

PLS-CADD^{TM} will generate a 3-Dimensional Terrain to facilitate graphical assessment of the line.

Wire Sags and Tensions: This step involves determining the design tensions of the conductors and ground wires selected. Selection of optimum wire tensions is the most important activity of a transmission line design; the chosen tensions impact structure loads at line angles, sags under various weather situations, which in turn affect clearances, vibrations and forces in guy wires and anchors. Appropriate conductor tensions are also important at locations of line crossings where one line crosses another directly above or below. The sag of the wires and the mandated clearances will govern the eventual heights of the structures.

Structure Design: Transmission structural design basically involves ensuring safety and integrity of the system when subjected to various wire and climactic loads. This

Figure 1.4b Steel Pole Deadend – Tangent and 90 degrees.

Figure 1.4c Steel Pole – Deadend Tangent.

process must adhere to and satisfy various code and industry regulations defining the performance of the structure. Structural materials may differ but the underlying design concepts are the same.

Foundation Design: Each structure must be securely embedded or anchored into the ground and facilitate safe transfer of structure loads to the ground strata below. To determine foundation requirements, the engineer must first evaluate the nature and condition of the soil in the vicinity of the structure as well as the variation of soil with depth and along the alignment. The choice of eventual foundation type will further depend on structure material and configuration and loads.

Design Drawings: The framing drawings show the full geometry of the structure, insulator and wire attachment points and heights. Individual hardware units and their

Figure 1.4d Drilled Shaft Foundation for a Steel Transmission Pole.

Figure 1.5 Lattice Tower.

Figure 1.6 Prestressed Concrete Pole.

Figure 1.7 Laminated Wood Pole.

Figure 1.8a Composite Pole with Cross Arms.

sub-components are shown in the assembly drawings. Finally, the Plan and Profile (P & P) drawings constitute the view of the entire line – plan and elevation – and showing all significant features defining the terrain.

1.5 SCOPE OF THIS BOOK

The knowledge related to the transmission line design is available in different design guides and manuals and not available in a simple form (i.e.) single source reference. The authors intend to fill that gap with this book, which will serve as a reference for information on lines, structures and foundations and are presented in a form useful as a textbook for students and educators at universities as well as practicing utility engineers.

As stated earlier, the sub-processes of design are shown in Figures 1.10, 1.11 and 1.12. These components define the scope of this book.

Transmission line and structure design in United States is governed by IEEE's *National Electrical Safety Code* (hereinafter called NESC); therefore, unless otherwise stated, all material in this book implicitly refers to NESC. The other important

Figure 1.8b Composite H-Frame (Courtesy: RS Technologies).

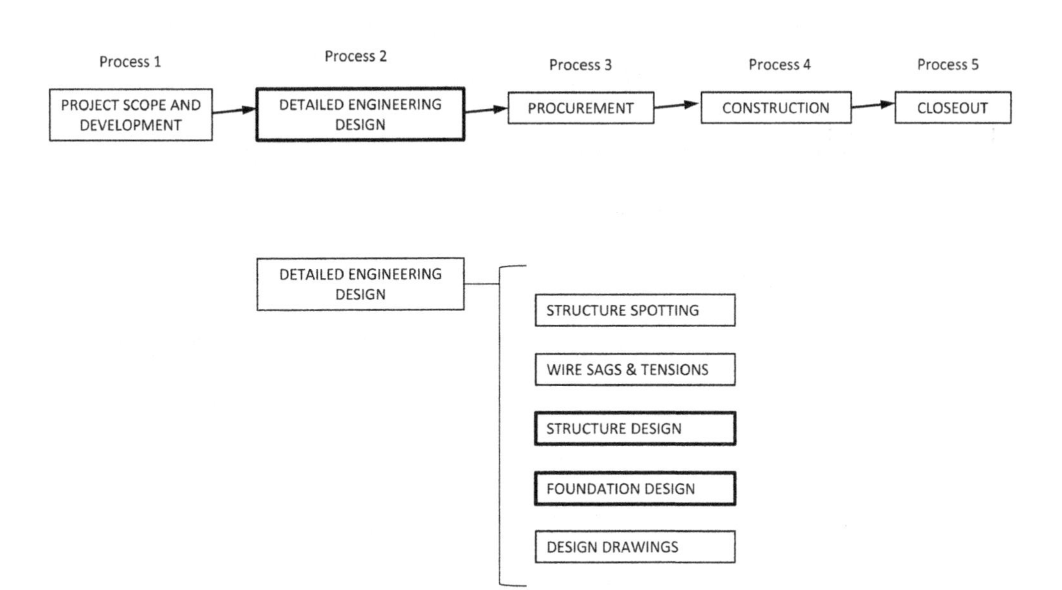

Figure 1.9 Transmission Line Design Process.

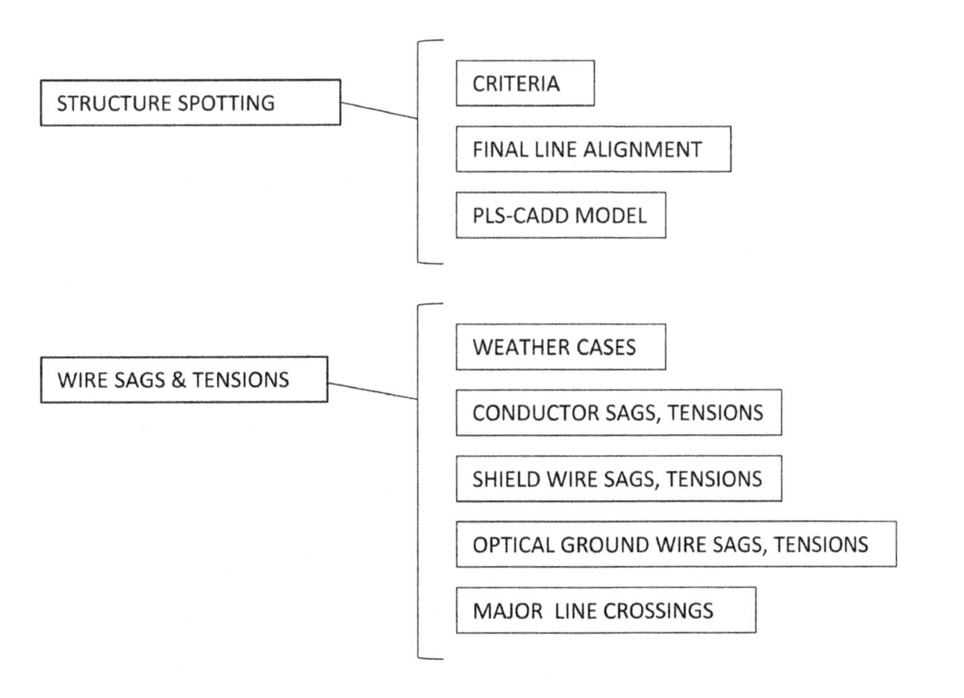

Figure 1.10 Transmission Line Design Process. (cont'd)

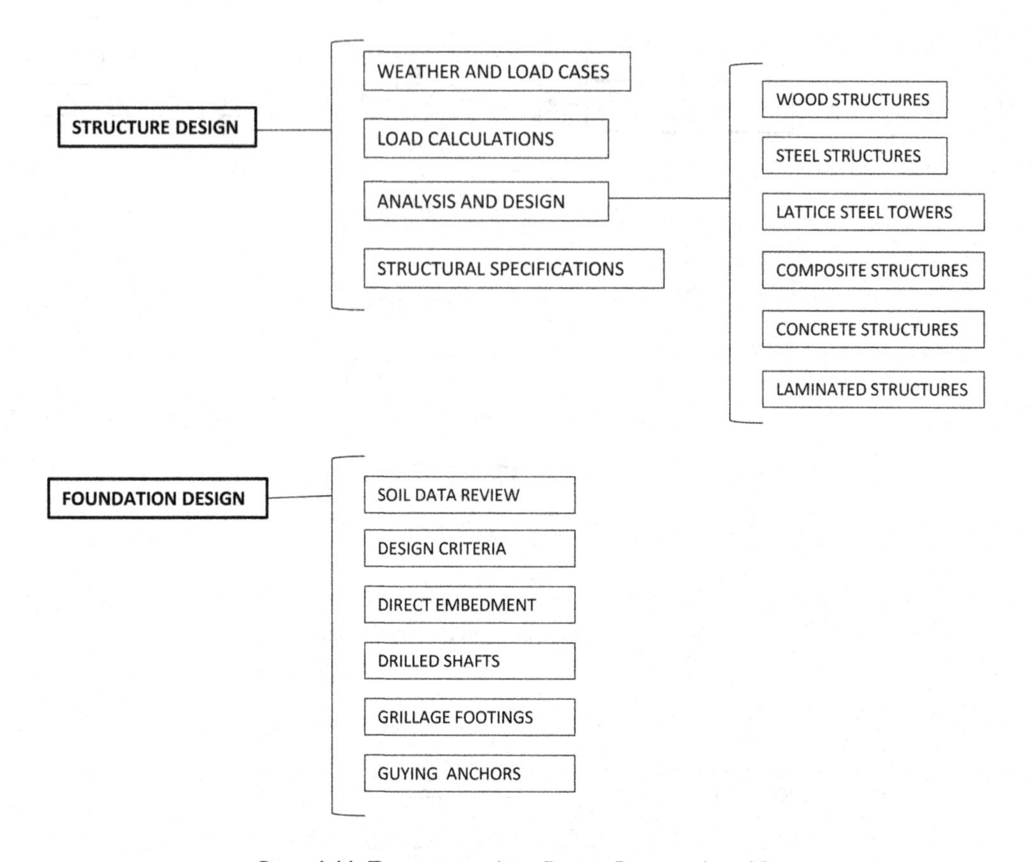

Figure 1.11 Transmission Line Design Process. (cont'd)

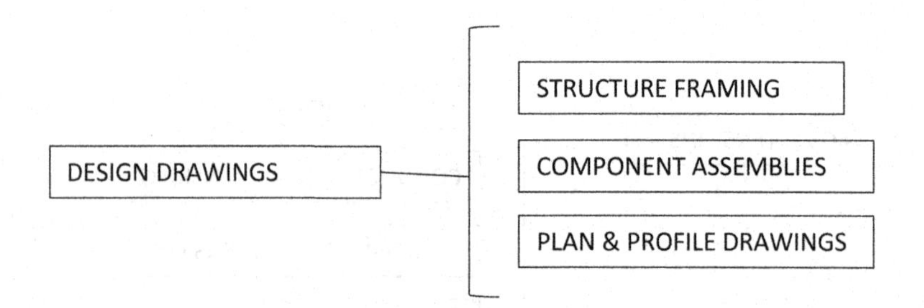

Figure 1.12 Transmission Line Design Process. (cont'd)

reference is the Rural Utilities Service (RUS)'s Bulletin 200 titled *Design Manual for High Voltage Transmission Lines*, which supplements NESC in various ways. Together, these two main references (along with many others as noted in the text) constitute the sources of design guidelines discussed in this textbook.

Chapter 2

General design criteria

The primary operation of overhead transmission lines involves transmittal of electrical current through the conductors, safely and reliably, through various terrains, climactic, environmental and ground conditions. This calls for satisfying design criteria related to both electrical components of the transmission system as well as the mechanical support system which includes steel and wood poles, frames, lattice towers and foundations. The design of transmission line structures requires involvement of civil, mechanical engineers and electrical engineers.

The basic design philosophy of all transmission lines anywhere in the world is the same: all structures and their associated foundations must safely withstand the imposed electrical and mechanical loads without excessive stresses and deformations. The conductors must also function safely without excessive sags and lateral movement. Although performance of a line is judged on the basis of electrical and structural behavior, other elements such as maintenance, inspection and repair also constitute an important segment of the overall line criteria.

In the United States, NESC, RUS Bulletin 1724E-200, IEEE-1724 and many others provide guidelines on the various parameters that are considered in design. Additionally, guidelines are specified by other organizations such as American Society of Civil Engineers (ASCE), Institute of Electrical and Electronics Engineers (IEEE), American Society for Testing of Materials (ASTM), American Concrete Institute (ACI) and American National Standards Institute (ANSI) through various standards. The scope, nature and application of the various guidelines vary from region to region within the country depending on climactic conditions. Additionally, local rules and federal regulations may also be applicable for a specific transmission line. However, the underlying intent is the same: to ensure reliable and safe transmittal of electrical current.

Table 2.1 summarizes the major design criteria associated with a typical high voltage overhead transmission line. This chapter takes a closer look at each of those criteria.

2.1 CLIMATE

The design criteria for a transmission line project should accurately reflect the various climatological conditions in the area. Regulatory codes often give weather-related loads in terms of temperature, ice, wind and a combination of ice and wind. Utilities,

Table 2.1 Typical Design Criteria for a Transmission Line.

Climate	*Electrical*	*Structural Analysis and Design*	*Constructability*
Extreme Wind	Regulatory Codes	Structure Spotting	Construction Considerations
Ice	Right-Of-Way	Ruling Spans	Environmental Constraints
Combined Ice and Wind	Clearances	Sags and Tensions – Galloping –Wire Tension Limits	Regulatory Issues
Extreme Ice with Concurrent Wind	Insulator Swing		Public Acceptance
High Intensity Winds	Shielding		
Pollution	Lightning Performance and Grounding	Insulators	
Temperature	Insulation Requirements	Hardware	
	Conductor Operating Temperature	Guy Wires and Anchors	
	Corona and Field Effects	Structural Analysis – Computer Programs – Loads and Strength Criteria – Grades of Construction – Structural Design Criteria –Weather Cases – Load Cases	
	EMF and Noise		
	Galloping		
	Ampacity		
		Foundation Design Criteria	

Note: Unless otherwise specified, all voltages in the following sections refer to phase to phase nominal alternating current (AC) voltage.

based on their own experiences and analysis, will often adopt a higher value of wind speed or ice thickness than the NESC minimums. The idea of considering climatological entities is based on the intent to determine a realistic set of weather cases for design. ASCE Manual 74 (2010) is one of the many sources of information pertaining to structural design of transmission lines. Climatological considerations also include incidence of lightning, temperature, humidity, pollution and corrosion-causing elements in the area.

2.1.1 Extreme wind

Wind in an area is defined in terms of the maximum "3-second gust" and must consider *both* the magnitude and frequency of wind speeds. Current American standards refer to a 50-year minimum return period (RP) which roughly translates to 2% probability of occurrence in any year. Transmission lines requiring higher reliability (example: extra high voltage (EHV) lines 345 kV and above) may use a 100 year return period which halves the probability of occurrence. Standards such as NESC, ASCE-7 (2010),

Table 2.2 Velocity Pressure Exposure Coefficients and Gust Response Factors for Wires.

Height of Wire at Structure In feet (meters)	*Velocity Pressure Exposure Coefficient* k_z	*Gust Response Factor* G_{RF} *for Various Span Lengths in feet (meters)*			
		251–500 (76.5 to 152.4)	*501–750 (152.7 to 228.6)*	*751–1000 (229 to 305)*	*1001–1500 (305.1 to 457.2)*
Up to 33 *(10.0)*	1.00	0.86	0.79	0.75	0.73
34–50 *(10.4 to 15.2)*	1.10	0.82	0.76	0.72	0.70
51–80 *(15.5 to 24.4)*	1.20	0.80	0.75	0.71	0.69
81–115 *(24.7 to 35.1)*	1.30	0.78	0.73	0.70	0.68
116–165 *(35.4 to 49.5)*	1.40	0.77	0.72	0.69	0.67

(Source: RUS/USDA)

ASCE-74 and RUS Bulletin 200 provide wind speed maps for the U.S. The extreme wind loading specified in NESC is also known as Rule 250C.

The equation used for computing wind pressure in *psf* corresponding to a given wind speed is:

$$p = 0.00256 * V^2 k_z G_{RF} C_d \mathrm{I} \tag{2.1a}$$

where:
V = Basic wind speed in miles per hour, 3-second gust measured at 33 ft above ground
k_z = Velocity Pressure Exposure Coefficient
G_{RF} = Gust Response Factor
C_d = Shape Factor (1.0 for round objects, 1.6 for flat surfaces)
I = Importance Factor

The wind pressure parameters in NESC are based on Exposure Category C which is open terrain with scattered obstructions.

If SI units are used:

$$p = 0.613 * V^2 k_z G_{RF} C_d \mathrm{I} \tag{2.1b}$$

where:
p = wind pressure in Pa (N/m^2)
V = Basic wind speed in meters per second, 3-second gust measured at 10 m above ground

Table 2.2 and 2.3 shows the values of k_z and G_{RF} for various spans and structure heights. For spans and heights outside of these ranges, formulae given in the NESC or RUS Bulletin 200 can be used. Shape Factors for shapes other than round and flat are:

Square or Rectangular Shapes 1.6
Hexagonal or Octagonal Poles 1.4
Dodecagonal (12-sided) Poles 1.0

Table 2.3 Velocity Pressure Exposure Coefficients and Gust Response Factors for Structures.

Height of Structure In feet (meters)	*Velocity Pressure Exposure Coefficient* k_z	*Gust Response Factor* G_{RF}
Up to 33 *(10.0)*	0.92	1.02
34–50 *(10.4 to 15.2)*	1.00	0.97
51–80 *(15.5 to 24.4)*	1.10	0.93
81–115 *(24.7 to 35.1)*	1.20	0.89
116–165 *(35.4 to 50.3)*	1.30	0.86

(Source: RUS/USDA)

Table 2.4a Adjustment Factors for Wind for Various Return Periods.

Load Return Period RP (years)	*Relative Reliability Factor*	*Wind Load Adjustment Factor*
25	0.50	0.85
50	*1.0*	*1.00*
100	2.0	1.15
200	4.0	1.30
400	8.0	1.45

(With permission from ASCE)

Table 2.4b Adjustment Factors for Ice and Concurrent Wind for Various Return Periods.

Load Return Period RP (years)	*Relative Reliability Factor*	*Ice Thickness Adjustment Factor*	*Concurrent Wind Load Factor*
25	0.50	0.80	1.00
50	*1.0*	*1.00*	*1.00*
100	2.0	1.25	1.00
200	4.0	1.50	1.00
400	8.0	1.85	1.00

(With permission from ASCE)

The Importance Factor is generally taken as 1.00 for utility structures.

In the US, it is common to use a basic wind speed of 90 mph to 110 mph (145 kmph to 177 kmph) as a typical design load case for extreme wind loading. Coastal winds often reach 150 mph (241 kmph). The minimum 50-year return period is considered as a baseline or reference; ASCE-74 provides adjustment factors in case other RP's for wind and ice are considered (Tables 2.4a and 2.4b).

Table 2.4c Ice, Wind, Temperature and Constants.

NESC Loading		*Design Temperature (°F)*	*Radial Ice Thickness (in)*	*Wind Loading (psf)*	*Constants (lbs/ft)*
Loading District	Heavy	0	0.50	4	0.30
	Medium	15	0.25	4	0.20
	Light	30	0	9	0.05
	Warm Islands (SL – 9000 ft)	50	0	9	0.05
	Warm Islands (above 9000 ft)	15	0.25	4	0.20
Extreme Wind		60	0	Variable*	N/A
Extreme Ice with Concurrent Wind		15	Variable*	Variable*	N/A

(Source: RUS/USDA)
SL = Sea Level
based on ASCE uniform ice thickness maps
Note: 1 psf = 47.88 Pa, 1 in = 25.4 mm, 1 ft = 0.3048 m, 1 lbs/ft = 14.594 N/m, deg C = (5/9)(deg F-32).

2.1.2 Combined ice and wind district loading

This is an empirical load case which was developed based on experience and expert judgement. The NESC categorizes areas in USA into loading "districts" based on specified thickness of glaze ice accumulation and wind pressure: Heavy, Medium, Light and Warm Islands. The design parameters applicable to these four zones are shown in Table 2.4c. The district loading specified in NESC is also known as Rule 250B.

2.1.3 Extreme ice with concurrent wind

This refers to the situation where ice build-up on a transmission wire is accompanied by small wind. The radial ice increases the projected area of the wire and thereby the transverse load due to wind. Current American design practices include a concurrent ice and wind load case with typical ice thickness varying from ¼ inch to 1 inch (6.35 mm to 25.4 mm) and a wind speed of 30 mph to 60 mph (48 kmph to 96.5 kmph), equivalent to a wind pressure of 2.3 psf to 9.2 psf (110 Pa to 440 Pa). The extreme ice with concurrent wind loading specified in NESC is also known as Rule 250D.

Again, the 50-year return period is considered as a baseline or reference. Table 2.4b provides adjustment factors in case other RP's are considered.

Extreme ice

Some areas experience heavy icing than required by NESC and utilities often develop in-house design standards for such cases with ice alone. This load case is known as *Extreme Ice*. The load case is used only with ice and no wind is considered. Some US utilities use 1 inch to 1½ inches of ice as their internal standard. A design temperature of 32°F is generally used for this load case.

2.1.4 High intensity winds

High Intensity Winds (HIW) such as tornadoes, downbursts and microbursts are localized events defined by very high wind velocity and are confined to a limited area. The frequency and magnitude of microbursts are not well understood to develop an acceptable design process referring to overhead transmission line structures. Research (ASCE-74) indicates 86% of tornadoes in USA are between F0 and F2 on the Fujita scale (wind speeds ranging from 40 mph to 157 mph) and with a maximum path width of 530 ft (161.5 m). It is not practical to design a transmission line to resist all tornadoes; however, economical designs can still be produced with existing wind criteria per ASCE-74 (Refer to Chapter 6 for a detailed discussion on this topic).

2.1.5 Pollution

Insulator contamination in salty environments in coastal or industrial areas should be considered while selecting insulators for a transmission line. The required insulation level for a specific transmission line must be calibrated with reference to anticipated pollution in the area. RUS Bulletin 200 provides guidance in this matter.

2.1.6 Temperature

Temperature is an important design parameter for transmission lines. All NESC loadings, wire sags and clearances refer to specific design temperatures.

2.2 ELECTRICAL DESIGN

Overhead transmission lines are designed to satisfy a wide range of local and regulatory requirements. While the demands of these requirements vary from location to location and are often line-specific, the underlying principle is one of legality, safety and reliability.

2.2.1 Regulatory codes

These relate to the local and national standards pertaining to both electrical operations as well as other land use considerations. For example, in areas close to runways in airports, U.S. Federal Aviation Administration (FAA) has laws limiting the height of a transmission structure to 200 feet (61 m). Protected lands such as wetlands do not permit or severely limit construction activities. In some areas, operating high-voltage transmission lines can be hazardous to raptors (eagles, hawks, falcons and other protected species of birds) and in such areas avian protection guidelines are specified. Other land use restrictions include wild life habitats and environmentally-sensitive locations.

From a design perspective, regulations refer to the NESC to which all designs adhere to. The other code is the RUS Bulletin 200 which is applicable to all rural high-voltage lines funded by the government. Most states in the USA have adopted NESC as governing code without any changes; but some states require supplemental standards via their administrative codes. In California and Hawaii, for example, electrical design is usually governed by GO-95 (2016) and GO-6 (1969), respectively. Supporting these codes are the various material and structural codes, design manuals and design guides

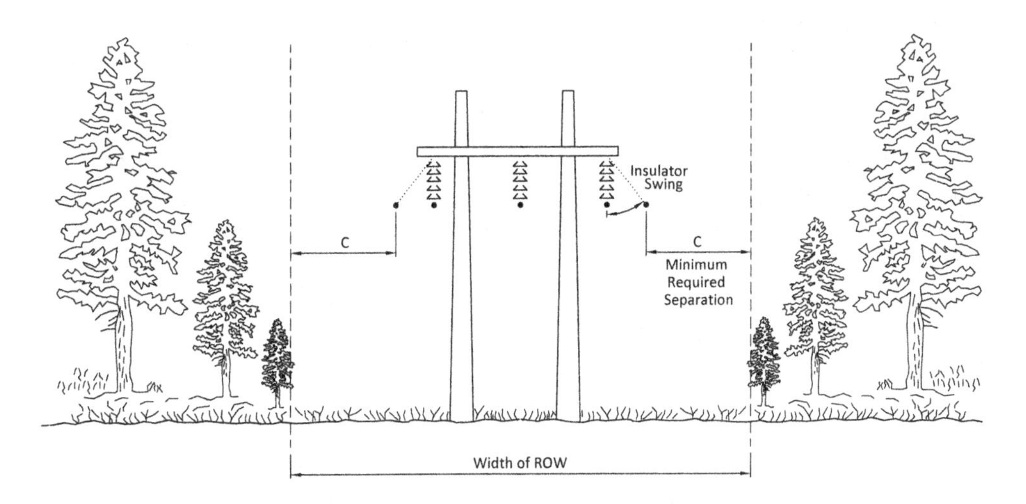

Figure 2.1 Right of Way Clearing Guide (Source: RUS/USDA).

(steel, concrete, lattice, foundations etc.). Many individual utilities in the US have their own in-house design standards which meet or exceed the NESC standards. Other international codes, standards and manuals will be briefly discussed in Section 2.7.

Another special situation is when transmission lines of one utility cross those of another or when transmission lines cross distribution circuits. Wire clearances in these cases must satisfy the applicable regulatory code. Lastly, overhead lines near highway or railroad crossings and underground pipelines in the vicinity of transmission lines are regulated by various state or local laws with special permitting and clearance requirements.

2.2.2 Right of way

The transmission line must be designed with adequate right-of-way (ROW) width to provide legal access to line repair and maintenance crews, vegetation management and to facilitate adequate distance from objects nearby. The actual width of line alignment depends on the voltage, number of circuits, type of structure adopted for the line and clearance from various objects required. Figure 2.1 shows the concept of ROW and how it relates to insulator swing and clearance to objects. Right of Way needs (referring to H-Frame type structures and various voltages) are shown in Table 2.5.

RUS Bulletin 200 provides a formula for calculating the required ROW width W for a typical pole structure given the required distance from conductors to a nearby object or such as building and other installation under moderate wind conditions. Referring to Figure 2.2:

$$W = A + 2(L_i + S_f)\sin \emptyset + 2\delta + 2C \tag{2.2}$$

where:
A = Horizontal Separation between the two suspension insulators
L_i = Length of the insulator string

Table 2.5 Typical Right of Way Widths in feet (*meters*).

Nominal Phase-to-Phase Voltage	*69 kV*	*115 kV*	*138 kV*	*161 kV*	*230 kV*	*345 kV*	*500 kV*
Single Circuit ROW	75 to 100 (*22.9 to 30.5*)	100 (*30.5*)	100 to 150 (*30.5 to 45.7*)	100 to 150 (*30.5 to 45.7*)	125 to 200 (*38.1 to 61*)	170 to 200 (*52 to 61*)*	200 to 300 (*61 to 91.4*)*

(Source: RUS/USDA)
**Commonly used values*

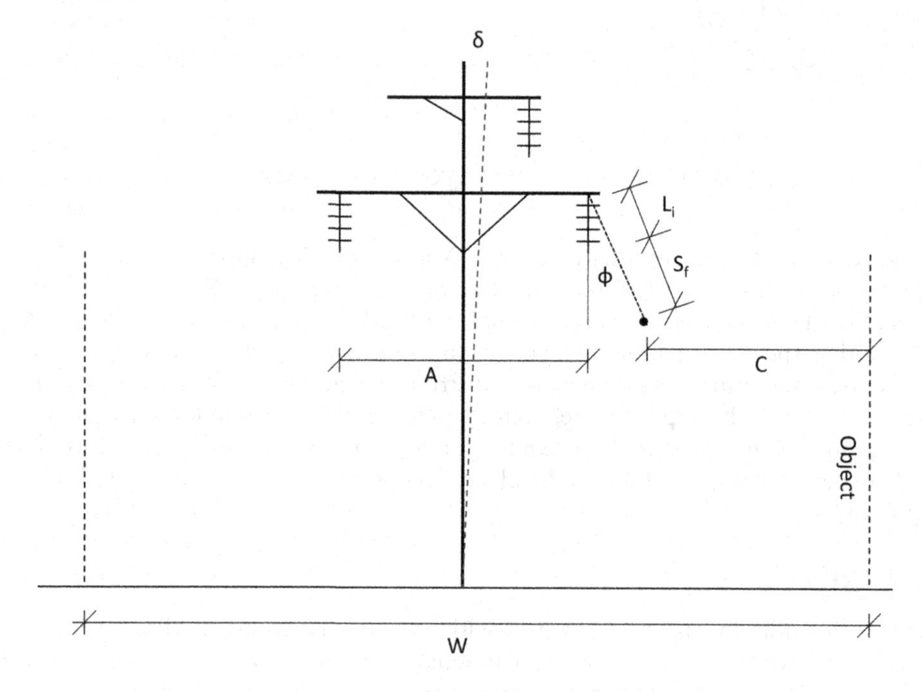

Figure 2.2 ROW Width Calculation for Single Pole Structures.

S_f = Conductor final sag at 60°F at 6 psf (290 Pa) wind
C = Required horizontal clearance between conductors and nearby objects
Ø = conductor/insulator swing angle under 6 psf (290 Pa) wind
δ = pole deflection under 6 psf (290 Pa) wind

In some cases the structure deflection is very small and negligible, especially for braced H-Frames. In such a case, the equation simplifies to:

$$W = A + 2(L_i + S_f)\sin Ø + 2C \quad (2.3)$$

It must be noted that weather conditions other than moderate 6 psf wind are also often considered. The required horizontal clearance C for objects such as buildings and other installations is provided by NESC. Without wind (i.e.) conductor at rest, the required

C values are higher; and therefore, one has to check the ROW for conductor at rest condition also. Note that some utilities allow a buffer of 2 ft (0.61 m) to be added to the required horizontal clearance.

Additionally, ROW width is also required to satisfy the horizontal and radial clearances to adjacent vegetation to prevent flashover between conductors and vegetation.

Example 2.1 A 69 kV line is planned on an available strip $W = 75$ ft (22.9 m) wide by an electric utility. The following data is given:

Horizontal distance between outermost insulators $= A = 9$ ft (2.74 m)
Suspension Insulator with 4 bells, Length $= L_i = 2$ ft-6 in (0.76 m)
Conductor Sag at 60°F with 6 psf wind $= S_f = 5$ ft (1.52 m)
The swing angle of the insulator under moderate 6 psf wind at 60°F $= \emptyset = 38°$
The required C per internal standards = 10 ft (3.05 m) away. Neglect pole deflections.
Determine if the strip provides adequate ROW for the line based on Equation 2.3.

Solution:

See Figure 2.2. From Eqn. 2.3, we have the required ROW width as

$$W = 9 + (2)(2.5 + 5)(\sin 38°) + (2)(10) = 38.23 \text{ ft } (11.7 \text{ m})$$

The 75 ft strip is adequate for ROW.

2.2.3 Clearances

Design clearances are one of the primary parameters of a transmission line design process. Clearances are primarily provided to safeguard public for activities reasonably anticipated in the vicinity of a transmission line. In general, the typical clearances required for a transmission line design are:

(a) Vertical Clearances of energized conductors above ground and other surfaces
(b) Clearance between Wires – Underbuild
(c) Clearance to Nearby Objects
(d) Clearance to Structure Surface
(e) Clearances between Wires – Phase and Ground Wires

Figure 2.3 shows the typical situations in which clearances are identified.

The NESC and RUS Bulletin 200 provide tables and formulas for *minimum recommended* clearances in each of the above categories.

One of the parameters influencing clearance calculations is the Maximum Phase-to-Ground Operating Voltage (MOV_{PG}) which is a function of Maximum Phase-to-Phase Operating Voltage (MOV_{PP}). In general:

$$MOV_{PP} = 1.05 \text{ Nominal Phase-to-Phase Voltage up to 230 kV, 345 kV and 765 kV} \quad (2.3a)$$

$$MOV_{PP} = 1.10 \text{ Nominal Phase-to-Phase Voltage kV for 500 kV} \quad (2.3b)$$

$$MOV_{PG} = MOV_{PP}/\text{sqrt}(3) \text{ or} = MOV_{PP}/1.732 \quad (2.3c)$$

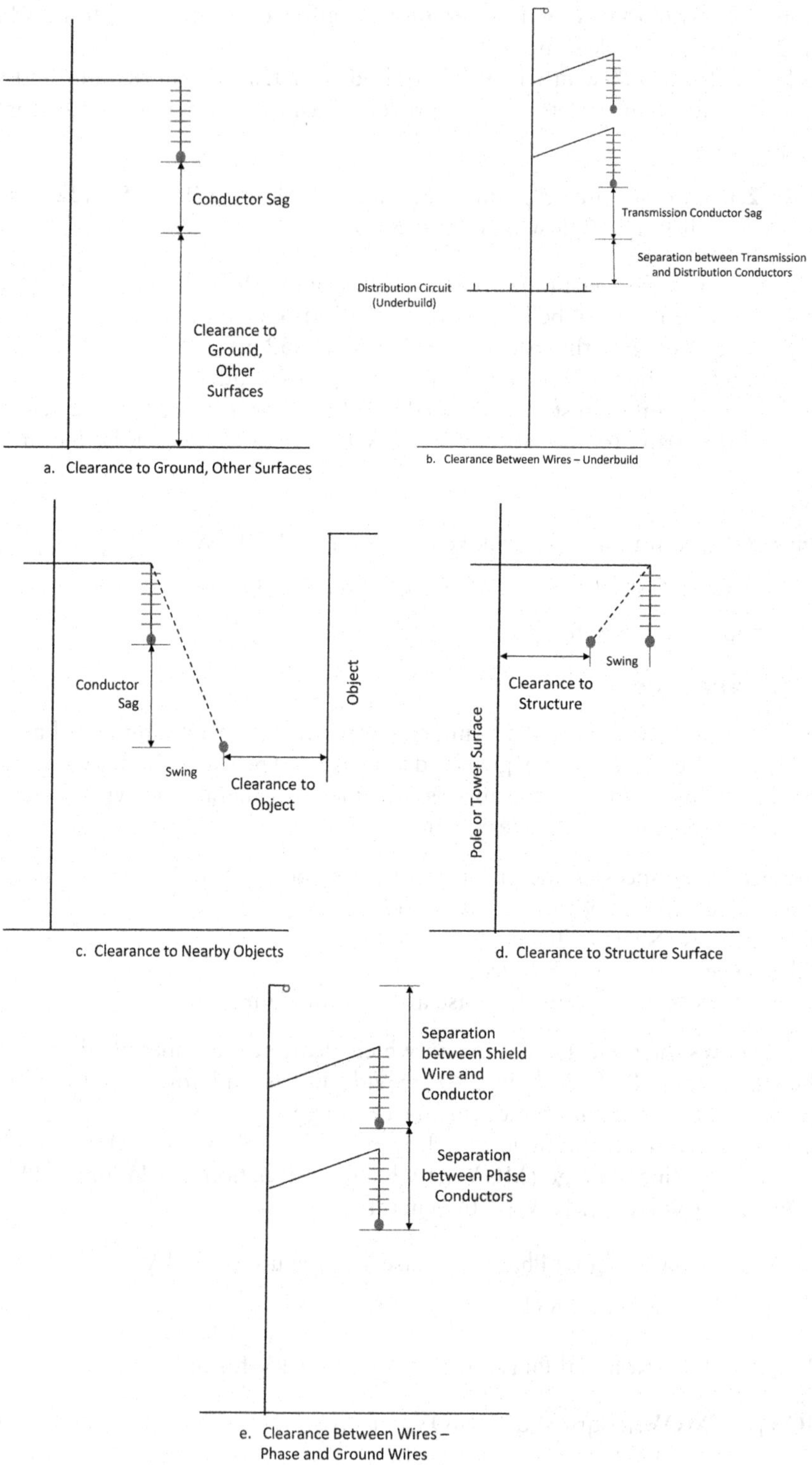

Conductor Sag

Clearance to
Ground,
Other
Surfaces

a. Clearance to Ground, Other Surfaces

Transmission Conductor Sag

Separation between Transmission
and Distribution Conductors

Distribution Circuit
(Underbuild)

b. Clearance Between Wires – Underbuild

Conductor
Sag

Swing

Clearance to
Object

Object

c. Clearance to Nearby Objects

Swing

Clearance to
Structure

Pole or Tower Surface

d. Clearance to Structure Surface

Separation
between Shield
Wire and
Conductor

Separation
between Phase
Conductors

e. Clearance Between Wires –
Phase and Ground Wires

Figure 2.3 Definitions of Various Clearances.

Clearances for voltages above 230 kV are more complex to evaluate considering the influence of several extra variables such as switching surge factors and electrostatic effects. For more information on electrostatic effects, the reader is referred to RUS Bulletin 62-4 (1976).

Several clearance tables are provided below. In these tables, the voltages are AC and nominal phase-to-phase unless specified otherwise. Altitude correction is not applied to these clearances. Additional buffers are mandatory over the minimum provided in these tables.

Vertical clearance of energized conductors to ground and other surfaces

The most important of all clearances is the vertical clearance of an energized conductor from the ground which governs structure heights. For voltages exceeding 98 kV AC to ground, Rule 232D of NESC also provides an alternative procedure for calculating vertical clearances of energized conductors above ground and other surfaces. The clearance is defined basically as a sum of a reference height and an electrical component accounting for maximum switching surge factor, non-standard atmospheric conditions and a margin of safety. This procedure also uses the Maximum *Crest-to-Ground* Operating Voltage (MOV_{CG}) which is defined as:

$$MOV_{CG} = Sqrt(2) * MOV_{PG} \tag{2.3d}$$

Tables 2.6a-1 and 2.6a-2 show the typical ground and other vertical clearances recommended by NESC and adopted in the US using Rule 232B and Rule 232C. For comparison, Table 2.6a-3 shows typical ground and other vertical clearances for EHV Lines computed per the alternative procedure (Rule 232D) using switching surge factors shown in the table. Utilities themselves also have their own in-house clearance standards, including buffers (adders), based on experience and consideration of various design and construction uncertainties. These standards meet and often exceed NESC values.

Clearance between wires – under-build

Dual-use structures (Figure 2.3b) often contain a distribution under-build or communication wires below the transmission circuits. The clearances between wires of the transmission circuit and those of the under-build are very important in determining the relative heights of the structures. Table 2.6b-1 shows these clearances. This table refers to clearances both at the structure as well as in mid-span and is a modified form of clearances from RUS Bulletin 200. The vertical clearances *at structure* apply regardless of horizontal separation between transmission and under-build conductors.

Minimum recommended vertical clearances *within span* apply to one of the following conditions which yields the least separation between the upper and lower conductors:

a. Upper conductor final sag at 32°F (0°C) with no wind and with radial ice thickness as applicable for the particular loading district;
b. Upper conductor final sag at 167°F (75°C)
c. Upper conductor final sag at maximum design temperature with no wind (usually 212°F or 100°C for ACSR conductors)

Table 2.6a-1 Typical Conductor Vertical Clearances* to Ground, Roadways, Rails or Water Surface in feet (*meters*) (U.S.).

	Nominal Phase-to-Phase Transmission Voltage (kV)					
Object Crossed	*34.5 to 46*	*69*	*115*	*138*	*161*	*230*
Maximum Operating Voltage (Phase-To-Phase)	–	*72.5*	*120.8*	*144.9*	*169.1*	*241.5*
Maximum Operating Voltage (Phase-To-Ground)	–	*41.8*	*69.7*	*83.7*	*97.6*	*139.4*
Roads, streets subject to truck traffic; Alleys, parking lots, driveways; other lands cultivated traversed by vehicles	18.7 *(5.7)*	19.2 *(5.9)*	20.1 *(6.1)*	20.6 *(6.3)*	21.0 *(6.4)*	22.4 *(6.8)*
Railroad Tracks	26.7 *(8.1)*	27.2 *(8.3)*	28.1 *(8.6)*	28.6 *(8.7)*	29.0 *(8.8)*	30.4 *(9.3)*
Spaces and ways accessible to pedestrians only	14.7 *(4.5)*	15.2 *(4.6)*	16.1 *(4.9)*	16.6 *(5.1)*	17.0 *(5.2)*	18.4 *(5.6)*
Water Areas – No Sail Boating	17.2 *(5.2)*	17.7 *(5.4)*	18.6 *(5.7)*	19.1 *(5.8)*	19.5 *(6.0)*	20.9 *(6.4)*
Water Areas – Sail Boating Suitable (Less than 20 acres)	20.7 *(6.3)*	21.2 *(6.5)*	22.1 *(6.7)*	22.6 *(6.9)*	23.0 *(7.0)*	24.4 *(7.4)*
Water Areas – Sail Boating Suitable (Over 2000 acres)	40.7 *(12.4)*	41.2 *(12.6)*	42.1 *(12.8)*	42.6 *(13.0)*	43.0 *(13.1)*	44.4 *(13.5)*

(Source: RUS/USDA)
**Does not include any buffer or adder.*

Table 2.6a-2 Typical Conductor Vertical Clearances* to Ground, Roadways, Rails or Water Surface in feet (*meters*) (U.S.) – EHV Lines.

	Nominal Phase-to-Phase EHV Transmission Voltage (kV)		
Object Crossed	*345*	*500*	*765*
Maximum Operating Voltage (Phase-To-Phase)	*362*	*550*	*803*
Maximum Operating Voltage (Phase-To-Ground)	*209.0*	*317.6*	*463.6*
Roads, streets subject to truck traffic; Alleys, parking lots, driveways; other lands cultivated traversed by vehicles	24.7 *(7.5)*	28.4 *(8.7)*	33.2 *(10.1)*
Railroad Tracks	32.7 *(10.0)*	36.4 *(11.1)*	41.2 *(12.6)*
Spaces and ways accessible to pedestrians only	20.7 *(6.3)*	24.4 *(7.4)*	29.2 (8.9)
Water Areas – No Sail Boating	23.2 *(7.1)*	26.9 *(8.2)*	31.7 *(9.7)*
Water Areas – Sail Boating Suitable (Less than 20 acres)	26.7 *(8.1)*	30.4 *(9.3)*	35.2 *(10.7)*
Water Areas – Sail Boating Suitable (Over 2000 acres)	46.7 *(14.2)*	50.4 *(15.4)*	55.2 *(16.8)*

**Does not include any buffer or adder.*

Table 2.6a-3 Typical Conductor Vertical Clearances* to Ground, Roadways, Rails or Water Surface in feet (*meters*) (U.S.) – EHV Lines by NESC Alternative Procedure**.

	Nominal Phase-to-Phase EHV Transmission Voltage (kV)		
Object Crossed	*345*	*500*	*765*
Maximum Operating Voltage (Phase-To-Phase)	*362*	*550*	*803*
Maximum Operating Voltage (Phase-To-Ground)	*209.0*	*317.6*	*463.6*
Maximum Operating Voltage (Crest-To-Ground)	*295.7*	*449.1*	*655.9*
Roads, streets subject to truck traffic; Alleys, parking lots, driveways; other lands cultivated traversed by vehicles	25.6 (*7.8*)	30.0 (*9.1*)	36.3 (*11.1*)
Railroad Tracks	33.6 (*10.2*)	38.0 (*11.6*)	44.3 (*13.5*)
Spaces and ways accessible to pedestrians only	21.6 (*6.6*)	26.0 (*7.9*)	32.3 (9.8)
Water Areas – No Sail Boating	24.1 (*7.3*)	28.5 (*8.7*)	34.8 (*10.6*)
Water Areas – Sail Boating Suitable (Less than 20 acres)	27.6 (*8.4*)	32.0 (*9.8*)	38.3 (*11.7*)
Water Areas – Sail Boating Suitable (Over 2000 acres)	47.6 (*14.5*)	52.0 (*15.9*)	58.3 (*17.8*)

**Does not include any buffer or adder.*
***Used Switching Surge Factors: 345 kV – 2.37, 500 kV – 2.08, 765 kV – 1.85 (based on NESC Alternate Method).*

Table 2.6b-1 Typical Conductor Vertical Clearances to Distribution Underbuild Conductors in feet (*meters*) (U.S.)*.

	Upper Level Conductors						
	Nominal Phase to Phase Transmission Voltage (kV)						
Object Crossed	*34.5*	*46*	*69*	*115*	*138*	*161*	*230*
Maximum Operating Voltage (Phase-To-Phase)	*36.2*	*48.3*	*72.5*	*120.8*	*144.9*	*169.1*	*241.5*
Maximum Operating Voltage (Phase-To-Ground)	20.2	*27.9*	*41.6*	*69.7*	*83.7*	*97.6*	*139.4*
Clearance at Support							
25 kV and Below**	4.2 (*1.28*)	4.5 (*1.37*)	4.9 (*1.5*)	5.9 (*1.8*)	6.3 (*1.92*)	6.8 (2.07)	8.2 (2.50)
34.5 kV	4.4 (*1.35*)	4.7 (*1.45*)	5.1 (*1.55*)	6.0 (*1.83*)	6.5 (2.0)	7.0 (2.13)	8.4 (2.56)
Clearance within Span							
25 kV and Below**	3.2 (0.97)	3.3 (*1.0*)	3.7 (*1.13*)	4.7 (*1.45*)	5.1 (*1.55*)	5.6 (*1.71*)	7.0 (2.13)
34.5 kV	3.3 (*1.0*)	3.5 (*1.1*)	3.8 (*1.16*)	4.9 (*1.5*)	5.3 (*1.65*)	5.8 (*1.77*)	7.2 (2.2)

(Source: RUS/USDA)
**Does not include any buffer or adder.*
***includes communication conductors.*

The sag of the under-build conductor to be used is the final sag, measured at the same ambient temperature as the upper conductor without electrical and ice loading.

Clearance between wires on different supporting structures

Transmission lines often cross another, usually a higher voltage (upper) line crossing a lower voltage line. The clearances between wires of the upper level circuit and those of the lower level are very important in determining the relative heights of the structures for the upper line. Table 2.6b-2 shows these clearances. This table refers to the situation where higher voltage wires cross lower voltage wires and where the

Table 2.6b-2 Typical Conductor Vertical Clearances to Crossing Conductors in feet (*meters*) (U.S.)*, ++, **.

	Upper Level Conductors							
	Nominal Phase-to-Phase Transmission Voltage (kV)						*Nominal Phase-to-Phase EHV Transmission Voltage (kV)****	
Object Crossed	*34.5 to 46*	*69*	*115*	*138*	*161*	*230*	*345*	*500*
Maximum Operating Voltage (Phase-To-Phase)	–	*72.5*	*120.8*	*144.9*	*169.1*	*241.5*	*362*	*550*
Maximum Operating Voltage (Phase-To-Ground)	–	*41.6*	*69.7*	*83.7*	*97.6*	*139.4*	*209.0*	*317.6*
Lower Level Conductors								
Communication	5.2 (*1.6*)	5.7 (*1.7*)	6.6 (*2.0*)	7.1 (*2.2*)	7.5 (*2.3*)	8.9 (*2.7*)	N/A	N/A
500 kV								22.0 (*6.7*)
345 kV							16.0 (*4.9*)	19.1 (*5.8*)
230 kV						9.8 (*3.0*)	13.7 (*4.2*)	16.8 (*5.1*)
161 kV					7.0 (*2.1*)	8.4 (*2.6*)	12.3 (*3.7*)	15.4 (*4.7*)
138 kV				6.1 (*1.9*)	6.6 (*2.0*)	8.0 (*2.4*)	11.8 (*3.6*)	14.9 (*4.5*)
115 kV			5.2 (*1.6*)	5.6 (*1.7*)	6.1 (*1.9*)	7.5 (*2.3*)	11.3 (*3.4*)	14.4 (*4.4*)
69 kV		3.3 (*1.0*)	4.1 (*1.2*)	4.7 (*1.4*)	5.2 (*1.6*)	6.6 (*2.0*)	10.5 (*3.2*)	13.5 (*4.1*)
46 kV and Below	2.3 (*0.7*)	2.8 (*0.9*)	3.7 (*1.1*)	4.2 (*1.3*)	4.7 (*1.4*)	6.1 (*1.9*)	9.8 (*3.0*)	12.8 (*3.9*)

(Source: RUS/USDA)
**Does not include any buffer or adder.*
++Higher Voltage should cross Lower Voltage.
***Applies only to lines with ground fault relaying.*
****Based on commonly adopted industry values.*

wires are on different supporting structures. Lower voltage wires crossing over higher voltage wires, though theoretically possible, is not operationally recommended. Note that Table 2.6b-2 is a modified form of RUS clearances from Bulletin 200.

Clearance to nearby objects

Horizontal clearances to various objects are given in Tables 2.6c-1 (modified RUS) and 2.6c-2. The tables provide clearances required for wires at rest and when wires are displaced under 6 psf (290 Pa) wind at 60 deg. F temperature. The horizontal clearance

Table 2.6c-1 Typical Conductor *Horizontal* Clearances to Objects in feet (*meters*) (U.S.)*.

	Nominal Phase-to-Phase Transmission Voltage (kV)					
	34.5 to 46	*69*	*115*	*138*	*161*	*230*
Maximum Operating Voltage (Phase-To-Phase)	–	*72.5*	*120.8*	*144.9*	*169.1*	*241.5*
Maximum Operating Voltage (Phase-To-Ground)	–	*41.6*	*69.7*	*83.7*	*97.6*	*139.4*
Object						
From buildings, walls, projections, guarded windows, areas accessible to pedestrians – At Rest	7.7 (*2.3*)	8.2 (*2.5*)	9.1 (*2.8*)	9.6 (*2.9*)	10.0 (*3.0*)	11.4 (*3.5*)
From buildings, walls, projections, guarded windows, areas accessible to pedestrians – Displaced by Wind++	4.7 (*1.4*)	5.2 (*1.6*)	6.1 (*1.9*)	6.6 (*2.0*)	7.0 (*2.1*)	8.4 (*2.6*)
From lighting support, traffic signal support or supporting structure of another line – At Rest	5.0 (*1.5*)	5.0 (*1.5*)	5.7 (*1.7*)	6.1 (*1.9*)	6.6 (*2.0*)	8.0 (*2.4*)
From lighting support, traffic signal support or supporting structure of another line – Displaced by Wind++	4.7 (*1.4*)	5.2 (*1.6*)	6.1 (*1.9*)	6.6 (*2.0*)	7.0 (*2.1*)	8.4 (*2.6*)
From signs, chimneys, billboards, radio and TV antennas, tanks and other installations – At Rest	7.7 (*2.3*)	8.2 (*2.5*)	9.1 (*2.8*)	9.6 (*2.9*)	10.0 (*3.0*)	11.4 (*3.5*)
From signs, chimneys, billboards, radio and TV antennas, tanks and other installations – Displaced by Wind++	4.7 (*1.4*)	5.2 (*1.6*)	6.1 (*1.9*)	6.6 (*2.0*)	7.0 (*2.1*)	8.4 (*2.6*)
From portions of bridges which are readily accessible and supporting structures not attached – At Rest	7.7 (*2.3*)	8.2 (*2.5*)	9.1 (*2.8*)	9.6 (*2.9*)	10.0 (*3.0*)	11.4 (*3.5*)
From portions of bridges which are readily accessible and supporting structures not attached – Displaced by Wind++	4.7 (*1.4*)	5.2 (*1.6*)	6.1 (*1.9*)	6.6 (*2.0*)	7.0 (*2.1*)	8.4 (*2.6*)
From railroad cars (only to lines parallel to tracks)	12.1 (*3.7*)	12.1 (*3.7*)	13.1 (*4.0*)	13.6 (*4.1*)	14.0 (*4.3*)	15.5 (*4.7*)

(Source: RUS/USDA).
**Does not include any buffer or adder.*
++*Wind refers to 6 psf (290 Pa) at 60 deg. F.*

Table 2.6c-2 Typical Conductor *Horizontal* Clearances to Objects in feet (*meters*) (U.S.) – EHV Lines*.

	*Nominal Phase-to-Phase EHV Transmission Voltage (kV)***	
	345	500
Maximum Operating Voltage (Phase-To-Phase)	*362*	*550*
Maximum Operating Voltage (Phase-To-Ground)	*209.0*	*317.6*
Object		
From buildings, walls, projections, guarded windows, areas accessible to pedestrians – At Rest	14.0 *(4.27)*	17.7 *(5.39)*
From buildings, walls, projections, guarded windows, areas accessible to pedestrians – Displaced by Wind++	11.0 *(3.35)*	14.7 *(4.48)*
From lighting support, traffic signal support or supporting structure of another line – At Rest	10.5 *(3.20)*	14.2 *(4.33)*
From lighting support, traffic signal support or supporting structure of another line – Displaced by Wind++	10.0 *(3.05)*	13.7 *(4.18)*
From signs, chimneys, billboards, radio and TV antennas, tanks and other installations – At Rest	14.0 *(4.27)*	17.7 *(5.39)*
From signs, chimneys, billboards, radio and TV antennas, tanks and other installations – Displaced by Wind++	11.0 *(3.35)*	14.7 *(4.48)*
From portions of bridges which are readily accessible and supporting structures not attached – At Rest	14.0 *(4.27)*	17.7 *(5.39)*
From portions of bridges which are readily accessible and supporting structures not attached – Displaced by Wind++	11.0 *(3.35)*	14.7 *(4.48)*
From railroad cars (only to lines parallel to tracks)	18.0 *(5.50)*	21.0 *(6.40)*

**Does not include any buffer or adder.*
***Based on commonly adopted industry values.*
++*Wind refers to 6 psf. (290 Pa) at 60 deg. F.*

covers what is commonly known as conductor displacement due to wind. However, the clearances do not cover the blow-out of the conductor (determined with a computer program such as PIS-CADD™). Vertical clearances to various objects are given in Tables 2.6d-1 (modified RUS) and 2.6d-2.

Buffers of 2.0 ft (0.61 m) to 5.0 ft (1.52 m) can be added to various clearances to provide additional safety. These adders are not absolute but are based on the judgement of the engineer. Some utilities in the US are known to use clearance buffers ranging up to 10.0 ft (3.05 m) on EHV lines.

Codes in various other countries more or less adopt similar clearances, quantitatively and qualitatively. Operating voltages often differ widely between the North American continent, Europe and Asia; so, any attempt to compare and contrast clearances can only be nominal at an informational level.

Clearances between wires – phase and ground wires

The separation between energized conductors of the same circuit and between conductors and ground wires should be adequate to prevent swinging contact or flashovers.

Table 2.6d-1 Typical Conductor *Vertical* Clearances to Objects in feet (*meters*) (U.S.)*.

	Nominal Phase-to-Phase Transmission Voltage (kV)					
	34.5 to 46	*69*	*115*	*138*	*161*	*230*
Maximum Operating Voltage (Phase-To-Phase)	–	*72.5*	*120.8*	*144.9*	*169.1*	*241.5*
Maximum Operating Voltage (Phase-To-Ground)	–	*41.6*	*69.7*	*83.7*	*97.6*	*139.4*
Nature of Object Underneath Wires						
Buildings not accessible to pedestrians	12.7 *(3.9)*	13.2 *(4.0)*	14.1 *(4.3)*	14.6 *(4.5)*	15.0 *(4.6)*	16.4 *(5.0)*
Buildings accessible to pedestrians and vehicles but not truck traffic	13.7 *(4.2)*	14.2 *(4.3)*	15.1 *(4.6)*	15.6 *(4.8)*	16.0 *(4.9)*	17.4 *(5.3)*
Lighting support, traffic signal support or supporting structure of a second line	5.5 *(1.7)*	5.5 *(1.7)*	6.2 *(1.9)*	6.6 *(2.0)*	7.1 *(2.2)*	8.8 *(2.7)*
Signs, chimneys, billboards, radio and TV antennas, tanks and other installations	8.2 *(2.5)*	8.7 *(2.7)*	9.6 *(2.9)*	10.1 *(3.1)*	10.5 *(3.2)*	11.9 *(3.6)*
Bridges – conductors not attached	12.7 *(3.9)*	13.2 *(4.0)*	14.1 *(4.3)*	14.6 *(4.5)*	15.0 *(4.6)*	16.4 *(5.0)*

(Source: RUS/USDA.)
*Does not include any buffer or adder.

Table 2.6d-2 Typical Conductor *Vertical* Clearances to Objects in feet (*meters*) (U.S.) – EHV Lines*.

	*Nominal Phase-to-Phase EHV Transmission Voltage (kV)***	
	345	*500*
Maximum Operating Voltage (Phase-To-Phase)	*362*	*550*
Maximum Operating Voltage (Phase-To-Ground)	*209.0*	*317.6*
Nature of Object Underneath Wires		
Buildings not accessible to pedestrians	19.0 *(5.79)*	22.0 *(6.71)*
Buildings accessible to pedestrians and vehicles but not truck traffic	20.0 *(6.10)*	23.0 *(7.01)*
Lighting support, traffic signal support or supporting structure of a second line	11.0 *(3.35)*	14.0 *(4.27)*
Signs, chimneys, billboards, radio and TV antennas, tanks and other installations	14.5 *(4.42)*	17.5 *(5.33)*
Bridges – Conductors not attached	19.0 *(5.79)*	22.0 *(6.71)*

**Does not include any buffer or adder.*
***Based on commonly adopted industry values, some values rounded to nearest half foot.*

Table 2.7 Typical Phase Separation at Structure*.

	Nominal Phase-to-Phase Transmission Voltage (kV)					
	34.5 to 46	69	115	138	161	230
Spacing between ...	Minimum Vertical Separation Required in feet (meters)[1,2]					
Phase Wires of Same Circuit	2.7 (0.82)	3.5 (1.07)	5.1 (1.55)	5.9 (1.80)	6.7 (2.04)	9.1 (2.77)
Phase Wires and Overhead Ground Wires (OHGW)	2.0 (0.61)	2.4 (0.73)	3.4 (1.04)	3.8 (1.16)	4.3 (1.31)	5.9 (1.80)

(Source: RUS/USDA.)
Does not include any buffer or adder.
[1]An additional 2.0 ft (0.61 m) should be added to the above clearances in areas of severe icing.
[2]Applicable for Standard RUS Structures only.

Table 2.8 Typical Phase Separation at Structure – EHV Lines*.

	Nominal Phase-to-Phase EHV Transmission Voltage (kV)		
	345	500	765
Spacing between ...	Minimum Vertical Separation Required in feet (meters)[1,2]		
Phase Wires of Same Circuit	15.0 (4.6)	22.0 (6.7)	32.0 (9.8)
Phase Wires and Overhead Ground Wires (OHGW)	8.0 (2.4)	12.0 (3.7)	19.0 (5.7)

Does not include any buffer or adder.
[1]An additional 2 ft to 5 ft (0.61 to 1.52 m) should be added to the above clearances in areas of severe icing.
[2]Based on values adopted by various utilities per industry standards.

NESC provides guidelines to determine the minimum required phase separation, both at the structure as well as mid-span. Tables 2.7 (modified RUS) and 2.8 show commonly used phase separation on structures. Clearances for circuits 345 kV and above included in Table 2.8 are based on data from actual lines built at those voltages. These values serve only as a guideline; the actual required clearances are a function of several variables including type of structure adopted, span lengths, voltage, terrain and possibility of Aeolian vibration and galloping. Longer spans are often associated with galloping and require larger separation between phase wires and between phase wires and overhead ground wires.

2.2.3.1 *Insulator swing*

Suspension insulator strings are usually free to swing about their points of attachment to the structure. It is, therefore, necessary to ensure that when the insulator strings do swing, minimum clearances are maintained to structure surfaces and guy wires. The amount of swing is a function of conductor tension, wind velocity, insulator weight, line angle etc. The RUS Bulletin 200 provides guidelines for determining minimum swing clearances (i.e.) separation from the structure, both in terms of distance as well

Table 2.9a Typical Wire Clearance from Structure Surface and Guy Wires.

	Nominal Transmission Voltage, Phase-to-Phase (kV)						
	34.5	*46*	*69*	*115*	*138*	*161*	*230*
Weather Condition	*Minimum Clearance Required in inches (meters)*						
No Wind Structure or Guy	19 *(0.48)*	19 *(0.48)*	25 *(0.64)*	42 *(1.07)*	48 *(1.22)*	60 *(1.52)*	71 *(1.80)*
Moderate Wind 6 psf (290 Pa)[1] to Structure	9 *(0.23)*	11 *(0.28)*	16 *(0.41)*	26 *(0.66)*	30 *(0.76)*	35 *(0.86)*	50 *(1.27)*
Moderate Wind 6 psf (290 Pa)[1] to Guy Wires	13 *(0.33)*	16 *(0.41)*	22 *(0.56)*	34 *(0.86)*	40 *(1.02)*	46 *(1.17)*	64 *(1.63)*
High Wind Structure or Guy[+]	3 *(0.076)*	3 *(0.076)*	5 *(0.13)*	10 *(0.25)*	12 *(0.30)*	14 *(0.36)*	20 *(0.51)*

(Source: RUS/USDA.)
[1]At 60 deg. F, final sag.
[+]Not NESC Requirement.

as maximum allowed swing angles. These clearances are generally evaluated at the three weather conditions shown below and are derived from the NESC for conditions (a) and (b).

(a) No Wind
(b) Light or Moderate Wind (6 psf minimum)
(c) High Wind

Swing clearance is aimed towards minimizing the possibility of a structure flashover during switching operations. Additionally, it helps protect electrical maintenance workers by providing safety clearances while working on the transmission structures. Table 2.9a shows safe clearances suggested by RUS for voltages up to 230 kV; Table 2.9b gives safe clearances suggested for EHV voltages of 345 and 500 kV. The values also indicate safe distance of energized conductors from guy wires (air gap).

The RUS Bulletin 200 also specifies the required *horizontal* separation between various wires on a transmission structure. This separation is mandated to prevent swinging contacts or flashovers between phases of the same or different circuit.

$$H = (0.025) * V + F_c\sqrt{S_f} + L_i \sin \emptyset \qquad (2.4)$$

where:
H = Horizontal Separation between the two conductors in feet
V = Maximum Phase-to-Phase voltage in kilovolts (1.05 times the nominal voltage)
L_i = Length of the insulator string (zero for post insulators), in feet
S_f = Conductor final sag at 60°F, no load, in feet (Note the difference in definition of S_f compared to Eqn. 2.2)
F_c = Experience Factor (1.15 to 1.25, depending on ice and wind loading intensity)
$\emptyset$ = Insulator swing angle under 6 psf at 60°F, (290 Pa) wind

Table 2.9b Typical Wire Clearance from Structure Surface and Guy Wires – EHV Lines.

	Nominal Transmission Voltage Phase-to-Phase (kV)	
	345	*500*
Weather Condition	*Minimum Separation Required in feet (meters)*[1,2]	
No Wind Structure or Guy	9.00 *(2.74)*	13.00 *(3.96)*
Moderate Wind 6 psf (290 Pa)[3] to Structure	6.10 *(1.86)*	9.30 *(2.83)*
Moderate Wind 6 psf (290 Pa)[3] to Guy Wires	8.40 *(2.56)*	12.30 *(3.75)*
High Wind Structure or Guy	2.50 *(0.76)*	3.60 *(1.10)*

[1]Clearances are based on commonly adopted industry values.
[2]For 765 kV, a clearance of 14 ft to 18 ft (4.3 m to 5.5 m) is adopted by some US utilities for the No Wind case.
[3]At 60 deg. F, final sag.

Elevation effect

The various clearances discussed so far are applicable for line locations at altitudes of 3300 ft or less. For locations at a higher altitude, NESC recommends altitude corrections to be added to the given clearances. These corrections include adding a specified amount of clearance for each 1000 ft in excess of 3300 ft.

Example 2.2 For the same data/line of the previous example E2.1, determine horizontal separation required between conductors. Assume $F_c = 1.20$, $S_f = 4.0$ ft (1.22 m).

Solution:

From Eqn. 2.4, we have:

$H = (0.025)(1.05)(69) + (1.20)(\text{sqrt}(4.0)) + (2.50)(\sin 38°) = 5.75$ ft (1.75 m)

Note: This S_f is defined differently than in Example 2.1.

2.2.4 Shielding

When lightning strikes a transmission line, it may hit either the overhead ground wire or a phase conductor. If a conductor is hit, there will certainly be a flashover of the insulation. To prevent and minimize such an occurrence, the overhead ground wire is used to intercept the lightning strike, "shielding" the conductors. To optimize this shielding, the shielding angle (see Figure 2.4) must be 30° or less. If a location is known to have an unusually high exposure to lightning strikes, and structure heights are over 90 ft (27.4 m), even smaller shielding angles should be used. Table 2.10 shows typical shielding angles relative to structure height as well as voltage. Note that for situations where the height and voltage give two different shielding angles, the smaller value shall be adopted.

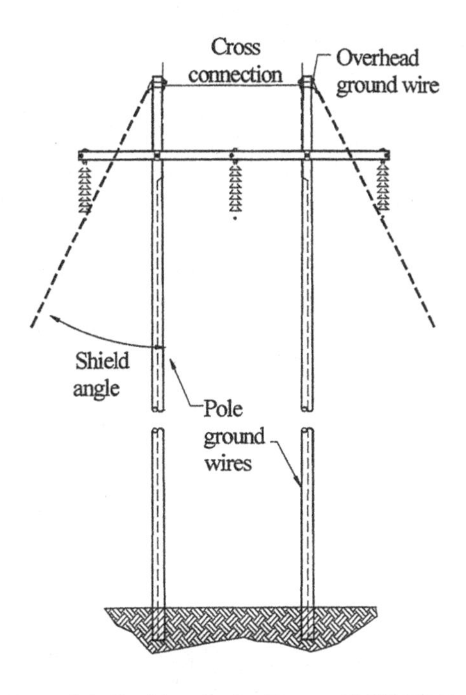

Figure 2.4 Shielding Angle (Source: RUS/USDA).

Table 2.10 Typical Shielding Angles.

Structure Height in feet (meters)	*Shielding Angle (degrees)*
92 *(28.0)*	30
99 *(30.2)*	26
116 *(35.4)*	21
Line Voltage in kV Phase-to-Phase	*Shielding Angle (degrees)**
69 to 138	30
161 to 230	25
345	20
500	15
765	10 to 12

(Courtesy: RUS/USDA.)
*common industry values.

For voltages exceeding 230 kV, the shielding angle often controls the required vertical and horizontal separation between the ground wire and conductor. For areas with a history of lightning strikes, the clearance between the overhead ground wire (OHGW) and the conductor shall be determined with specific reference to the Isokeraunic levels.

2.2.5 Lightning performance and grounding

Design criteria should include the line's target Isokeraunic Level of the area. This level is usually given in terms of number of thunderstorms in the area per year. An overhead ground wire (OHGW) should be provided at all places where the Isokeraunic Level is over 20 per RUS Bulletin 200. This wire should be grounded at every structure by way of the structure ground wire. For H-Frame structures with two (2) ground wires, the OHGW must each be connected to the structure ground wire and to one another. This will be beneficial in the event that if one structure ground wire breaks, both overhead ground wires will still remain grounded. RUS Bulletin 200 recommends that lightning outages of 1 to 4 per 100 miles per year is acceptable for lines in the 161 to 230 kV range.

All circuits must be adequately grounded to facilitate protection against lightning strikes. RUS recommends that individual transmission structures should have a footing resistance of less than 25 ohms, measured in dry soil conditions, especially within ½ mile (0.8 km) of a substation in high Isokeraunic areas. Outside of this distance, a resistance of less than 40 ohms is allowed by some utilities at voltages upto 230 kV. Ground rods or counterpoise wires may be used to obtain a footing resistance below the maximum allowable value.

A map of Isokeraunic levels at various locations in the US is given in RUS Bulletin 200.

2.2.6 Insulation requirements

Insulators are needed to provide both a mechanical means to hold the line conductors as well as electrical isolation required to withstand impulse voltage events such as lightning impulses and switching impulses. Insulators are basically a physical means of providing an "air gap" between an energized conductor and the grounded (earthed) portion of the structure. Toughened glass, porcelain and non-ceramic (polymer) insulators are commonly used on overhead lines. Recommended insulation levels are generally in terms of number of "bells" (porcelain) or "sheds" (polymer) in an insulator string. The number of bells or sheds also determines the so-called "leakage distance" or "dry-arcing distance" – the distance measured along the insulating surface from the ground point to energized point.

Tables 2.11, 2.12 and 2.13 provide guidelines for suggested minimum insulation levels in suspension and horizontal post insulators (See also Appendix 11 for more insulator data). Additional insulation may be warranted when required due to higher altitudes, contamination and high soil resistance. Other insulation considerations include critical impulse flashover, switching surge factor flashover and strength requirements. For deadends, where the insulator string is in line with the conductor, two extra bells should be provided relative to the tangent string.

It must be noted that for voltages up to 230 kV, the most severe stress on insulators is generally due to lightning strikes; therefore, the most important characteristic is the impulse flashover. RUS Bulletin 200 recommendations for insulation levels include both positive and negative flashover values.

2.2.7 Conductor operating temperature

Clearances from energized conductors to ground are generally evaluated at the wire's specified Maximum Operating Temperature (MOT). ACSR (Aluminum Conductor

Table 2.11 Minimum Insulation Levels Needed for Suspension Insulators*,**.

Nominal Voltage (between phases) (kV)	*Rated Dry Flashover Voltage (kV)*
34.5	100
46	125
69	175
115	315
138	390
161	445
230	640
345	830
500	965
765	1145

*Based on ANSI C29.
**NESC.
(Courtesy: IEEE NESC ®C2-2012)
National Electrical Safety Code.
Reprinted With Permission from IEEE.

Table 2.12 Recommended Insulation Levels for Suspension Insulators at Sea Level*,**,***.

Nominal Line-to-Line Voltage kV	*60 Hz Low Freq. Flashover kV*		*Impulse Flashover kV*		*No. of Bells***	*Total Leakage Distance, ft (m)*
	Dry	*Wet*	*Positive*	*Negative*		
34.5 to 46	215	130	355	340	3	2.88 *(0.88)*
69	270	170	440	415	4	3.83 *(1.17)*
115	435	295	695	670	7	6.71 *(2.05)*
138	485	335	780	760	8	7.66 *(2.33)*
161	590	415	945	930	10	9.58 *(2.92)*
230	690	490	1105	1105	12	11.5 *(3.51)*

Note: 1 in = 25.4 mm, 1 ft = 0.3048 m
*Based on RUS Bulletin 200.
**5-3/4 in × 10 in bells.
***Wood Tangent or small angle structures.

Table 2.13 Recommended Insulation Levels for Horizontal Post Insulators at Sea Level*,**.

Nominal Line-to-Line Voltage kV	*60 Hz Low Freq. Flashover kV*		*Impulse Flashover kV*		*Total Leakage Distance, ft (m)*
	Dry	*Wet*	*Positive*	*Negative*	
34.5	100	70	210	260	1.83 *(0.56)*
46	125	95	255	344	2.42 *(0.74)*
69	180	150	330	425	4.42 *(1.35)*
115	380	330	610	760	8.33 *(2.54)*
138	430	390	690	870	9.17 *(2.80)*

*Based on RUS Bulletin 200.
**Tangent or small angle structures.

Steel Reinforced) wires function well even at an elevated temperature of 100°C (212°F) without any significant loss of strength. High performance wires such as ACSS (Aluminum Conductors Steel Supported) can sustain temperatures up to 200°C (392°F). ACCR (Aluminum Conductor Composite Reinforced) conductors often used in long span crossings can withstand temperatures over 200°C (392°F).

Nominal wire clearances to ground are usually given on P & P (Plan and Profile) drawings in terms of MOT. This in turn will help determine the required structure heights for that location. Conductor operating temperature also influences line ratings (see Ampacity subsection below).

2.2.8 Corona and field effects

Corona is the ionization of the air that occurs at the surface of the conductor and hardware due to high electric field strength at the surface of the metal. Field effects are the secondary voltages and currents that may be induced in nearby objects. Corona is a function of the voltage of the line, conductor diameter and the condition of the conductor and may also result in radio and television interference, light and ozone production.

2.2.9 EMF and noise

Operation of power lines produce electric and magnetic fields commonly referred to as EMF. Noise on transmission lines is also due to the effect of corona. The EMF produced by alternating current in the USA has a frequency of 60 Hz. Electric field strength is directly proportional to the line's voltage; the higher the voltage, the stronger the electric field. But this field is also inversely proportional to the distance from the conductors. That is, the electric field strength decreases as the distance from the conductor increases.

2.2.10 Galloping

Galloping is a phenomenon where transmission conductors vibrate with large amplitudes and usually occurs when steady, moderate wind blows over a conductor covered with ice. Ice build-up makes the conductor slightly out-of-round irregular in shape leading to aerodynamic lift and conductor movement. Such movement of conductors results in:

(a) contact between phase conductors or between phase conductors and ground wires resulting in electrical short circuits
(b) conductor failure at support point due to violent dynamic stress caused by galloping

During galloping, conductors oscillate elliptically at frequencies of 1 Hz or less. Shorter spans, usually less than 600 ft (183 m), are anticipated to gallop in a single loop configuration; longer spans are expected to gallop in double loops. Overlapping of these ellipses means possible conductor contact which must be avoided. Therefore, adequate clearance must be maintained between phase wires and between phase and ground wires, to prevent loop contact. Another way to address galloping problems is to adjust the wire tensions to an optimum level so that sag and lateral movement are minimal. Shorter line spans or advanced conductors such as T2 (Twisted Pair) may also help in reducing galloping effects. Anti-galloping devices are also used on existing lines to mitigate galloping issues.

2.2.11 Ampacity

Ampacity of a conductor is the maximum current in amperes the wire can carry at its maximum design temperature. This rating is a measure of the conductor's electrical performance and thermal capability. The maximum conductor design temperature for sags and clearances is also the line's maximum operating temperature (MOT) for Ampacity. Although the MOT is the primary criteria governing these ratings, wind, ambient temperature and sun (solar heating) conditions are also considered in the calculation of Ampacity. IEEE Standard 738 (2013) provides excellent guidance for determining Ampacity Ratings for conductors.

For example, according to RUS Bulletin 200, the Ampacity ratings for Drake ACSR conductor (795 kcmil 26/7), which is very popular in North America, are: 972 Amps (summer) and 1257 Amps (winter), both calculated at an MOT of 212°F (100°C) and with a wind speed of 2 fps (0.61 meters/sec) and at ambient temperatures of 104°F (summer) and 32°F (winter).

2.3 STRUCTURAL DESIGN OF TRANSMISSION LINES

2.3.1 Structure spotting

Prior to any design, the engineer must establish the alignment of the transmission line taking into consideration the various criteria. These criteria include property boundary issues, soil conditions, road and other clearances, regulatory issues, environmental concerns, costs, impact on public property, aesthetics and construction access. Potential structure locations are identified along the chosen alignment using structure spotting process. Structure spotting is the design process which determines the height, location and type of consecutive structures on the Plan and Profile (P & P) sheets. The potential points are identified based on terrain, expected access issues and land use. However, structure's height and strength requirements are not considered as primary criteria. Survey coordinates of the points will be input into computer programs such as PLS-CADD™ (2012) to generate the terrain showing graphical view of the transmission line. Structures with known strength ratings (i.e.) maximum allowable span for a given configuration and height are used alternatively to 'spot' structures to make best use of their height and strength.

Among the key factors that impact structure spotting are vertical and horizontal clearances, structure capacity, insulator swing, conductor separation, galloping and uplift. Both manual and computerized spotting processes are available. In the case of manual spotting, structure strength is expressed in terms of maximum allowable Weight and Wind Spans.

The weight span (also called vertical span) is the horizontal distance between lowest points on the sag curve of two adjacent spans. The wind span (sometimes

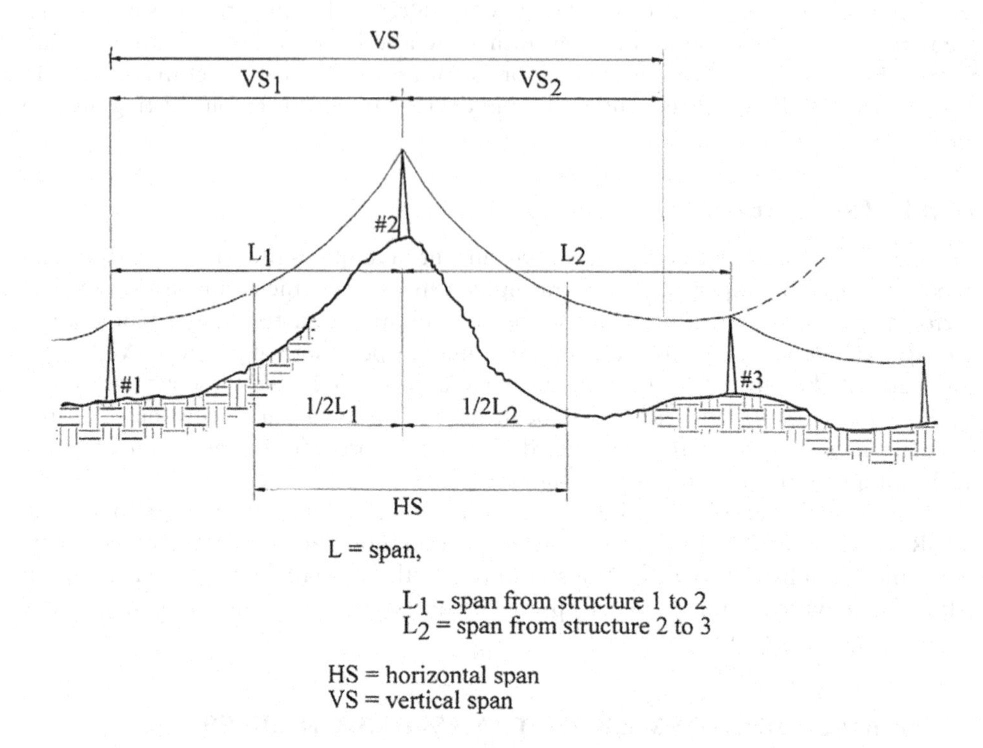

Figure 2.5 Definition of Wind and Weight Spans (Source: RUS/USDA).

called horizontal span) is the horizontal distance between the geometrical midpoints of the adjacent spans. Figure 2.5 shows the definitions of wind and weight spans.

2.3.2 Ruling spans

The concept of Ruling Span (RS) is used in the design and construction of a transmission line to provide a uniform span length representative of the various lengths of spans within a tension section (i.e.) between deadends. A ruling span is an assumed design span which approximates the mechanical performance of given tension section. This span allows sags and tensions to be defined for structure spotting and conductor stringing.

Mathematically, the Ruling Span is defined as:

$$\mathrm{RS} = \sqrt{[L_1^3 + L_2^3 + L_3^3 + \cdots + L_n^3]} / \sqrt{L_1 + L_2 + L_3 + \cdots + L_n} \tag{2.5}$$

where:

$L_1, L_2, L_3, \ldots, L_n$ = lengths of different spans in the tension section.

Example 2.3 A tension section between two deadends of a 161 kV transmission line contains the following spans: 920 ft (280.4 m), 1060 ft (323.1 m), 1010 ft (307.8 m) and 870 ft (265.2 m). Determine the Ruling Span.

Solution:

$RS = \{[920^3 + 1060^3 + 1010^3 + 870^3]/[920 + 1060 + 1010 + 870]\}^{0.5}$
$= 973.55$ ft (296.7 m)

2.3.3 Sags and tensions

Sags and tensions of wires are interdependent. Sags depend on the span length, tension, loading condition, type of wire (conductor, overhead ground wire or optical ground wire) and tensions in turn depend on history of stressing and wire temperature relative to a given weather condition. Long-term stress behavior includes creep which in turn affects tension. Aeolian vibrations of wires induce dynamic stresses. Extreme operating temperatures soften the conductor and induce higher sags. Therefore, the determination of conductor and ground wire sags and corresponding tensions – as a function of temperature and loading – is of fundamental importance in transmission line design. Wire stringing charts showing these sags and tensions are critical to field technicians during installation process.

Wire sags and tensions are generally determined using well known computer program SAG10™ from Southwire (2014). Alternatively, wire sags and tensions can also be calculated within PLS-CADD™.

The maximum anticipated sag of a conductor – usually at its highest operating temperature or heavy ice loads also helps determine the required height of a pole or structure at a location to satisfy mandated ground clearance.

Example 2.4 The maximum sag of a conductor at a particular level span in a 230 kV transmission line is calculated as 13 ft (4 m). Determine the nominal height above ground of a pole required for that location. If 11% of the total pole is embedded into ground, what is the total pole length needed? Assume a ground clearance buffer of 2.5 ft (0.76 m).

Solution:

This problem serves to illustrate the use of clearance tables discussed in Section 2.2.3.
From Table 2.7 – Minimum phase-to-phase separation for a 230 kV line = 9.1 ft (2.77 m)
Minimum separation between OHGW and phase wire = 5.9 ft (1.80 m)
From Table 2.6a-1 – Minimum Ground Clearance required = 22.4 ft (6.8 m)
With buffer: Design Ground Clearance = 22.4 + 2.5 = 24.9 ft or 25 ft (7.62 m)
Height of pole required = 5.9 + (2)(9.1) + 13 + 25 = 62.1 ft (18.93 m).
Total Length of the Pole needed = 62.1/(1.00 – 0.11) = 69.8 or 70 ft (21.34 m)
Use 70 ft (21.34 m)

Additionally, the following Sag-Tension relationships will be useful in many situations:

$$\frac{S_S}{S_R} = \frac{L_S^2}{L_R^2} \tag{2.6}$$

where:
S_S = Sag of a specific span
S_R = Sag of ruling span
L_s = Specific Span
L_R = Ruling Span

For a level span:

$$T_H = \frac{wL_s^2}{8S_m} \tag{2.7}$$

where:
S_m = Sag at mid-span
T_H = Horizontal wire tension at support, at constant temperature
w = Weight of wire, including wind, or wire per foot

Vertical separation of phase wires is a function of sags which in turn are related to spans. RUS Bulletin 200 provides an equation relating *vertical* separation at mid-span to maximum allowable span.

$$L_{mcs} = L_R * \sqrt{(V_R - V_S)}/\sqrt{(S_L - S_U)} \tag{2.8}$$

where:
L_R = ruling span in feet
L_{mcs} = maximum possible span limited by conductor separation in feet
V_R = required vertical separation at mid-span in feet
V_S = vertical separation at support in feet
S_L = ruling span sag of lower conductor without ice in feet
S_U = ruling span sag of upper conductor with ice in feet

The requirement of using iced upper and un-iced lower conductors is related to the issue of differential ice accumulation. Ice on overhead ground wires is unlikely to melt whereas ice on phase wires is prone for melting given the heat generated by passing of current. This unbalanced ice is also a condition to check separation between ground wires and energized phase wires.

Equations 2.4 and 2.6 can be combined to determine maximum allowable horizontal spans for tangent structures based on *horizontal* separation of conductors. The example below illustrates the process for an H-frame.

Example 2.5 The following data apply to a 69 kV line with a ruling span of 800 ft (243.8 m). The suspension insulators each carry 4 bells for a string length of 2.5 ft (0.76 m). Experience Factor is taken as 1.00.

Determine the maximum horizontal span limited by conductor horizontal separation.
H = Horizontal separation of phase wires = 10.5 ft (3.2 m)
Ruling Span Sag = 20.4 ft (6.22 m) at 60°F final
ϕ = Swing angle under 6 psf (290 Pa) wind = 59.4 deg. ($\sin\phi = 0.861$)

Solution:

Substituting these values in Equation 2.4:

$10.5 = (0.025)(1.05)(69) + (1)(\sqrt{S_f}) + (2.5)(0.861)$ or $S_f = 42.72$ ft (13.02 m)

Using this value in Equation 2.6:
$(L_{max}/800)^2 = (42.72/20.4) = 2.09$ or $L_{max} = 1157$ ft (352.9 m)

Example 2.6 Given a 138 kV structure with a phase separation at pole of 9.64 ft (2.94 m), a ruling span of 800 ft (243.8 m) with the following additional data:

Sag of lower conductor at without ice = 18.7 ft (5.7 m)
Sag of upper conductor at with ice = 22.4 ft (6.83 m)
Required separation at mid-span = 5.7 ft (1.74 m)

Determine maximum span limited by conductor vertical separation.

Solution:

From Equation 2.8:

$L_{mcs} = (800)(\text{sqrt}[(5.7–9.64)/(18.7–22.4)]) = 825.5$ ft (251.6 m)

2.3.3.1 *Galloping*

The issue of galloping was discussed in Section 2.2.10 where wire tension control is noted as one of the methods of controlling galloping vibrations. Usually, for long spans in excess of 600 ft (183 m), the optimum wire tension is often governed by galloping movement defined by double loop ellipses. This tension is determined in a trial-and-error procedure by adjusting the tension so that the ellipses do not touch each other. At angles and deadends, a factored value of this wire tension is used to design the structure itself; therefore, it is important to evaluate the impact of galloping while designing transmission lines with large spans.

2.3.3.2 *Tension limits*

Throughout the life of a transmission line, conductors are subject to a variety of mechanical and climactic loading situations including wind, ice, snow and temperature variations. NESC and RUS Bulletin 200 therefore specify tension limits for conductors and shield wires which refer to state of loading, temperature and climactic parameters. Table 2.14 shows these combined limits.

These limits generally refer to ACSR conductors. For other types of conductors, manufacturer's guidelines must be followed.

Table 2.14 Recommended Wire Tension Limits.

	Tension Limit (% of Breaking Strength)			
Tension Condition	Conductors	Overhead Ground Wires (HS)	Overhead Ground Wires (EHS)	Notes
1. Maximum Initial Unloaded	33.3[1]	25	20	Must be met at the design temperature specified for the loading district
2. Maximum Final Unloaded	25	25	20	
3. Standard Loaded (usually NESC District Loads with a load factor of 1.0)	50[2]	50[2]	50[2]	
4. Maximum Extreme Wind	70[3,4]	80	80	Usually taken as 60°F
5. Maximum Extreme Ice	70[4]	80	80	Usually taken as 15°F
6. Extreme Ice with Concurrent Wind	70[3,4]	80	80	Usually taken as 15°F

(Courtesy: RUS/USDA).
[1]35% per NESC
[2]60% per NESC
[3]80% per NESC
[4]For ACSR conductors only.

Initial Unloaded Tension refers to the conductor as it is strung initially before any ice or wind is applied.

Final Unloaded Tension refers to the state of the wire after it has experienced ice and wind loads, long term creep and permanent inelastic deformation.

Standard Loaded situation refers to the conductor state when it is loaded with simultaneous ice and wind per NESC loading districts as defined in Section 2.1.2.

Extreme Wind Tension is the tension when wind is acting on the conductor as defined in Section 2.1.1. No ice is allowed on the wire for this condition.

Extreme Ice Tension is the tension when the conductor is loaded with specified amount of radial ice as defined in Section 2.1.3. No wind is allowed on the wire for this condition. This load case is outside of NESC requirements.

Extreme Ice with Concurrent Wind refers to the situation where extreme ice on the wire is accompanied by a moderate amount of wind as defined in Section 2.1.3.

Tension in the wires is also a criterion to determine whether vibration dampers are needed or not (see Section 5.1.1.12 for details).

2.3.4 Insulators

Insulators are an integral component of the mechanical system defining a transmission structure and are employed in several basic configurations: post or suspension, angle and deadend. All insulators specified for a structure must consider both mechanical strength as well as electrical characteristics.

Post insulators are used where insulator swing is not permitted and are usually subject to cantilever forces. Suspension (and angle) insulators are subject to vertical loads due to wire weight and tensile loads due to line angles and are rated by their tensile strength. Deadend insulators carry direct wire tensions and are selected on the basis of tensile strength. NESC recommends the following allowable percentages of strength ratings for line post insulators:

Cantilever 40% (Ceramic and Toughened glass) & 50% (Non Ceremic)
Tension 50%
Compression 50%

For suspension type insulators, NESC recommends allowable strength rating of 50% of combined mechanical and electrical strength in case of ceramic and toughened glass insulators. For non-ceramic insulators such as polymer insulators, it is 50% of the specified mechanical load. This recommendation applies only for the case of combined ice and wind district loading (Rule 250B) with load factors of unity.

These percentages are applied to the insulator manufacturer ratings which are different for ceramic and non-ceramic insulators. For post and braced post insulators, manufacturers also provide interaction strength diagrams showing combined effects of simultaneous application of vertical, transverse and longitudinal loads. For porcelain or glass suspension insulators, the ultimate strength rating is usually denoted as "Combined Mechanical and Electrical" whereas for polymer insulators the terms "Specified Mechanical/Cantilever/Tensile Load" are used, as appropriate.

2.3.5 Hardware

All components defining a transmission structure also contain various connecting hardware and must meet or match the strength ratings of the main connected part. For example, the anchor shackles and/or yoke plates used to connect deadend insulator strings to the structure must withstand the loads imposed via the insulator string. This also applies to all other associated accessories such as conductor and ground wire splices, suspension and deadend clamps and other hardware.

2.3.6 Guy wires and anchors

The design of guyed structures involves both guy wires as well as anchors that transfer wire load into the ground. Guy wires are generally galvanized steel (often the same stranding and rating as the overhead ground wire) or aluminum-clad steel. Strength ratings apart, corrosion resistance is also often considered while selecting a guy wire. A wide variety of anchoring systems are available to choose from: log anchors, plate anchors, rock anchors, helical screw anchors etc., each with a specific set of usage and design criteria. The nature of soil/rock at the location plays an important role in anchor design.

As with insulators, limits are placed on allowable loads on guy wires and anchors. Typical allowable percentages of strength ratings for guys and anchors are:

Guy Wire 90% (NESC)
Anchor 100% (NESC)
Guy Wire and Anchor 65% (RUS)

In situations where the guy wire may be in close proximity to energized conductors, a guy insulator is often used on the guy wire to provide additional protection. For visual safety, the bottom portion of all guy wires are typically enclosed in colored PVC markers, usually yellow, to render them visible during night time.

2.4 STRUCTURAL ANALYSIS

The utility industry now uses tubular steel poles as well as steel lattice towers, spun pre-stressed concrete poles, wood poles and composite poles. Simple suspension (tangent) and small-angle single pole structures can be quickly analyzed using spreadsheets. The advent of powerful digital computers and software such as PLS-POLE™, TOWER™ and PLS-CADD™, now enables accurate modeling and analysis of not only individual structures of a transmission line but also the *entire transmission line* in a single modeling session. POLE and TOWER also have provisions for input of foundation capacities to facilitate fuller representation of the structure. Plan and Profile drawings can now be digitally processed, printed and saved in various formats and sizes.

2.4.1 PLS-POLE™

PLS-POLE™ is a powerful structural analysis and design program for transmission structures. The program is capable of handling wood, laminated wood, steel, concrete and composite structures and performs design checks of structures under specified loads. It can also calculate maximum allowable wind and weight spans to aid in structure spotting. Virtually any transmission, substation or communications structure can be modeled, including single poles, H-Frames and A-Frames. These models can be rapidly built from components such as pole shafts, davit arms, cross arms, guy wires, X- and V-braces and all types of insulator configurations.

Pole shaft databases for standard classes of steel poles from various suppliers are built into the program and cover both galvanized and weathering steel poles. The component databases that are used with PLS-POLE™ include various steel shapes (round and multi-sided pole shafts), wood, laminated wood, concrete and composite poles of various classes, guy wires and insulators etc. Custom elements can be added by the user as needed.

PLS-POLE™ is capable of performing both linear and nonlinear analyses. Nonlinear analysis allows 2nd order or P-Delta effects, helps to detect instabilities and to perform accurate buckling checks. The program can perform steel pole design checks based on ASCE 48-11 or other specified codes. The current version of the program has built-in checks for various international codes. For poles used as communication

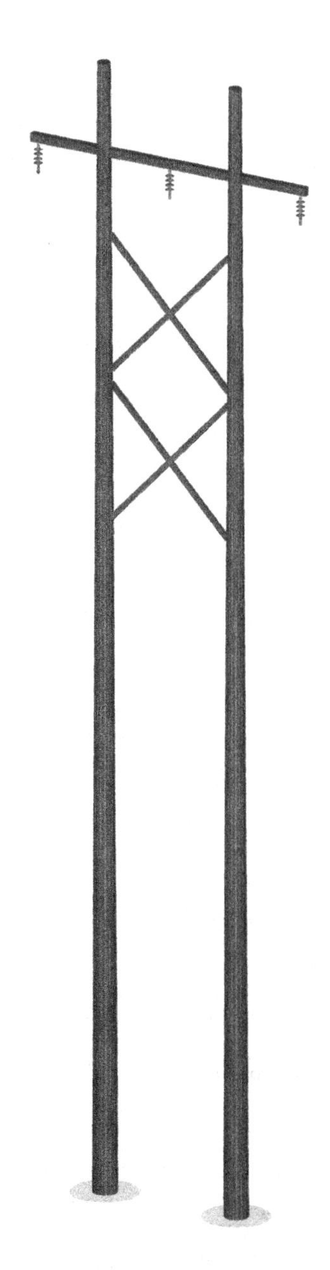

Figure 2.6a PLS-POLE Model of a Braced H-Frame.

structures carrying antennae, the program also has a design check option using ANSI TIA-222 (2006) standards.

Structures assembled in PLS-POLE™ can be exported to PLS-CADD™ and installed in the line's 3-dimensional model. Figures 2.6a, 6b and 6c show renderings of typical structure models developed with PLS-POLE™.

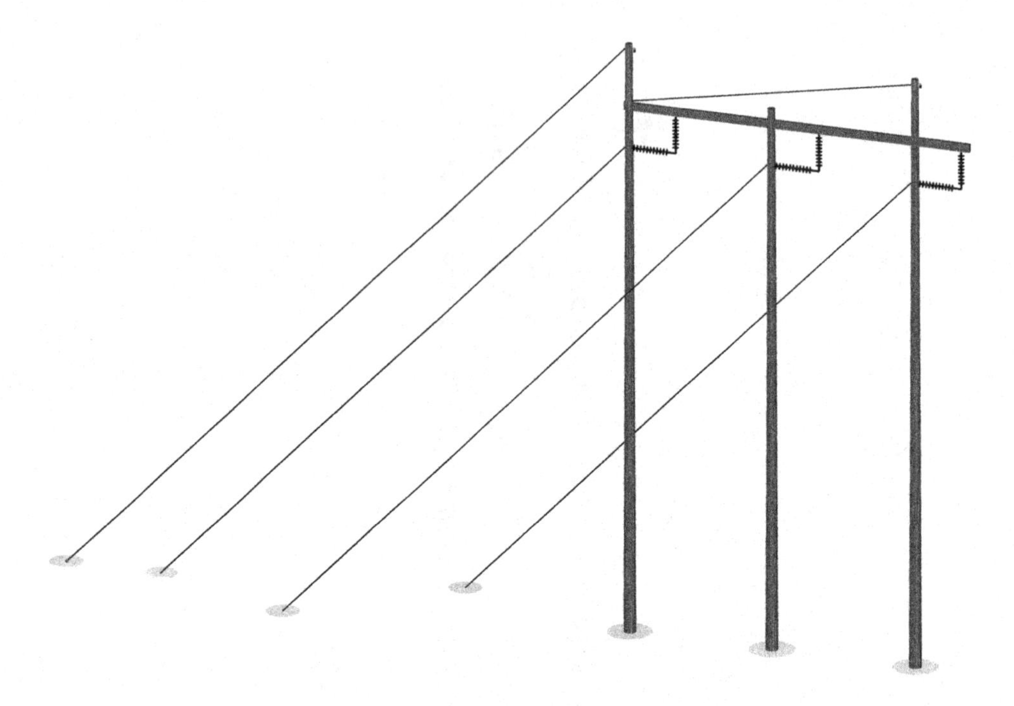

Figure 2.6b PLS-POLE Model of a 3-Pole Deadend.

2.4.2 TOWER™

TOWER™ is a powerful structural analysis and design program from Power Line Systems (PLS) for lattice steel transmission towers based on a 3-dimensional finite element analysis scheme for axially-loaded trusses. Both self-supported and guyed towers can be modeled. The program performs design checks of structures under user specified loads. For a given lattice tower structure, it can also calculate maximum allowable wind and weight spans and interaction diagrams between different ratios of allowable wind and weight spans.

The program is capable of performing both linear and non-linear analysis and facilitates detailed input (angle members, bolts, bolt configurations, member slenderness ratios, connection eccentricity and restraint etc.) It calculates the forces in the members and components and compares them against calculated capacities for the selected code or standard. Overstressed members easily identified graphically and in the TOWER™ output reports.

TOWER™ is capable of performing design checks based on various world standards: ASCE Standard 10-15, ANSI/TIA-222, Canadian Standard CSA S37, ECCS, CENELEC, AS 3995, BS 8100 and others.

The component databases that are used with TOWER™ include various steel angles (equal and unequal legs, single and double angles), tower bolts of various specifications and insulators etc. Unique shapes (flat bars, channels, T-sections for example)

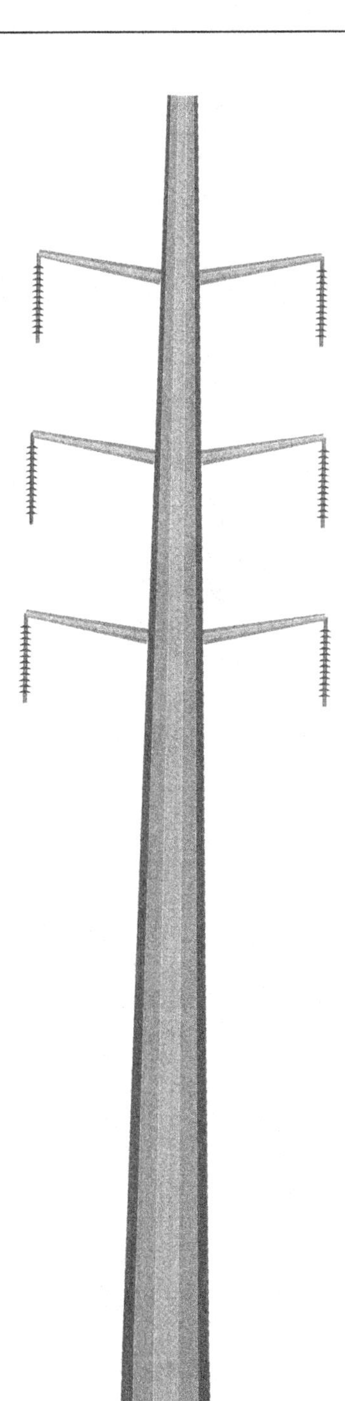

Figure 2.6c PLS-POLE Model of a Steel Pole.

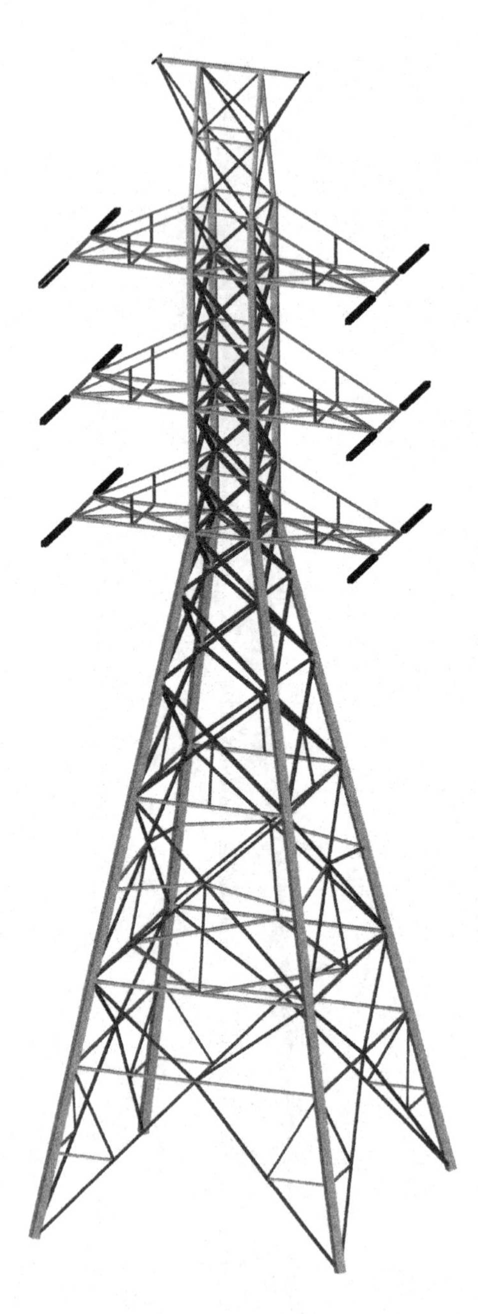

Figure 2.7 TOWER Model of a Lattice Tower.

can be created where necessary. Structures assembled in TOWER™ can be exported to PLS-CADD™ and installed in the line's 3-dimensional model.

Figure 2.7 shows rendering of a double-circuit deadend tower developed with TOWER™.

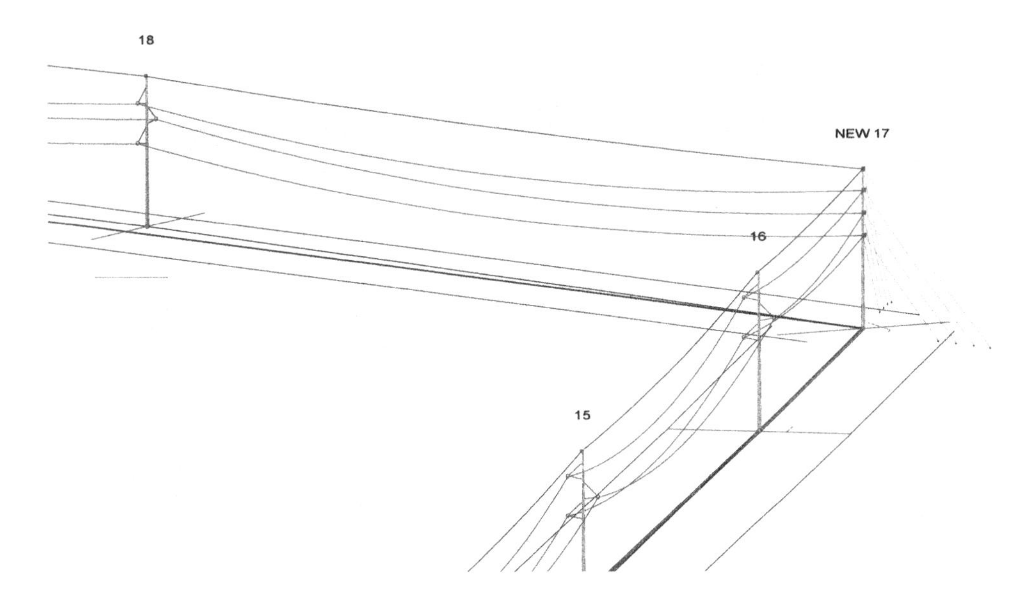

Figure 2.8 PLS-CADD Model of a Transmission Line.

2.4.3 PLS-CADD™

PLS-CADD™ is a line modeling, analysis and design program for transmission lines and structures. This program in integrated with its sister modules PLS-POLE and TOWER. A typical 3-dimensional view of a line segment showing single poles with braced post insulators is shown in Figure 2.8. A Plan and Profile drawing generated using the software is shown in Figure 2.9.

2.4.4 Load and strength criteria

The fundamental design philosophy behind current transmission structure design is that all structures shall be designed and detailed in such a way so as to sustain imposed factored design loads without excessive deformations and stresses. The objective is to design a structure with resistance exceeding the maximum anticipated load during its lifetime and to produce a structure with an acceptable level of safety and reliability. Mathematically,

$$(\emptyset)(R) \geq (\gamma)(P) \tag{2.9}$$

where:
$\emptyset$ = Strength Factor (SF)
R = Strength Rating of Component
γ = Load Factor (LF) for specified load
P = Applied Load

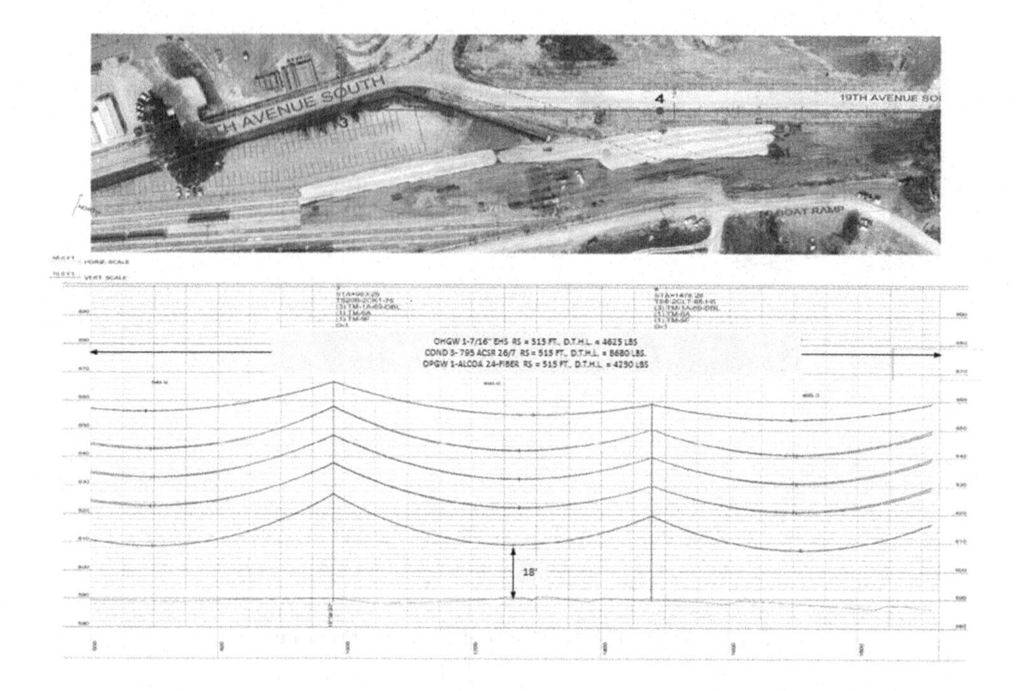

Figure 2.9 Typical Plan and Profile of a Transmission Line.

The strength factor 'Ø' limits the resistance 'R' and accounts for the variability of the resistance property. The load factor LF accounts for the uncertainty of the given load and/or simplifying assumptions made in the analysis.

This concept is similar to the LRFD (Load and Resistance Factor Design) used in general structural design where the erstwhile deterministic approach is replaced with a more robust probabilistic approach, matching strengths with loads of various types and incorporating statistical variations associated with different load categories and strength of components.

Strength and Load factors suggested by RUS and NESC are shown in Tables 2.15a, 2.15b and 2.15c. All values refer to new, Grade B Construction which is the highest grade associated with safety and reliability per NESC. Both NESC and RUS also provide guidelines for strength factors for replaced or rehabilitated structures. The strength factors of RUS are somewhat conservative and are often adopted by utilities in rural areas and whose lines and structures are designed and built according to RUS/USDA guidelines.

Grades of construction

NESC defines three grades of construction, namely, Grades B, C and N. Grade B is the highest or the type of construction with stringent strength standards while N is the lowest grade. Grade B standards are typically used by US utilities for transmission lines.

Table 2.15a RUS Strength and Load Factors (Grade B New Construction).

Structure or Component with Rule 250B	*Strength Factor* (ϕ) (Note 3)
Steel and Prestressed Concrete Structures	1.00
Wood and Reinforced Concrete Poles (Note 4)	0.65
Wood Crossarms and Cross Braces (Note 4)	0.50
Fiber-Reinforced Polymer Poles, Cross Arms and Braces	1.00
Guy Wire Assemblies (Note 1)	0.65
Guy Anchors and Foundations	0.65
Guy Attachment Assemblies (Note 2)	0.65
Conductor Support Hardware (Note 6)	1.00
Structure or Component with Rule 250C and 250D	*Strength Factor* (ϕ) (Note 3)
Steel and Prestressed Concrete Structures	1.00
Wood and Reinforced Concrete Poles (Note 4)	0.75
Wood Crossarms and Cross Braces (Note 4)	0.65
Fiber-Reinforced Polymer Poles, Cross Arms and Braces	1.00
Guy Wire Assemblies (Note 1)	0.65
Guy Anchors and Foundations	0.65
Guy Attachment Assemblies (Note 2)	0.65
Conductor Support Hardware (Note 6)	0.80
Loading Component with Rule 250B	*Load Factor (LF)*
Vertical Loads	1.50
Transverse Loads	
Wind	2.50
Wire Tension	1.65
Longitudinal Loads @ Crossings and Elsewhere	
General	1.33
Deadends	1.65
Loading Component with Rule 250C and 250D	*Load Factor (LF)*
Vertical Loads	1.10
Transverse Loads	
Wind	1.10
Wire Tension	1.00
Longitudinal Loads @ Crossings and Elsewhere	
General	1.00
Deadends	1.10

Note: Rule 250B refers to NESC District Loading (Light, Medium, Heavy or Warm Islands)
Rule 250C refers to NESC Extreme Wind Loading
Rule 250D refers to NESC Extreme Ice and Concurrent Wind Loading
1. A value different than 0.65 may be used, but should not exceed 0.90.
2. This strength factor of 0.65 may be increased for steel and prestressed concrete poles.
3. It is recognized that structures will experience some level of deterioration after installation. These strength factors are for new construction.
4. For wood structures, when the deterioration reduces the structure strength to 2/3 of that required when installed, the wood structure should be replaced or rehabilitated. If the structure or structure component is replaced, the structure or structure component needs to meet the strength for the original grade of construction. The rehabilitated portions of the structures have to be greater than 2/3 of that required when installed for the life of the line.
5. When calculating the additional moment due to deflection, deflections should be calculated using loads prior to application of the load factor.
6. Conductor Support Hardware is any hardware not a part of the structure, guy assembly, or guy attachment. Conductor support hardware may be splices, extension links, insulator string yokes, y-clevis balls, ball hooks, deadend clamps, etc.
(Source: RUS/USDA Bulletin 200).

Table 2.15b NESC Load Factors[1] (Grade B New Construction).

Loading Component with Rule 250B	*Load Factor (LF)*
Vertical Loads[2]	1.50
Transverse Loads	
Wind	2.50
Wire Tension	1.65[1]
Longitudinal Loads	
In General	1.10
At Deadends	1.65[1]
Loading Component with Rule 250C and 250D	*Load Factor (LF)*
250C Wind Loads	1.00
All Other Loads	1.00
250D All Loads	1.00

Note: Rule 250B refers to NESC District Loading (Light, Medium, Heavy or Warm Islands)
Rule 250C refers to NESC Extreme Wind Loading
Rule 250D refers to NESC Extreme Ice and Concurrent Wind Loading
[1] For guys and anchors associated with structures supporting communication conductors and cables only, this factor may be reduced to 1.33.
[2] Where vertical loads significantly reduce stress in a structure member, a vertical load factor of 1.00 should be used for the design of such member. Such member shall be designed for the worst case loading.
(Courtesy: IEEE NESC ®C2-2012) National Electrical Safety Code

2.4.4.1 *Structural design criteria*

Overhead transmission lines are continuous structural systems with supporting elements (poles, towers, frames) and cable elements (conductors and shield wires). To ensure safe and reliable service, these are typically designed by considering the various load conditions listed in Table 2.16.

2.4.4.2 *Weather cases*

A comprehensive analysis and design of a transmission line involves various weather cases related to both the performance of the line itself (conductors and insulators) as well as the supporting structures (poles, frames and towers). A collection of these weather cases is called "Criteria File" which is a critical input to line design software PLS-CADD^{TM}. Tables 2.17a and 2.17b show the full list of weather cases that are typically considered in transmission design in the United States. Several parameters such as wind pressures and ice thickness magnitudes vary from utility to utility and the numbers provided in the table are for illustration purpose only. The case labeled Warm Islands is applicable only for islands located from 0 to 25 degrees latitude, north or south. This zone covers Hawaii, Guam, Puerto Rico, Virgin Islands and American Samoa. The '*k*' factor is explained below.

2.4.4.3 *Load cases*

Table 2.18 gives a typical list of load cases needed for transmission structure analysis and design. These loads are applied on the structures, supported wires, hardware,

Table 2.15c NESC Strength Factors (Grade B New Construction).

Structure[1] or Component with Rule 250B	*Strength Factor (φ)*
Metal and Prestressed Concrete Structures, Crossarms and Braces[6]	1.00
Wood and Reinforced Concrete Structures, Crossarms and Braces[2,4]	0.65
Fiber-reinforced Polymer Structures, Crossarms and Braces[6]	1.00
Support Hardware	1.00
Guy Wire[5,6]	0.90
Guy Anchors and Foundation[6]	1.00
Structure[1] or Component with Rule 250C and 250D	*Strength Factor (φ)*
Metal and Prestressed Concrete Structures, Crossarms and Braces[6]	1.00
Wood and Reinforced Concrete Structures, Crossarms and Braces[3,4]	0.75
Fiber-reinforced Polymer Structures, Crossarms and Braces[6]	1.00
Support Hardware	1.00
Guy Wire[5,6]	0.90
Guy Anchors and Foundation[6]	1.00

Note: Rule 250B refers to NESC District Loading (Light, Medium, Heavy or Warm Islands)
Rule 250C refers to NESC Extreme Wind Loading
Rule 250D refers to NESC Extreme Ice and Concurrent Wind Loading
[1]Includes Pole
[2]Wood and reinforced structures shall be replaced or rehabilitated when deterioration reduces structure strength to 2/3 of that required when installed. When new or changed facilities modify loads on existing structures, the required strength shall be based on revised loadings. If a structure of component is replaced, it shall meet the strength required by this table. If a structure or component is rehabilitated, the rehabilitated portions of the structure shall have strength greater than 2/3 of that required when installed.
[3]Wood and reinforced structures shall be replaced or rehabilitated when deterioration reduces structure strength to 3/4 of that required when installed. When new or changed facilities modify loads on existing structures, the required strength shall be based on revised loadings. If a structure of component is replaced, it shall meet the strength required by this table. If a structure or component is rehabilitated, the rehabilitated portions of the structure shall have strength greater than 3/4 of that required when installed.
[4]Where a wood or reinforced concrete structure is built for temporary service, the structure strength may be reduced to values as low as those permitted by footnotes 2 and 3 provided the structure strength does not decrease below the minimum required during the planned life of the structure.
[5]For guy insulator requirements, see Rule 279 of NESC.
[6]Deterioration during service shall not reduce strength capability below required strength.

insulators and other equipment. The philosophy behind each of the load cases is explained below. The reader must note the additional NESC Factor '*k*' applied for the NESC district load cases only. This is an arbitrary factor added to the resultant of the vertical and horizontal loads as shown in the figure at the bottom of the Table 2.18.

NESC district loadings

These 4 cases – also called District Loadings – are included to meet the requirements of Rule 250B of NESC and include the mandated climactic and load factors. The 'k' factor is an additional load item applied to the resultant of the conductor vertical and wind load components while determining the sag and tension of the wire. Only the structural design of angle and deadend structures and tangent structures located on

Table 2.16 Load Conditions Considered in Design of a Transmission Line.

No.	*Type of Loads*	*Load Case Description*
1	Weather	Extreme Wind in Any Direction
2		Extreme Ice Combined with Reduced Wind
3		Unbalanced Ice without Wind
4		Reduced Ice Combined with Substantial Wind
5	Failure Containment	Broken Conductors or Ground Wires
6	Construction and Maintenance	Stringing of Wires
7		Structure Erection
8	Legislated	NESC-mandated Cases

(with permission from ASCE)

Table 2.17a Weather Cases for a Typical PLS Criteria File.

No.	*Description*	*Wire Temp. (°F)*	*Wind Pressure (psf)*	*Ice Thickness (in)*	*Constant k (lb/ft)*	*Remarks*
1a	NESC Heavy with *k*	0	4	½	0.30	Basic Load Cases for Structure Design
1b	NESC Medium with *k*	15	4	¼	0.20	
1c	NESC Light	30	9	0	0.05	
1d	NESC Warm Islands	50	9	0	0.05	
1e		15	4	¼	0.20	
2	NESC Extreme Wind[1]	60	20.7[2] (for 90 mph)	0	0	
3	Extreme Ice	32	0	1	0	
4	NESC Extreme Ice with Concurrent Wind	15	4	1	0	
5a	Insulator Swing – No wind	60	0	0	0	Weather Cases for Insulator Swing
5b	Insulator Swing – Moderate Wind	0	6	0	0	
5c	Insulator Swing – High Wind	60	20.7	0	0	
6	Galloping (Swing)	32	2	½	0	Weather Cases for Galloping Checks
7	Galloping (Sag)	32	0	½	0	

Note: 1 psf = 47.88 Pa, 1 inch = 25.4 mm, 1 mph = 1.609 kmph, 1 lb/ft = 14.594 N/m, deg C = (5/9)(deg F-32)
[1]All structures, including those below 18 m (60 ft) in height, shall be checked for Extreme Wind condition without any conductors with wind acting in any direction.
[2]Wind pressure must be calculated using Equation 2.1a and including all applicable parameters.

Table 2.17b Weather Cases for a Typical PLS Criteria File (cont'd).

No.	*Description*	*Wire Temp. (°F)*	*Wind Pressure (psf)*	*Ice Thickness (in)*	*Constant k (lb/ft)*	*Remarks*
8	Low Temp	−10	0	0	0	Uplift Cases
9	Low Temp	−20	0	0	0	
10	T-0	0	0	0	0	
11	T-32	32	0	½	0	Conductor Separation; Differential Ice Loading
12	T-50	50	0	0	0	Various Wire
13	T-60	60	0	0	0	Installation
14	T-70	70	0	0	0	Temperatures
15	T-80	80	0	0	0	
16	T-90	90	0	0	0	
17	T-100	100	0	0	0	
18	T-120	120	0	0	0	
19	T-167	167	0	0	0	High Temp.
20	T-212	212	0	0	0	Transmission Sag
21	T-392	392	0	0	0	(Clearance Purposes)

Note: 1 psf = 47.88 Pa, 1 in = 25.4 mm, 1 mph = 1.609 kmph, 1 lb/ft = 14.594 N/m, deg C = (5/9)(deg F-32)

the line angle are impacted by the 'k' factor because the tension of the wire is affected by the factor.

The designations of *Heavy*, *Medium*, *Light* and *Warm Islands* are based on ice and wind expected in those areas. The user may refer to NESC for geographic boundaries of these 4 zones. For a given transmission line, only one of the four zones is applicable unless the line crosses more than one zone.

Extreme wind

The purpose of this NESC load is to ensure that the structure is capable of withstanding high winds that may occur within the geographic territory. As mentioned earlier, the wind force used is based on the fastest 3-second gust at 33 ft (10 m) above ground. Per NESC, this load is currently applied to structures over 60 ft (18 m) in height. However, RUS Bulletin 200 recommends all structures be checked for Extreme Wind regardless of height. Maps showing design wind speeds are provided by NESC as well as RUS Bulletin 200. NESC also requires all structures (irrespective of height) to be checked for extreme wind applied in any direction on the structure without conductors.

Extreme ice

This load case considers the possibility of extreme ice storm or a storm that develops icing conditions. Usually this case is defined by accumulation of radial ice of 1 in (25.4 mm) thickness on conductors; but often utilities located in icing regions adopt a higher value of 1.5 in (38.1 mm) or more.

Table 2.18 Typical Load Cases for Analysis and Design.

Load Case	*Wind Speed, mph (kmph)*	*Wind Pressure, psf (Pa)*	*NESC Constant 'k' (lb/ft)*	*Temperature (°F)*	*Ice Thickness, Inches (mm)*
NESC Heavy	40 (64)	4 (190)	0.30	0	½ (12.5)
NESC Medium	40 (64)	4 (190)	0.20	15	¼ (6.5)
NESC Light	60 (96)	9 (430)	0.05	30	0
NESC Warm Islands	60 (96)	9 (430)	0.05	50	0
	40 (64)	4 (190)	0.20	15	¼ (6.5)
NESC Extreme Wind	90 (144)[1]	20.7 (991)[1]	0	60	0
Extreme Ice	0	0	0	32	1.0 (25)
NESC Extreme ice with Concurrent Wind	*40 (64)*	4 *(190)*	0	15	1.0″ *(25)*
Construction	*28 (44.8)*	2 *(96)*	0	60	0
Broken Wires	40 *(64)*	4 *(190)*	0	0	½ *(12.5)*
Failure Containment[2]	40 *(64)*	4 *(190)*	0	0	½ *(12.5)*
Uplift	0	0	0	−10 to −20[3]	0
Deflection	*28 (44.8)*	2 *(96)*	0	60	0

[1]Values shown are typical. For other wind speeds, use appropriate values.
[2] All wires are cut on one side of the structure.
[3] Cold curve.

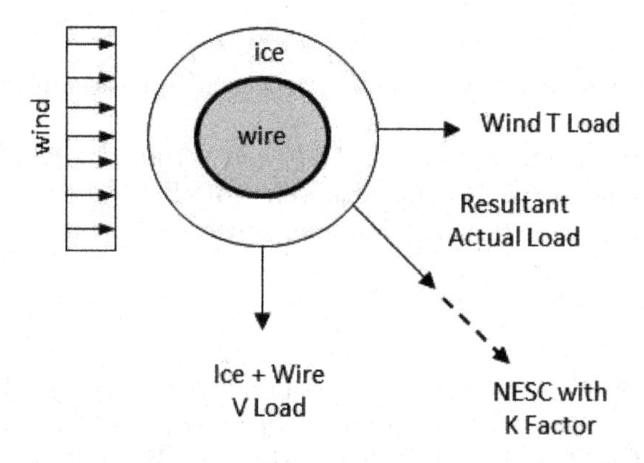

Definition of NESC Constant *'k'*.

Extreme ice with concurrent wind

The intent of this NESC load case is to design a structure for extreme ice accompanied by wind. Per NESC, this load is currently applied to structures over 60 feet (18 m) in height. Since ice can stay on conductors for 4 to 5 days and may see subsequent wind, a 40 mph (64 kmph) wind at 4 psf (190 Pa) wind is shown in Table 2.18; this is used to satisfy ASCE Manual 74 requirement that wind supplementing ice must be equal to about 40% of extreme wind case. Maps showing uniform ice thickness with concurrent wind speeds are provided in NESC as well as RUS Bulletin 200. These maps show uniform ice thickness typically ranging from ¼ inch (6.35 mm) to 1 inch (25.4 mm) and concurrent wind speeds ranging from 30 mph (48.3 kmph) to 60 mph (96.5 kmph), equivalent to a wind pressure of 2.3 psf to 9.2 psf (110 Pa to 440 Pa).

In all NESC load cases in which wind is included, the horizontal wind pressure is applied at right angles to the direction of the line, except where wind is applied in all directions without wires. Also, NESC does not consider ice on structures and wind-exposed surface areas.

Construction

This load case is to ensure the structural integrity of not only the main structure but also the arm or steel vang supporting the insulator/stringing block and arm strength during wire tensioning. This is because one of the worst loadings that a given arm or conductor attachment point will endure is during the stringing of the conductors. A small wind is also considered in this case. This load is applied as additional vertical and horizontal loads to the phase that induces the highest structural stresses with all conductors installed and stringing in the last conductor. A typical example will be tensioning at the structure tensioner down slope (1:1).

Broken wires

The idea behind including this case is to ensure that in the event any phase wire or overhead ground wire fails, the failure does not cause any additional damage to the structure or lead to a cascading type of line failure. Another motivation is related to the cost and availability of replacement structures and the long lead times for fabrication. Designing a structure for potential broken wire cases and the slight cost associated with it is well worth considering that removal of a line from service, even temporarily, is avoided. This is critical for lattice transmission towers carrying HV and EHV circuits.

Broken wire loads are generally applied at selected wire location (conductor or shield wire) that induces the greatest stress in the structure. More than one load case may be needed if the highest stress location is not readily apparent. ASCE Manual 74 provides more information on design longitudinal loading on structures, historically called everyday wire tension or broken wire load.

Failure containment

This load case is designed to reduce potential for catastrophic failures. Severe climactic conditions often produce a cascading type of failure where one structure fails and collapses completely. This subjects adjoining structures with severe unbalanced loads – much more than what they were designed for – upon which they collapse in a domino-like sequence. Deadends and heavy angles, and in some cases tangent structures too, are subjected to such a load criteria. Containment loading is characterized by the absence of ALL wires on one side of the structure. The conditions that define this load case are usually set by the utility based on their judgment.

ASCE Manual 74 also provides guidelines for longitudinal loading on structures including a procedure based on Residual Static Load (RSL), which is a final effective static tension in a wire after all the dynamic effects of a wire breakage have subsided.

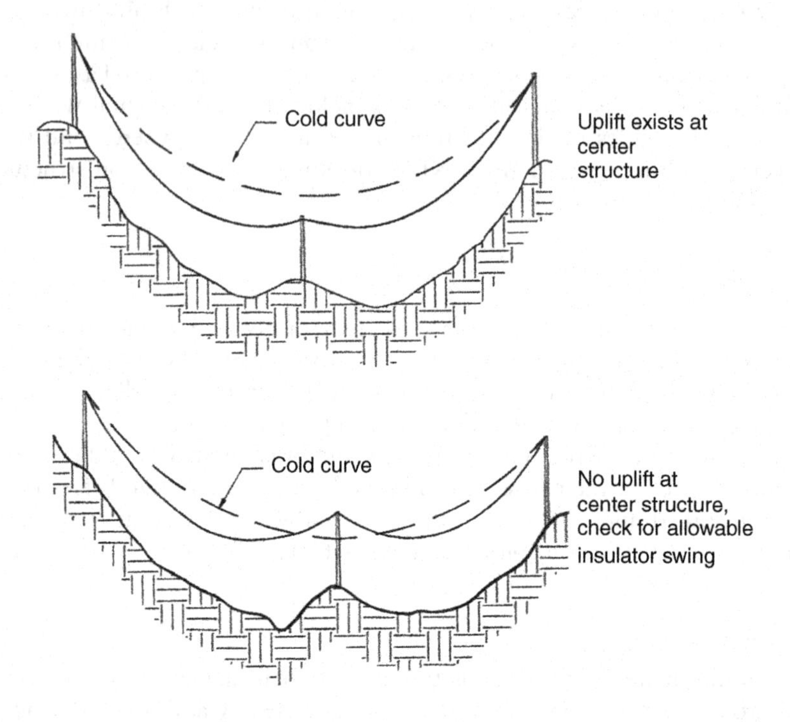

Figure 2.10 Uplift Condition (Source: RUS/USDA).

Uplift

Uplift is defined as negative vertical span. Figure 2.10 depicts the situation where uplift occurs in a transmission line. On steeply inclined spans in hilly terrains, when the cold sag curve shows the low point to be above the lower support structure, the conductors in the uphill span exert upward forces on the lower structure. The magnitude of this force at each attachment point is related to the weight of the loaded conductor from the lower support to the low point of sag. This uplift force is more pronounced at colder sub-zero temperatures. Uplift must be avoided for suspension, pin-type and post insulators. A quick check for uplift can be made using a sag template (see Chapter 5).

Deflection

In addition to the above loading cases, designers also often check situations such as Deflection Loading where tangent poles and frames are designed to limit the pole top deflection to a specified level, usually to 1% to 2% pole height above ground for a given loading. This ensures adequate stiffness of the structure to limit flexural deformations and thereby help keep insulators plumb and clearances intact. The climactic conditions for this load case usually include the average annual ambient temperature for the structure geographical location.

2.5 FOUNDATION DESIGN CRITERIA

Each structure must be securely embedded or anchored into the ground and facilitate safe transfer of structure loads to the ground strata below. To determine foundation requirements, the engineer must first evaluate the nature and condition of the soil in the vicinity of the structure. The choice of eventual foundation type will further depend on geotechnical characteristics of strata underneath, structure material, configuration, loads, constructability and economy. A majority of tangent poles are directly-embedded into ground; but systems defined by large lateral forces and moments require concrete drilled shafts whose design is a bit more labor-intensive.

Design criteria for foundations depend on type of soil and loads imposed. Design loads are usually factored reactions obtained from structural analysis from computer programs. In lattice towers, the loads transmitted are primarily compression and uplift loads. For single pole structures, the loads transmitted are overturning moment, shear (lateral) and axial loads. Where moment foundations such as drilled shafts are required, it is important to specify allowable deflection or rotation criteria, both elastic and non-recoverable. Chapter 4 will contain more information on this issue.

As with structures, foundations are also designed for given Strength Factors shown in Tables 2.15a and c. However, unlike structures, guidance in the area of foundation design is not well laid out by the codes and standards. Therefore it is common practice to use internal design criteria for foundations. These criteria vary from utility to utility. The structural design of all reinforced concrete systems is in general governed by the ACI 318 (2011). For drilled shafts with a diameter greater than 30 in (0.76 m), ACI-336-3R (1998) is recommended.

Computer programs such as CAISSON™ (2011), LPILE™ (2015) or MFAD™ (2015) are used to quickly size a drilled shaft or directly-embedded pole under moment loads. Drilled shafts under moment loads are primarily designed as laterally-loaded piles. Guyed systems require anchors to transmit guy wire loads to the ground. Anchor types range from classical (log) to grouted (rock) to helical (screw) with widely-varying holding strengths depending on site-specific soils. If soil data is available, helical anchors are evaluated using HeliCAP™ (2007).

Geotechnical properties of ground strata are an integral part of the design criteria. In general, properties such as allowable bearing capacity, unit weight (dry, moist and submerged), friction angle, cohesion, subgrade modulus and others are required. Chapter 4 will discuss these issues in greater detail.

2.6 CONSTRUCTABILITY

Constructability refers to the "readiness to be built" and to the process of reviewing and ensuring that a particular transmission line project *can be* actually built as designed, given the numerous technical and non-technical parameters that control the construction phase. This "review" looks at the construction constraints, environmental barriers, political issues, risks and acceptance to local community.

2.6.1 Construction considerations

The primary construction-related issue is availability of access roads, both temporary and permanent. These roads are used to move crew, vehicles, materials and equipment in and out of the construction site. Their secondary purpose is to provide maintenance access to the structures for repair and to the ROW in general. Access needs also vary depending upon the type of structure (pole, frame or lattice tower), function (tangent, angle or deadend) and foundation (grillage, direct embedment, concrete shafts or guy anchors). These considerations in turn will dictate the type and number of vehicles and equipment that must reach the work site. Use of cranes and helicopters for heavy construction impose additional challenges.

The work areas for conductor pulling and tensioning equipment must be integrated into access road plans. However, these activities are performed only during construction and therefore the access is temporary.

Constructability assessment also involves evaluating the process of hauling poles to the site. Concrete poles are about four to five times heavier than steel structures are difficult to handle and move. Lattice steel towers, which are typically set on concrete shaft foundations, require the most concrete at each tower site. If unguyed poles are used at line angles, this mandates drilled shafts due to large base forces.

Most utilities will have in-house procedures for inspection, assessment and maintenance and these are likely to dictate their choice of a structure type for a specific line. Steel or concrete structures require minimal maintenance while wood systems need frequent inspections and condition monitoring.

Because of difficulty procuring ROW or easements, and obtaining permits for new lines, many utilities strive to improve their future options by selecting structure types for current projects that will permit easy upgrading or uprating initiatives.

2.6.2 Environmental constraints

Topographic characteristics of the terrain often provide the most difficult construction challenges. Areas with steep slopes, erodible soils, sensitive streams, wetlands, restricted lands and habitats impose constraints on the process. Construction activities in the neighborhood of federally-protected lands and wildlife refuges require a wide range of government permits and adherence to agency-imposed work schedules.

2.6.3 Regulatory issues

Any potential crossing of public (state and federal) lands invariably leads to regulatory requirements, consultations and approvals. Prior to construction, the project may require several permits at various levels. For example, road and highway departments require permits for transmission line construction where the line crosses a highway or expressway. Areas attached to sensitive natural resources and recreational usage demand additional regulatory compliances.

2.6.4 Public acceptance

Any project involving transmission line construction must take into account its impact on the residents of the area it covers and possible public opposition to the presence

Table 2.19 Selected World Design Codes, Standards and Manuals.

Title	*Published By**	*Country*
ASCE Manual 74 Guidelines for Electrical Transmission Line Structural Loading	ASCE	USA
National Electrical Safety Code ANSI C2	IEEE/ANSI	
Standard 738 for Calculating the Current-Temperature Relationship of Bare Overhead Conductors	IEEE	
ASCE Standard 48-11 (previously 72) Design of Steel Transmission Pole Structures	ASCE	
ANSI O5.1 Specifications and Dimensions for Wood Poles	ANSI	
ASCE Standard 10-15 Design of Latticed Steel Transmission Structures	ASCE	
ASCE Manual 123 Prestressed Concrete Transmission Pole Structures	ASCE	
Manual of Steel Construction – Allowable Stress Design	AISC	
Bulletin 1724E-200 Design Manual for High Voltage Transmission Lines	RUS/USDA	
ASCE Manual 91 Design of Guyed Electrical Transmission Structures	ASCE	
Standard 691 Guide for Transmission Structure Foundation Design and Testing	IEEE	
Standard 524 Guide to the Installation of Overhead Transmission Line Conductors	IEEE	
ECCS Recommendation for Angles in Lattice Transmission Towers	ECCS	Europe
EN 10025 European Structural Steel Standard	CCI	UK
BS EN 50341 Overhead Electrical Lines exceeding AC 1 kV – Part 1 General Requirements, Common Specifications	ECES	UK
Eurocode 3 – Design of Steel Structures	ECCS	Europe
CIGRE Technical Brochure 207 Thermal Behaviour of Overhead Conductors	CIGRE	France
British Standard 8100 – Lattice Towers and Masts Code of Practice for Strength Assessment of Members of Lattice Towers and Masts	BS	UK
AS 3995 Design of Steel Lattice Towers and Masts	SA	Australia
AS/NZS 7000 Overhead Line Design – Detailed Procedures	SNZ	New Zealand
ESAA C(b) 1 Guidelines for Design and Maintenance of Overhead Distributions and Transmission Lines	ESAA	Australia
ENATS 43-125 Design Guide and Technical Specification for Overhead Lines Above 45 kV	ENA	UK
Technical Specifications 43-8, Overhead Line Clearances		UK
TC-07 Electricity Transmission Code	ESCSA	Australia
TP.DL 12.01 Transmission Line Loading Code	TP	New Zealand
IEC 60826 Design Criteria of Overhead Transmission Lines	IEC	Switzerland
IEC 61284 Overhead Lines – Requirements and Tests for Fittings	IEC	Switzerland
PN-90 B-03200 Konstrukcje Stalowe Obliczenia Statyczne I Projektowanie (Steel Structures Design Rules)	PS	Poland
CSA S37 Antennas, Towers and Antenna-Supporting Structures	CSA	Canada
CSA C22.3 1-15 Overhead Systems		
IS 802 Use of Structural Steel in Overhead Transmission Line Towers – Code of Practice	Bureau of INS	India
IS 800 Code of Practice for General Construction in Steel		
IS 4091 Code of Practice for Design and Construction of Foundations for Transmission Line Towers and Poles		
IS 398 Aluminum Conductor for Overhead Transmission Purposes – Specification		

*See *List of Abbreviations.*

of large structures, noisy construction equipment as well as potential disruption to regular community activities. Most utilities conduct meetings to inform the public of their intended projects and seek their input and cooperation.

2.7 CODES AND STANDARDS FOR LINE DESIGN

The design of transmission line structures involves both analysis of the structural system – wires, supports and foundations – as well as design checks ensuring compliance with established norms and guidelines. Worldwide, most countries have institutions and standards devoted to design, construction and safe operation of utility structures. These standards or codes cover a variety of parameters such as loading, conductors, materials (steel, wood and concrete), hardware, clearances, performance and other issues relevant to the system. Table 2.19 shows a selected, partial listing of some of the codes, manuals and design standards used in various countries. As mentioned in the Preface, the basic principles of transmission line structural analysis are more or less the same all over the world; but different regions impose different rules and regulations, mostly associated with their local experience, climate, economy and safety and reliability requirements. Despite that, these codes and standards have a universal purpose: to ensure that transmission structures and foundations are designed for safe and reliable operation during their lifetime.

As mentioned in the previous chapter, transmission line and structure design in United States is primarily governed by the NESC supported by RUS Bulletin 200 in various ways. It must be noted that the NESC is primarily a safety code which is not intended as a design specification or an instruction manual. Along with many other standards noted in the text at various locations, the above two references constitute the main sources of design guidelines discussed in this book.

PROBLEMS

P2.1 Determine adequate ROW for a 230 kV line for the following data.
Horizontal distance between insulators = 28 ft (8.53 m)
Suspension Insulator Length = 8 ft (2.44 m)
Sag at 60°F at 6 psf (290 Pa) wind = 10 ft (3.05 m)
The swing angle of the insulator under 6 psf wind = 45°
The required C per internal standards = 13 ft (3.96 m) away. Neglect pole deflections.

P2.2 Re-do the problem in P 2.1 assuming horizontal post insulators. Use A = 14 ft (4.27 m). What is the basic difference between the two cases? Discuss.

P2.3 A tension section of the above 230 kV transmission line contains the following spans: 1020 ft (310.9 m), 1090 ft (332.2 m), 1130 ft (344.4 m) and 1070 ft (326.1 m). Determine the Ruling Span.

P2.4 For a tangent structure in a 138 kV line, assuming maximum conductor sag of 9 ft (2.74 m), determine the minimum height of a pole required if 13% of the total length is used for ground embedment. Assume a ground clearance buffer of 3.5 ft (1.06 m).

P2.5 Re-do the problem in Example E2.5 assuming horizontal phase separation of 12.5 ft (3.81 m). What is your basic inference from the two cases? Discuss.

P2.6 Re-do the problem in Example E2.5 assuming a 161 kV H-frame structure with a horizontal phase separation of 15.5 ft (4.72 m). Assume ruling span of 1150 ft (350.5 m) and corresponding sag of 31.6 ft (9.63 m). Insulator string has 10 bells with a total length of 5.33 ft (1.62 m).

P2.7 Repeat Example E2.6 for a 69 kV structure. Use $V_R = 3.3$ ft (1.0 m) and $V_S = 7.6$ ft (2.32 m). Ruling span is 800 ft (243.8 m) but sags are now as follows: lower wire's 20.4 ft (6.22 m) and upper wire's 24.7 ft (7.53 m).

Chapter 3

Structural analysis and design

Transmission structures, whether wood, steel, lattice, concrete or composite, are one of the most visible elements of an electrical power transmission system. Regardless of material, they serve a single purpose: supporting the insulators, wires and equipment. The choice of a particular type of structure – material and configuration – for a transmission line, however, depends on several factors such as electrical, spatial, structural and economy.

Electrical factors include voltage, number of circuits (single or multiple, including distribution under-build), conductor bundles (single or multiple), communication cables or optical ground wires (OPGW), insulation, lightning protection and grounding.

Spatial constraints involve allowable spans, ROW (Right-of-Way) issues as well as vertical and horizontal clearances mandated by codes for the project. ROW is a function of voltage; so are the clearances required for wires above ground and above other specified wires. The adopted conductor configuration – horizontal, vertical or Delta – influences the nature of clearances. Conductor separation also plays a critical part in determining phase spacing and therefore, pole or structure height.

Structural factors involve strength and stiffness of the structure, its components and foundations in resisting the loads applied on them without excessive stress and deformation. Structure strength is traditionally indicated by horizontal and vertical span (HS and VS) limits, which are in turn dependent on strength of pole material, knee braces, cross braces, cross arms and uplift. Additionally, the ratio of HS/VS is limited by insulator swing (for tangent structures with suspension insulators). Finally, majority of transmission structures in USA must, as a minimum, be designed to the applicable NESC (National Electrical Safety Code) or RUS (Rural Utilities Service) loading guidelines (or utility-specified norms exceeding NESC) and these include climactic loads due to ice, wind and temperature. Some utilities are also specifying special HIW (High Intensity Wind) loading for lattice towers in areas of hurricanes, tornadoes and downbursts.

Economy is associated with the issues of constructability, erection techniques, inspection, assessment and maintenance. As discussed in Chapter 2, accessibility for construction must be considered while deciding on structure types. Mountainous or swampy terrains make access difficult for construction vehicles and use of specialized equipment or a helicopter may be required. Guyed structures also create construction difficulties since a wider area must be accessed to install guys and anchors often impacting the ROW and land use restrictions.

This chapter looks at various types and configurations of structural systems adopted for transmission lines in North America, materials, computer modeling, analysis and design, hardware and development of design drawings. Also discussed are the processes of selecting structural components and compliance with different code regulations for ensuring safety, integrity and strength of structures.

3.1 STRUCTURE MATERIALS

This section covers the structural and material properties of various elements that comprise a typical transmission structure. The three basic construction materials for transmission structures are wood, steel and concrete, although fiber glass (composite) poles and cross arms are being increasingly used for both transmission and distribution lines. Composite cross arms are also becoming popular for distribution under-build circuits on transmission poles given the ease of installation and structural strength.

Structure costs usually account for 30% to 40% of the total cost of a transmission line. Therefore, selecting an optimum structure and material becomes an important part of a cost-effective design. A structure study is usually performed to determine the most suitable structure configuration and material based on cost, construction and maintenance considerations, along with electric and magnetic field effects. Indirect considerations include material availability, environmental issues and other local constraints.

Key factors to consider when evaluating structure configuration:

- Depending on voltage and other factors, a horizontal phase configuration often gives the lowest structure cost.
- If ROW costs are high or if the width of ROW is limited, a vertical configuration may give a lower line cost.
- Horizontal configurations, however, may require more tree clearing.
- Vertical configurations may have a narrower ROW but structures are generally taller and may have aesthetic objections from public. Also, foundations may be costlier.
- From electric and magnetic field perspectives, a vertical configuration will have lower field strengths at the edge of ROW than horizontal configurations. Delta circuits will have lowest single-circuit field strengths.
- If H-Frames are considered as an option, both single and double X-braced configurations must be evaluated as they sustain larger spans and thus help reduce the number of structures per mile of line.

Key factors to consider when evaluating material choice:

- Pole type structures (wood, concrete, composite or steel) are generally used for voltages less than 230 kV; steel poles and lattice towers are preferred for higher voltages.
- For relatively smaller spans and lower voltages, wood structures are economical.
- In areas subject to severe climactic loads or on lines with bundled conductors, wood and concrete are uneconomical; steel structures provide a cost-effective option.

- If large, unbalanced longitudinal loads are specified as part of the loading scheme, guyed pole systems or lattice towers are better suited for this application.
- For voltages under 230 kV, wood H-Frames will usually give the lowest initial installed cost when larger level spans (up to 1000 ft or 300 m) are used.

The choice of a particular structure material often depends on the design spans of the line. Figure 3.1 shows the relationship between common transmission structure materials and design spans. It must be noted that the word 'poles' includes both single poles as well as multiple-pole systems such as H-Frames.

Figures 3.2, 3.3 and 3.4 show the classifications of structure types based on materials, function and support configuration. The advantages and disadvantages of various materials and structure types are shown in Table 3.1.

3.1.1 Wood

Wood poles used for transmission purposes are generally made of Douglas Fir or Southern Yellow Pine material with a designated fiber (bending) strength or Modulus of Rupture (MOR) of 8000 psi (55.2 MPa). They are directly embedded into the ground to a specified depth. In single poles, design is governed by bending at ground line and setting depth needed to resist lateral overturning forces. For wood cross arms, bending stress is generally limited to 7400 psi (51 MPa). Modulus of Elasticity 'E' usually varies

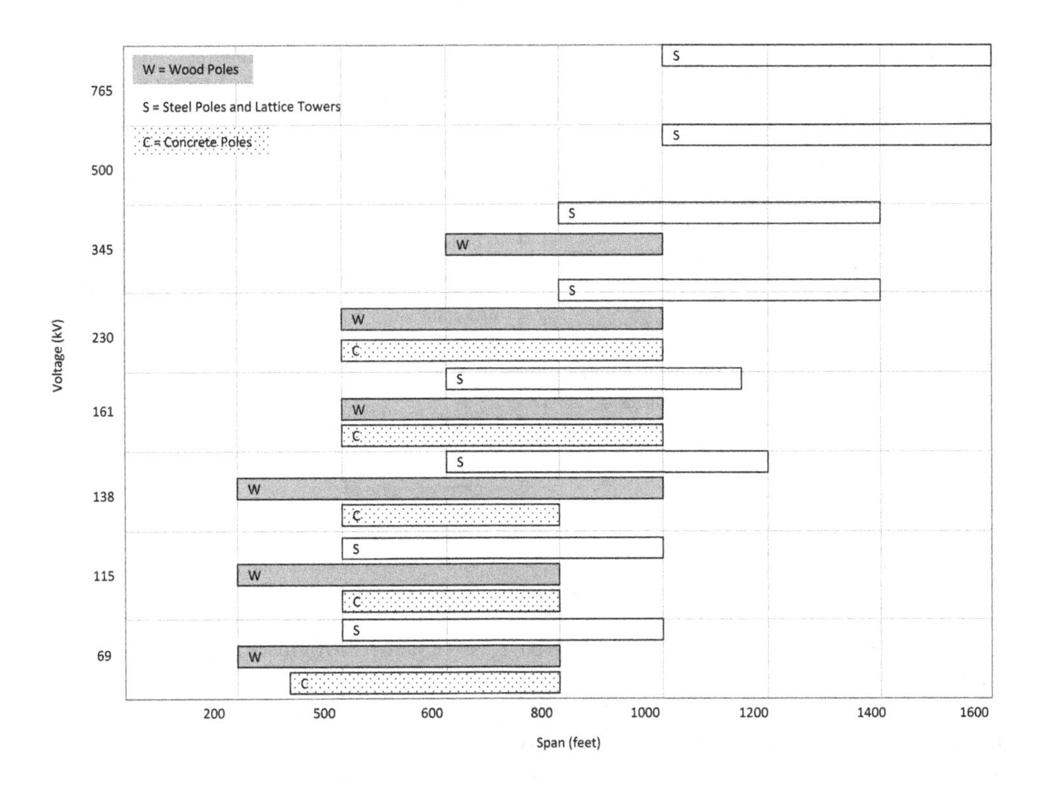

Figure 3.1 Span Lengths and Structure Materials.

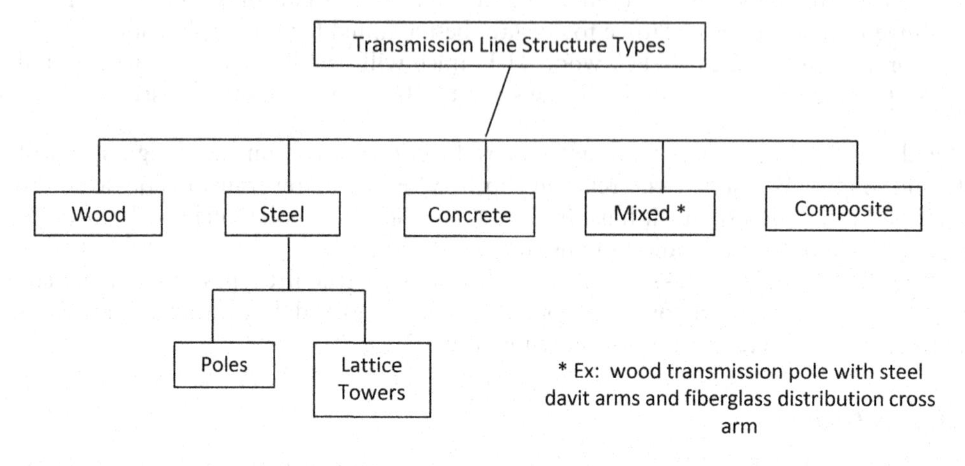

Figure 3.2 Structure Types – Based on Material.

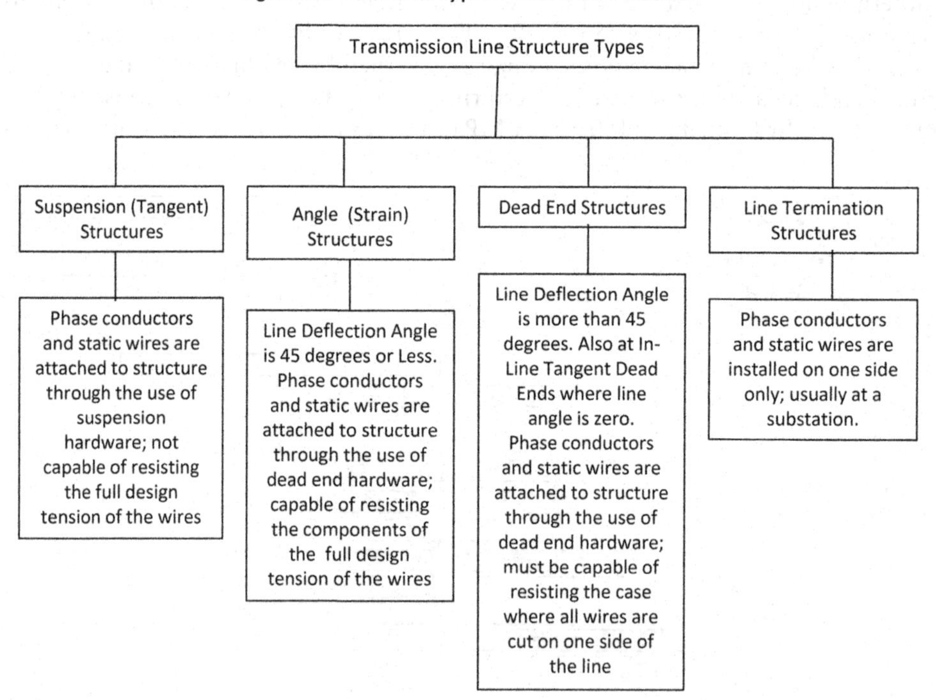

Figure 3.3 Structure Types – Based on Function.

from 1,800 ksi for Southern Yellow Pine to 1,920 ksi for Douglas Fir (12.40 GPa to 13.23 GPa).

Wood poles are used in single-pole, two-pole (H-Frames) and three-pole (angles and deadends) configurations, in both guyed and unguyed applications. Common transmission voltage range is 69 kV to 230 kV although they have been also used in 345 kV systems.

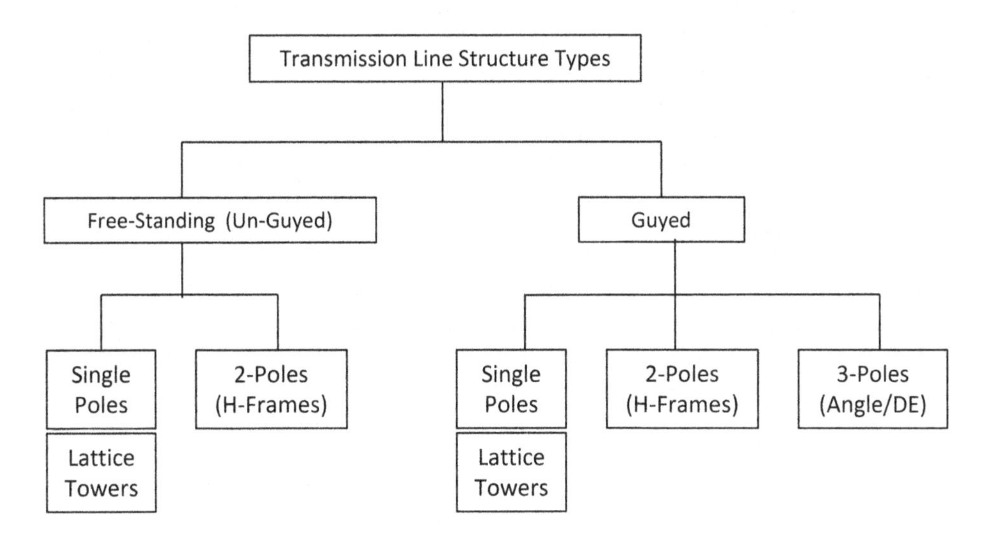

Figure 3.4 Structure Types – Based on Configuration.

Table 3.1 Structure Type Comparison.

Structure Type	*Advantages*	*Disadvantages*
Steel Lattice Towers	Most Versatile, Most Economical†	Aesthetic objections, needs large base area
Steel Pole Structures	Minimum Ground Area, Aesthetically Appealing	Angle poles must be often guyed or will be very heavy with large foundations
Wood Pole Structures	Low Initial Cost, Flexible Application	Short Life, High Maintenance, Limited in Strength
Concrete Pole Structures	Low Maintenance, Non-Corrosive	Very Heavy, Difficult to transport over large distances
Composite Structures	Low Maintenance, Non-Corrosive, Light Weight, Easy to Install	Higher Initial Cost**
Guyed Structures*	Most economical, particularly for higher voltages	Conflicts with edge of ROW

**Guying can occur in wood, steel, concrete pole structures as well as lattice towers.*
***Relatively new; long-term performance not yet known.*
†At extra high voltage.

Structurally, ANSI O5.1 (2015) categorizes wood poles into various classes in terms of a single lateral load applied 2 ft. (61 cm) below the top of the pole (See Table A2.1, Appendix 2). These standards cover 15 pole classes and lengths up to 125 ft. (38.1 m). Class 2 and higher (H-1, H-2 etc.) wood poles with tapered circular cross section are currently being used for small (30 ft. or 9.14 m) to large heights (100 ft. or 30.5 m). Wood pole dimensions based on class and height are also shown in Table A2.4 of Appendix 2. Note that the table refers to Douglas Fir and Southern

Yellow Pine poles. The diameters at pole top and 6 ft. (1.83 m) from the bottom are shown and diameters at GL (ground line) can be interpolated from these values.

The NESC (2012), in addition to specifying design loadings, also requires all utility structures to satisfy other strength-related norms. RUS Bulletins 200 (2015) and 700 (2011) also stipulate guidelines for wood pole preservative treatment and coatings. Since wood is a bio-degradable material, it degrades and deteriorates with time. Chemical treatment of poles with preservative coatings will increase durability; treated wood poles typically last up to 40 years. From a structural perspective, strength factors are normally used in design to account for this variation and decrease in wood strength. Special care is taken with protective coatings of the embedded or buried portion of the wood poles since soil contains corrosive elements which eat away the pole material. Woodpeckers are also known to attack wood poles by punching holes and many utilities are known to specify a protective nylon mesh wrap around transmission poles. In warmer climates, termites are also a problem.

For guyed wood poles, buckling under vertical loads must be checked as well as need for special bearing pads below pole butt to distribute large axial loads over the soil layer.

The performance of wood poles as a structural material is also dependent on various other issues such as moisture content, knots, decay, time in service etc. For more information on biological characteristics of wood, the reader is referred to ANSI O5.1, textbooks on wood structures, AITC Standard 109 (2007) and RUS Bulletins 700, 701 and 702 (2011).

3.1.1.1 *Laminated wood*

Laminated wood poles are manufactured using ¾ in. (19 mm) to 1 in. (25.4 mm) thick laminations of Douglas Fir or Southern Pine, glued using a high strength adhesive and cured in controlled atmospheric conditions. The pole can be tapered in both the transverse and longitudinal directions but usually the taper is provided in one direction only. All poles have a rectangular cross section. All design and preservation standards that apply to regular wood systems also apply to laminated wood; however, ANSI O5.2 (2012), RUS Bulletin 701 (2011) and AITC Standards 109, 110 and 111 (2007, 2001 and 2005) etc. are also additionally specified. Engineered laminated (E-Lam) wood poles are currently available in size from 30 ft. to 135 ft. (9.1 m to 41.1 m) and in various classes (Laminated Wood Systems, 2012). E-Lam poles have been successfully installed on transmission and distribution lines in single pole, H-Frame, tangent, angle, deadend, guyed and unguyed configurations.

Foundations for E-Lam poles can be direct-embedded with aggregate backfill, expansive foam or concrete. In soft soils and soils with low ground water table, a corrugated steel culvert with standard backfill is used. Foundation reinforcing systems include steel bearing angles affixed to the E-Lam pole in the embedded portion to provide added lateral resistance.

3.1.2 Steel

Tapered steel poles with various cross sections (round, 8-sided or 12-sided) are currently being used for moderate to large heights 50 ft. to 150 ft. (15 m to 45.7 m).

The poles are generally made of ASTM A572 Grade 65 (galvanized) or ASTM A871 Grade 65 (weathering) steel material with a stress rating of 65,000 psi (448.2 MPa). They are generally attached to concrete pier foundations via a base plate or directly embedded into the ground to a specified depth. As with wood poles, design is governed by bending at ground line and setting depth needed to resist lateral overturning forces. Since steel is not a biodegradable material and statistical strength variation is much less than wood, no strength reduction is required. Therefore, a strength factor of 1.00 is used. For multi-sided poles, local buckling must be checked in terms of the applicable flat width-to-thickness ratios. For round poles, local buckling is referred to the diameter-to-thickness ratio D/t, where 'D' is the average diameter of the pole shaft.

The length of steel pole shafts is generally limited by several parameters such as handling size at the manufacturing facility, length of the galvanizing tank, length of the flat-bed truck required for transport and other local constraints including weight limits on highway bridges. The average piece length is about 50 ft. to 60 ft. (15.24 m to 18.3 m). Taller poles are assembled by piecing together pole sections of various lengths either by means of splices (circumferential welds or overlap) or flange plates. The overlap length in case of a slip joint is generally about 1.5 times the bottom diameter of the upper piece. If flanges are used, these flange plates are welded to the pole segment and bolted using high strength structural bolts.

An important component of a steel pole is the weld connecting the pole shaft to the flange or base plate. For strength and structural integrity, these welds are generally recommended to be full penetration welds. Specifications for steel poles and structures are given in RUS Bulletins 204 (2008), 224 (2007) and ASCE Standard 48-11 (2011). Tables A3.1 (*a* and *b*), A3.2 and A3.3 of Appendix 3 shows various steel types and fasteners used in USA and elsewhere. Table A3.4 depicts the shapes used for transmission structures along with their geometric properties useful for design checks.

Tangent poles are directly embedded into the ground with a butt plate or shoe to help transfer axial load over a larger area and thereby minimize soil stress. Large angle and deadend poles are generally attached to concrete foundations by means of base plates and anchor bolts. Typical arrangements are shown in Chapter 4. The design basically involves verifying if the depth of embedment (setting depth) is adequate for resisting the ground line moments, and shears, in case of self-supported single poles. For concrete piers, soil data is needed to check pier ground line rotations and deflections, resistance to lateral loads and skin (side) friction resistance for resisting uplift and compressive loads.

Anchor bolts transfer tensile, compressive and shear loads from the structure to the concrete shaft. Threaded re-bars are the most common type of anchor bolt material. Table A3.13 of Appendix 3 gives material data for various anchor bolt steels. Bolt diameters range from 5/8" (16 mm) to 1¾" (44.5 mm); for larger piers, #18J rebar (2¼" or 57.2 mm diameter) meeting ASTM A615 Grade 75 is used. Current standard for anchor bolts is unified under ASTM F1554, with three grades, namely, 36, 55 and 105. The tensile strength of these threaded rods range from 60 ksi to 125 ksi (414 MPa to 862 MPa). Common diameters specified for transmission-level poles are 1½" to 2¼" (38 mm to 57.2 mm). The minimum embedment (development) length of anchor bolts into concrete is discussed in Section 3.5.1.2.

3.1.2.1 Wood equivalent steel poles

Manufacturers offer steel poles which are structurally 'equivalent' to wood poles – and referring to a given equivalency ratio – to enable designers specify quick replacements for damaged wood poles. These WES (wood-equivalent-steel) poles are available in both round and 12-sided shapes as well as light and heavy duty applications. However, caution should be exercised in using and specifying such 'equivalent' poles since it is impossible to equate the steel pole and wood pole at all points along the length. Also, the differences in material and section properties will result in differences in buckling capacity, deflections, secondary moments etc. (RUS Bulletin 214, 2009).

ANSI O5.1 (2015) defines wood pole classes with reference to tip load applied at 2 ft from pole tip. Steel pole equivalencies with ANSI wood poles are based on the "Equivalency Factor" (EF) which is defined as:

$$EF = WSF/SSF \tag{3.1}$$

where:
SSF = Steel Strength Factor (usually 1.0)
WSF = Wood Strength Factor (usually 0.65 to 0.75)
For NESC District (Rule 250B) Transverse Wind loading: $EF = 0.65/1.0 = 0.65$
For NESC Extreme Wind (Rule 250C) loading: $EF = 0.75/1.0 = 0.75$
(Note that the equivalency is valid since NESC load factors are the same for both wood and steel poles).

The WES poles originally developed by steel pole manufacturers are based on the ratio of *overload factors* for wood and steel used in the older versions of NESC. For Rule 250B loading, for example, the overload factors are 2.50 (wood) and 4.0 (steel). The equivalency factor is therefore 2.50/4.0 = 0.625. The associated strength factor is 1.0 for both wood and steel.

Standard class steel poles by RUS

The classification of standard steel poles based on the Equivalency Factor is shown in Table A3.5. These classes are defined per RUS Bulletin 214 in terms of a single lateral load applied 2 ft. (61 cm) below the tip of the pole. This strength requirement of RUS steel poles also includes a specified moment capacity that must be available 5 ft. (1.52 m) from the top. Additionally, RUS assumes that the point of fixity is located at a distance of 7% of pole length measured from the bottom; the pole must develop ultimate moment capacity at this location.

Tables A3.6 to A3.12 give various design data for these standard steel poles. These poles, when specified, reduce lead times involved in bidding, design, drawing preparation and ordering of material. The reader must keep in mind that the dimensions of standard class steel poles vary slightly from manufacturer to manufacturer; values shown in the above tables are typical to the referenced fabricator.

Example 3.1 Determine the RUS Standard Class designation for a Class H1 wood pole for (a) Transverse wind load and (b) Extreme wind load.

Solution:

Class H1 wood pole is defined as a pole rated for a horizontal lateral load of 5,400 lbs. (24.03 kN) applied at 2 ft from pole tip.

(a) Equivalency Factor for transverse wind = 0.65
Required horizontal load capacity of steel pole = 0.65 * 5,400 = 3,510 lbs. (15.62 kN)
Referring to Table A3.5, the pole class that is closest to this load is S-03.5 (3,510 lbs.).
(b) Equivalency Factor for extreme wind = 0.75
Required horizontal load capacity of steel pole = 0.75 * 5,400 = 4,050 lbs. (18.02 kN)
Referring to Table A3.5, the pole class that is closest to this load is S-04.2 (4,160 lbs.).

It must be noted that absolute point-to-point wood-to-steel equivalency does not exist and the engineer is cautioned to exercise sound judgment while determining equivalency. The reader is also referred to RUS Bulletin 214 and ASCE 48-11 which explain the issues and limitations related to various equivalencies in detail. ANSI O5.1 also limits the maximum wood pole class to H6; however, several manufacturers developed WES poles even for higher classes of H7 to H10.

For higher voltages and heavier loads, standard class poles are often not adequate; in such cases, steel poles are custom-designed with larger diameters and thicknesses (See Example in Appendix 1).

3.1.3 Concrete

Spun, prestressed, high strength concrete poles are currently used as transmission poles for heights ranging from 50 ft. to 120 ft. (15.2 m to 36.6 m). Square sections are typically used for distribution lines and street lighting poles; circular sections are preferred for transmission structures. The nominal taper of the concrete pole shall not exceed 0.216 in/ft. (1.8 cm/m). Pole dimensions vary from manufacturer to manufacturer, but typical size ranges are as follows:

Top outside diameter	7 in. to 16 in. (17.8 cm to 40.6 cm)
Bottom outside diameter	18 in. to 42 in. (45.7 cm to 106.7 cm)

These hollow poles are usually directly embedded into the ground to a specified depth. For single poles, design is governed by bending at ground line and setting depth needed to resist lateral overturning forces (bending and shear) without cracking of concrete or excessive bearing pressure on soil below the pole due to pole weight. Specifications for concrete poles and structures are given in RUS Bulletins 206 (2008), 216 (2009), 226 (2007) and ASCE Manual 123 (2012).

Concrete poles are also classified into various classes in terms of a single lateral load applied 2 ft. (61 cm) below the top of the pole. Table A4.1 of Appendix 4 shows various standard concrete pole classes per RUS Bulletin 216 as well as data on common prestressing steels (Table A4.2). RUS strength requirements for concrete poles include a minimum ultimate moment capacity at 5 ft. (1.52 m) from the pole top to ensure adequate bending strength at locations of high stress. Additionally, RUS assumes that the point of fixity is located at a distance of 7% of pole length measured from the bottom; the pole must develop ultimate moment capacity at this location.

Commercial manufacturer catalogs such as StressCrete (2009) also provide important information on the geometry and sizes of spun concrete poles. Special situations demand custom-designed poles of larger heights and diameters.

The suggested minimum concrete cover for steel is ¾ in. (19 mm). The 28-day compressive strength of concrete shall not be less than $f_c' = 8500$ psi (58.6 MPa). The range of concrete strengths employed in transmission pole manufacture is 8500 psi to 12000 psi (58.6 to 82.7 MPa).

The Modulus of Elasticity (in psi) of concrete as computed from ACI 318 (2014) is $57000\sqrt{f_c'}$ for 3 ksi (20.7 MPa) $\leq f_c' \leq$ 12 ksi (82.8 MPa). Since cracking is a design issue, two moduli are generally defined: cracked and un-cracked. Nominal values often used are 6000 ksi (41.4 GPa) for un-cracked and 2000 ksi (13.78 GPa) for cracked concrete. The modulus of rupture f_r is defined as $7.5\sqrt{f_c'}$ (in psi).

Prestressing steel strands are generally 3/8 in. to ½ in. dia. high-strength galvanized wires, Grade 250 ksi to Grade 270 ksi (1722 MPa to 1860 MPa), with tensile capacity ranging from 20 kips (89 kN) to 41 kips (182 kN). For other information, the reader is referred to ASCE-PCI Guides for Concrete Poles 257 and 412 (1987 and 1997). Confinement is provided by spiral wire sizes ranging from No. 5 to No. 11 ($\frac{1}{4}$ in. to $\frac{3}{8}$ in. dia.) and minimum spacing between spirals is 1 in. (25.4 mm). Spacing should not exceed 4 in. (10.2 cm) under any circumstances. Closer spacing may be required at the pole tip and butt segments where large radial stresses occur during load transfer from strands to concrete.

3.1.4 Lattice towers

Lattice Transmission Towers with steel angle members are commonly used as line support structures for heights ranging from 50 ft. to 300 ft. (15.2 m to 91.4 m) and for spans up to 3,500 ft. (1,067 m) and more. These towers can be self-supporting or guyed, usually with square bases. The towers are generally made of ASTM A36 or A572 Grade 50 steel members, connected by bolts (via gusset plate or direct member-to-member) or rivets. Angle sizes and geometrical properties are listed in AISC Manuals, ASD (1989) or LRFD (1995) or the latest combined version of the steel manual (2013).

The types of structural steel generally used on lattice towers and bolts used in tower joints are shown in Tables A3.1a and A3.2 in Appendix 3. The commonly used fastener specifications for latticed steel towers are ASTM A394 for bolts and A563 for nuts (See Appendix 13). The value of "E", the Modulus of Elasticity of steel, is taken as 29,000 ksi (200 GPa) for all steels. All tower bolts come with a washer, nut and a lock nut.

In contrast to wood or steel pole design, lattice tower design is governed by *individual* member behavior, often involving buckling (compression) and yielding (tension) or connection failures. Member slenderness ratios play an important role in lattice tower analyses along with amount of restraint offered at member ends which depends on the number of bolts (or rivets) used in the joint. Given the 3-dimensional nature of a lattice tower, structural stability is a critical issue checked during computer modeling.

End restraints are difficult to quantify and idealize; but are critical in defining the slenderness ratio of the angles idealized as beam-columns. This issue will be discussed in greater detail in Section 3.4.5.

3.1.5 Composite

Although fiberglass composite poles are becoming increasingly popular for a variety of reasons, one product that is now increasingly used are the composite cross arms in both transmission and distribution structures. Composite cross arms are widely used on distribution structures, both in tangent as well as deadend applications. The components are also integrated into PLS-POLE™ library for ready use.

Composite materials in general are non-isotropic and their elastic properties vary based on the direction and orientation of the constituent fibers with reference to applied loads. They are also dependent on type of epoxy bonding materials used in construction. This requires that the non-isotropic nature be considered in structural analysis. To facilitate easier analysis, some simplifications are made. One such simplification is the use of "bulk" material properties which represent the global response of the structure to a given loading. These bulk properties are determined through testing and theoretical calculations.

Appendix A14 contains information on composite poles as well as material properties useful for computer modeling. Care must be used in adopting values for computer models.

3.2 STRUCTURE FAMILIES

All transmission lines consist of structures mostly of the same material but of different sizes, insulator, line angle and loading configurations. These can be grouped into a "family" of structures specific for the project. Once the basic structure type (usually tangent) has been established, the "family" can be set up by adding systems of other configurations, specifically angle and deadend structures. In other words, the structure family basically contains tangent, angle (small, medium, large) and deadend structures.

For example, if a single wood pole with horizontal post insulators is chosen as the primary tangent structure for a specific voltage, then the extended "family" would consist of the following guyed structures to complement the tangent system:

1. Small Angle Structure for line angles 0° to 20°
2. Medium Angle Structure for line angles 21° to 45°
3. Large Angle/Deadend Structures for line angles 46° to 90°

Similarly, if a wood H-Frame with suspension insulators is chosen as the primary tangent structure, then the extended "family" would consist of the following guyed structures which complement the tangent system:

1. 3-Pole Small Angle Structure for line angles 0° to 20°
2. 3-Pole Medium Angle Structure for line angles 21° to 45°
3. 3-Pole Large Angle/Deadend Structures for line angles 46° to 90°

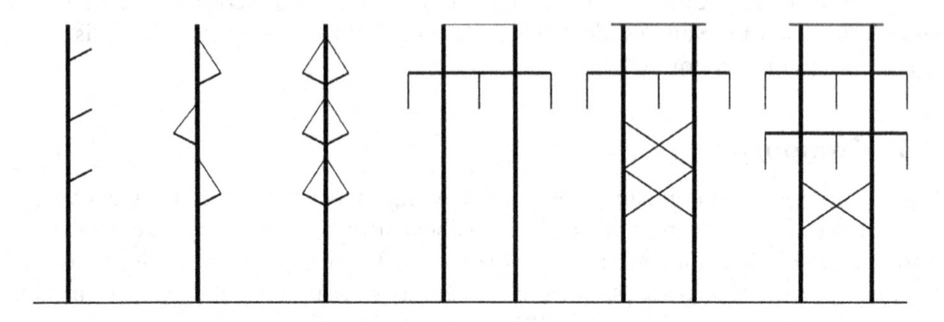

(a) 69 kV to 230 kV Single & Double Circuit Tangent Structures

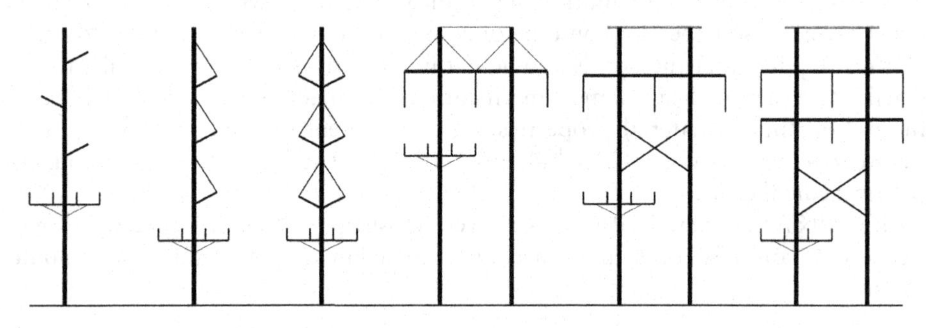

(b) 69 kV to 230 kV Single & Double Circuit Tangent Structures – With Distribution Under-Build

Figure 3.5 Structure Families – Based on Usage and Wire Configuration (Tangent).

PLS-POLE™ contains a library of standard RUS structure families and models for ready use. These families cover single pole systems, H-Frames and 3-pole systems for various voltages. The structures refer to RUS/USDA Bulletin 810 (1998) up to 69 kV and Bulletin 811 (1998) up to 230 kV. The user can also adopt a particular RUS model as a base and modify it to suit the requirements of the project, for voltages above 230 kV.

Structure families based on usage and wire configuration are shown in Figures 3.5 and 3.6 while Figure 3.7 depicts a typical family of H-Frames wood structures designed for various voltages up to 345 kV. Note that the footprint of the structure (i.e.) width of ROW needed, increases with increasing voltage. Figure 3.8 shows two- and three-pole guyed structures employed at running angles and deadends.

3.2.1 Structure models

In the US, most transmission structures are currently modeled on PLS family of programs. Models can be created to the level of detail required by the engineer. Where available, assembly units can be directly incorporated into the model (example: standard cross arms, pole top angle, davit arms etc.). For realistic representation, the nature of the connection of the unit to the pole must be correctly modeled. For instance, cross

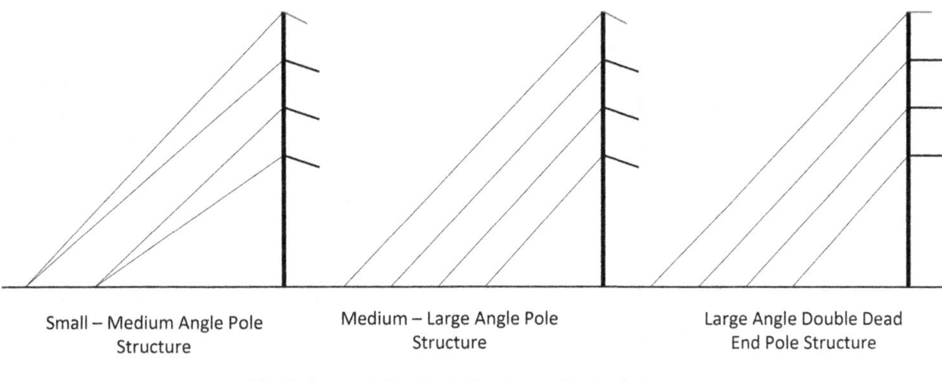

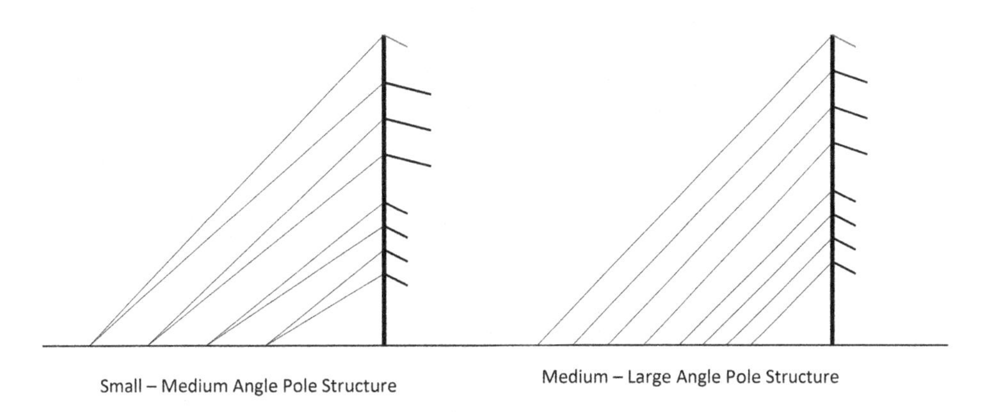

Figure 3.6 Structure Families – Based on Usage and Wire Configuration (Angle/DE).

arms are affixed to the poles of an H-Frame with through bolts, which offers partial fixity.

If the PLS Family of programs are used, models can be created on PLS-POLE™ or TOWER™ for analysis as a single independent structure or for future export to the PLS-CADD™ program. For the latter option, the set and phase numbers assigned to each wire/insulator must be clearly defined.

3.2.1.1 *Insulator attachment to structure*

The attachment of insulators to various structures depends on the structure configuration as well as material. Hardware associated with insulator connections is a function of pole material. Also, the end fittings of the insulators depend on what type of attachment is being sought. As mentioned earlier in Chapter 2, the effective length of the

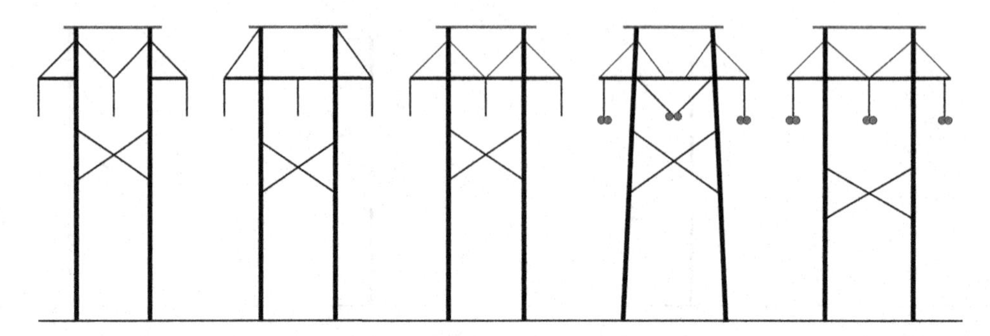

(a) 69 kV to 345 kV Single and Double Conductor Tangent Structures

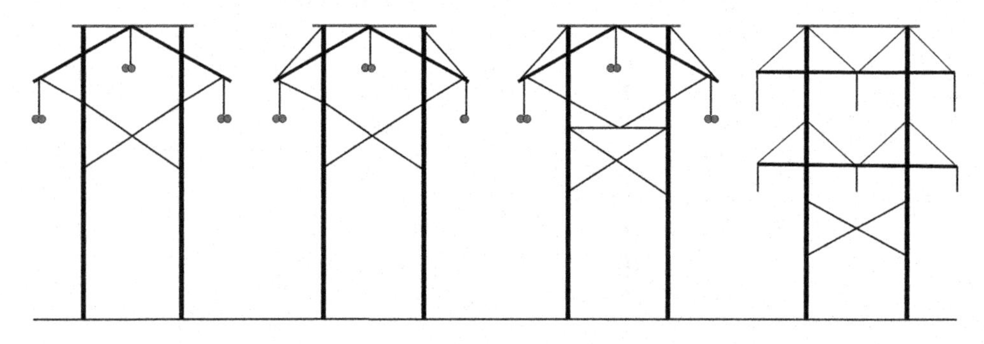

(b) 230 kV to 345 kV Single and Double Circuit Tangent Structures
Single or Bundled Conductors

Figure 3.7 Structure Families Involving H-Frames (Courtesy Hughes Brothers).

insulator string is critical in determining the sag-clearance in the span as well as the swing. This in turn will affect the required height of the pole/structure.

Attachment to Steel Structure: Suspension insulators are usually attached to the structure via steel tubular davit arms (tapered) or horizontal posts. In steel H-Frames, the insulators are typically attached via vertical vangs welded to the steel cross arm. In sub-station structures, tubular steel cross arms are generally used to support the deadend/terminal wire loads, in which case the vangs are horizontal. Angle and strain insulators are attached to the pole via welded vang plates. Where moderate line angles pose a challenge along with narrow ROW, the insulators are attached to the davit arms via swing brackets. Horizontal posts are attached to the pole via support brackets, flat base or gain base.

Attachment to Wood Structure: Suspension insulators are attached to the structure via davit arms or horizontal posts. In H-Frames, the insulators are attached via proper end fittings to the wood cross arm. Angle and strain insulators are attached to the pole directly via eye bolts or guying tees. Where moderate line angles pose a challenge along with a narrow ROW, the insulators are attached to the davit arms via swing brackets. Horizontal posts are attached to the pole via support brackets, flat base or gain base.

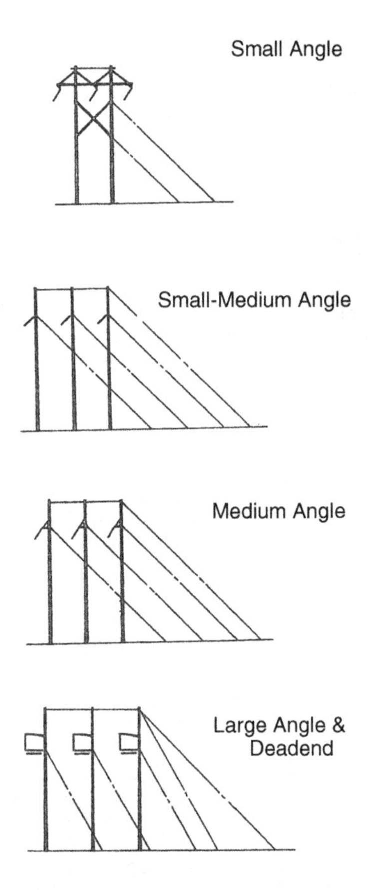

Figure 3.8 Typical 2- and 3-Pole Guyed Structures (Courtesy: Hughes brothers).

Figures 3.9a, b and c show various insulator attachments (braced line post, cross arm and davit arms).

Attachment to Concrete Structure: Insulator attachment to concrete is a bit more complex than other structures since the poles are prestressed (i.e.) contain stressed tendons inside the core wall. Holes to attach insulator hardware are pre-drilled carefully. For example, insulators are usually attached to the structure via steel brackets supporting horizontal posts (Figure 3.10). Angle and strain insulators are attached to the pole via bolted tees. Where moderate line angles pose a challenge along with narrow ROW, the insulators are attached to the davit arms via swing brackets. Horizontal posts are attached to the pole via support brackets, usually with a gain base inclined at 12 degree angle.

Attachment to Lattice Towers: Suspension insulators are usually attached to the structure via proper end fittings at the ends of the lattice arms. To control insulator swing and to maintain wire-structure surface clearances, 2-part insulators are often used. Angle and strain insulators are attached to the tower via vangs welded at the end of the lattice arms.

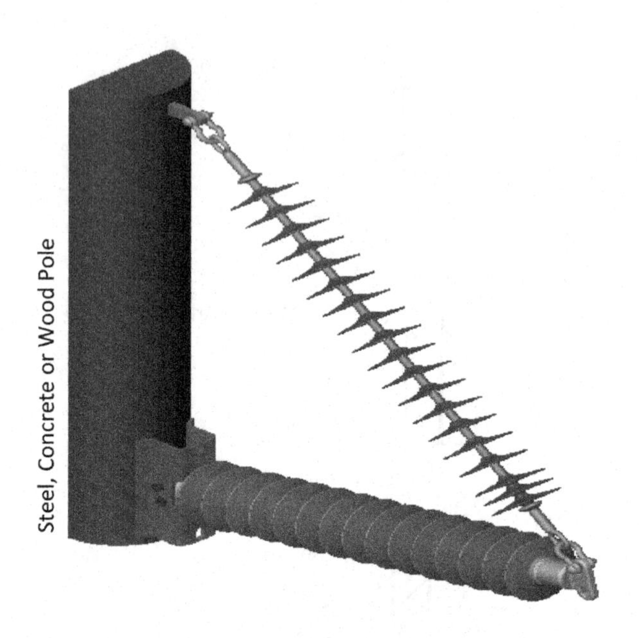

(a) Braced Line Post Insulators

Figure 3.9a Insulator Attachment to Structure – Braced Line Post Insulators.

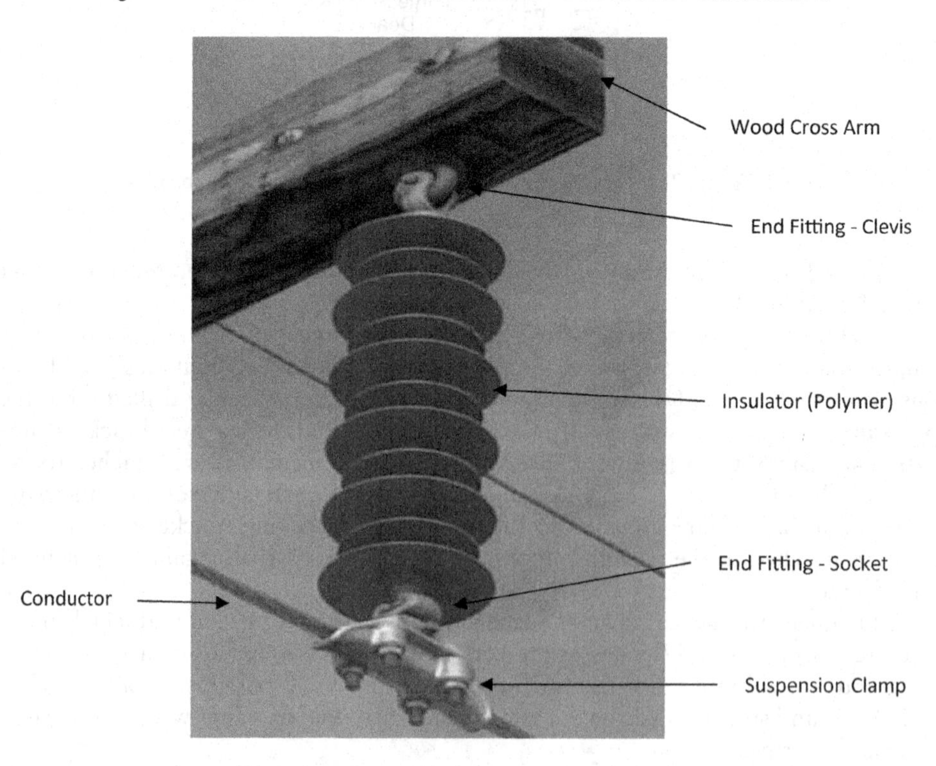

(b) Suspension Insulators on Cross Arms

Figure 3.9b Insulator Attachment to Structure – Suspension Insulators on Cross Arms.

(c) Suspension Insulators on Steel Davit Arms

Figure 3.9c Insulator Attachment to Structure – Suspension Insulators on Steel Davit Arms.

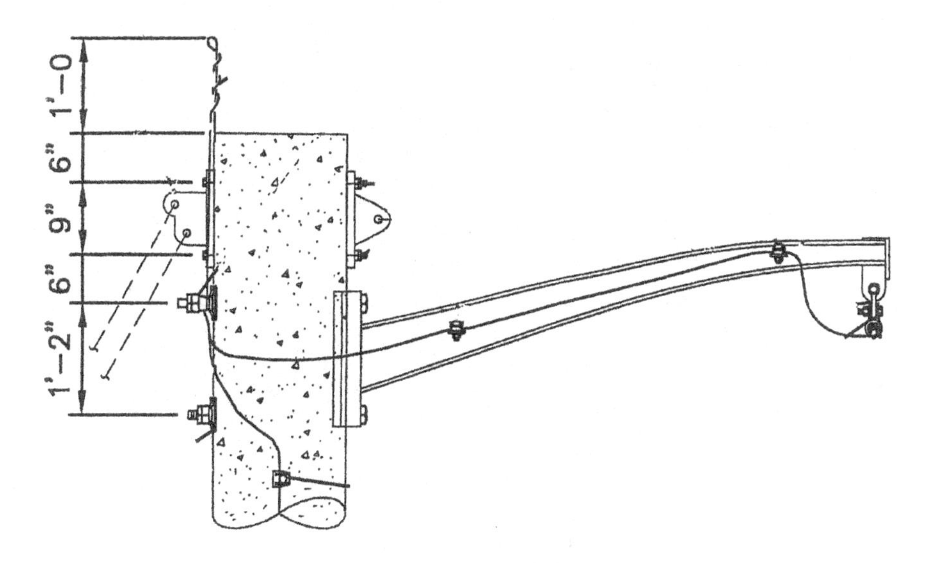

Figure 3.10 Insulator Attachment to Concrete (Source: Georgia Transmission).

The engineer is referred to various manufacturer catalogs (Ohio Brass, NGK Locke, MacLean, Hubbell etc.) for more information on insulators, end fittings and hardware.

Figures 3.11 to 3.13 show insulator attachment in typical structural systems.

Another situation where insulator attachment pattern is important is at locations where the wires transition from a vertical to a horizontal configuration or vice-versa. One example is from a vertical angle or deadend to a H-Frame. Figure 3.14 indicates a preferred way of connecting the phases so that potential for short circuiting (phase wire contact) is minimized.

3.2.2 Structure types

Transmission structures are divided into 4 functional categories for defining *strength requirements* and based on the manner in which the wire loads are resisted (See Figure 3.3).

Suspension or Tangent Structure: where all wires are attached to the structure using suspension insulators and clamps not capable of resisting tension on the wires.

Strain Structure: primarily used at running angles where all wires are attached to the structure using suspension or strain insulators and clamps where the transverse forces resulting from wire tensions are resisted by guy wires and anchors (or an unguyed system if steel poles are used).

Figure 3.11 Insulator Attachment to Lattice Tower.

Figure 3.12 Insulator Attachment to H-Frame with Double Cross Arms.

Figure 3.13 Insulator Attachment to Steel Pole.

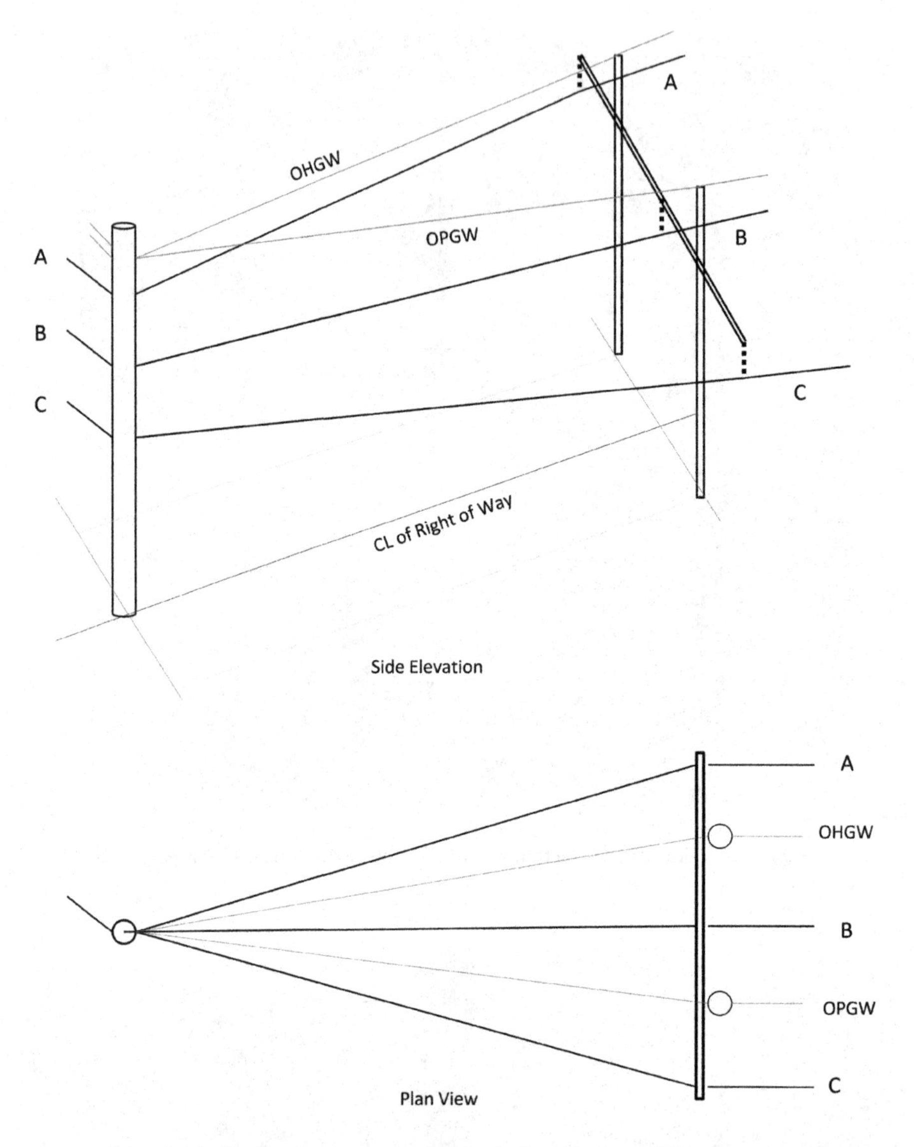

Figure 3.14 Phasing Arrangement for Vertical to Horizontal Construction.

Deadend Structure: primarily used at large angles and deadends where all wires are attached to the structure using strain insulators and botted deadend clamps (or compression deadend connectors) where the structure must have the ability to safely resist a situation where all wires are broken on one side, in addition to loading from intact wires.

Terminal Structure: where all wires are attached to the structure using strain insulators on one side only. This situation usually occurs at substation frames where wires are installed at a reduced tension on the spans coming into the substation.

Configuration-wise, the most basic structure type is the single pole system which is extensively employed for tangent, angle and deadend applications, in wood, steel, concrete and composite. Apart from lattice-type systems, the only other unique configuration popularly used is the 2-Pole H-Frame.

3.2.2.1 H-Frames

H-Frame structures are commonly used in 69 kV to 230 kV (and above) single or double-circuit high voltage transmission lines. They are often used in situations where spans are relatively moderate and ROW adequate. Design with H-Frames is generally performed in terms of "Allowable Spans" where the maximum allowable horizontal (and vertical) spans are determined as a function of several variables. Spans are often limited by X-brace and cross arm strengths, insulator swings or uplift. Design is often governed by the setting depth needed to resist lateral overturning forces in case of unbraced structures. Wood is the predominant material in most H-Frames although steel and composites are also being increasingly employed. Since asymmetrical bending is often involved, factors like backfill material often control the overturning resistance of the structure at ground line. Also, if the ratio of Vertical Span/Horizontal Span is less than 1.0 (excessive elevation difference), then the effects of the vee/knee braces also become predominant.

Cross Arms connect the two (or three) poles of the H-Frame and provide locations for attaching insulators. A double cross arm is often used to resist large vertical loads due to large spans or when the frame is a tangent deadend. Cross arm lengths range from 12 ft. to 40 ft. (3.7 m to 12.2 m) depending on voltage, phase separation etc.

X-bracing in H-Frames helps increase the allowable horizontal spans by increasing the structure strength. They also help in enhancing the lateral stiffness of the structure to resist transverse deflections. Design strength of typical RUS braces ranges from 20,000 lbs to 40,000 lbs (89 kN to 178 kN) in either tension/compression. All wood cross arms and braces used in RUS standard H-Frames are typical RUS units, defined by the following pole separations:

69 kV – 10½ ft. (3.2 m)
115 kV – 12½ ft. (3.8 m)
161 kV – 15½ ft. (4.7 m)
230 kV – 19½ ft. (6.0 m)

The reader is referred to Example 3.6 showing situations where various H-Frame types are chosen.

For 3-pole systems, the arm lengths vary from 25 ft. to 35 ft. (7.6 m to 10.7 m). For other pole spacing, the axial capacity of the X-braces or minimum brace size can be found using the catalogs from various manufacturers such as Hughes Brothers (2012).

3.2.2.2 Guyed structures

Wood structures at running angles and deadends are characterized by strain insulators and guy wires linked to an anchor. In case of single poles (vertical angles), the guys are usually "bi-sector" guys (i.e.) they are oriented along a line bisecting the line angle. For larger 3-pole angle systems, the guys and anchors are located on either side of the structure. Anchors can be individual (one anchor per guy wire) or combined (one anchor for two guy wires). At line locations where there is a change in wire tension, in-line guying is adopted.

For poles stabilized by guy wires, the wires are considered an integral part of the structural system. Design specifications include guy type, size, modulus of elasticity, rated tensile strength (RTS), allowable load (often as a % of RTS), installation tension (usually as a % of RTS) and location of attachment on pole and anchor on ground (guy slope or angle). The recommended guy angle to pole is 45°. Utilities specify several sizes of storm guys for wood poles, namely, 3/8 in., 7/16 in., ½ in. etc. up to ¾ in. (9, 11, 13 mm up to 19 mm) with ultimate tensile strengths from 10.8 kips to 58 kips (48 kN to 258 kN), respectively.

Anchors come in a variety of sizes and configurations (single log, double log, plate), helical screw and rock anchors. Virtually all guy-anchor systems provide means for grounding the overhead ground wire by connecting it to the anchor and therefore embedded in the ground.

At locations where guying at a pole is prevented for various reasons (lack of space, for instance), the system is guyed by means of a stub pole usually installed across the street or road. The guying here includes overhead wires from pole to stub pole and then the anchor guys from the stub pole to the ground.

From analysis perspectives, any structural system with a cable element (i.e.) a guy wire is predominantly a non-linear system. Therefore, such systems when analyzed on any computer program (such as PLS-POLE^{TM}) must use the non-linear option.

3.3 STRUCTURE LOADS

Determination of analysis loads on transmission structures involves the following:

1. Wire Loads – vertical, transverse and longitudinal loads on wires due to ice, wind, temperature
2. Structure Loads – vertical, transverse and longitudinal loads on structure due to wires, attachments and hardware

Secondary loads include those due to P-Delta (2nd Order) effects which will be discussed below.

3.3.1 Load cases and parameters

The weather conditions considered for structural design were discussed earlier in Chapter 2 in Sections 2.1 and 2.4. Also, Table 2.16, Table 2.17 (*a, b*) and Table 2.18 contain the required load case information. Though climate-related loading parameters were covered in Chapter 2, they are briefly summarized here.

Wind: Wind pressure on transmission structures is defined as force resulting from exposure of structure surface to wind. These surfaces include both the surface of the wires as well as structural system (steel or wood or concrete poles or lattice towers).

Radial Ice: This is the thickness of ice applied about the circumference of conductors and ground wires. In routine transmission line design, ice is only applied to conductors and ground wires but not to the surface of the structure, insulators and other hardware.

Temperature: This is a design parameter needed for calculating conductor and static wire sags and tensions. For example: Maximum sags (and corresponding clearances) are often evaluated at a conductor temperature of 100°C (212°F) for ACSR wires. Uplift situations are generally referred to a "cold weather case" for low temperatures in the range of 0°F to –20°F.

3.3.2 Load and strength factors

The Load and Strength (Reduction) Factors – from both NESC and RUS – are given earlier in Chapter 2 in Section 2.4.4 and in Tables 2.15a, 2.15b, and 2.15c.

3.3.3 Point loads

Transverse Load: This load is defined as force or pressure acting perpendicular to the direction of the line. For tangent structures, wind forces are usually applied as transverse loads on the structures. In angle structures and deadends located in angles, the transverse direction is parallel to the bi-sector of the line angle and the component of the wire tension acts in the transverse direction. All transverse loads are usually factored (i.e.) contain applicable Load Factors.

Vertical Load: This load is defined as force acting vertically due to gravity. For all structures, vertical forces usually include factored weight of wires (iced and non-iced), insulators and hardware, along with the weight of various components defining the system. All vertical loads are multiplied by a specified load factor to obtain the design load. Uplift loads, which occur due to uneven terrain and cold temperatures, are another form of vertical loads, acting against gravity.

Longitudinal Load: This load is defined as force or pressure acting parallel to the direction of the line. In angle structures and deadends located in angles, the longitudinal direction is perpendicular to the bi-sector of the line angle and the component of the wire tension acts in the longitudinal direction. For deadends located in zero line angles, with wires on one side, this load is simply the wire tension with the appropriate tension load factor.

Figure 3.15 shows the wire scheme used to illustrate the calculation of point loads V, T and L for various line angles. The loads are computed with the equations given below:

$$\text{Total Vertical V (lbs.)} = \{[(2 * t_{ri} + d_w)^2 - d_w^2] * 0.3109 + w_{bw}\} * LF_v * S_{wt} \quad (3.2a)$$

$$\text{Total Transverse T (lbs.)} = (2 * t_{ri} + d_w) * S_{wd} * \left(\frac{p_w}{12}\right) * LF_t + 2\sin\left(\frac{\theta}{2}\right) * T_w * LF_{wt} \quad (3.2b)$$

$$\text{Longitudinal L (lbs.)} = \cos\left(\frac{\theta}{2}\right) * T_w * LF_{wt} + \cos\left(\frac{\theta}{2}\right) * (-T_w) * LF_{wt} \quad (3.2c)$$

where:
t_{ri} = thickness of radial ice on wire (in)
d_w = diameter of bare wire (in)

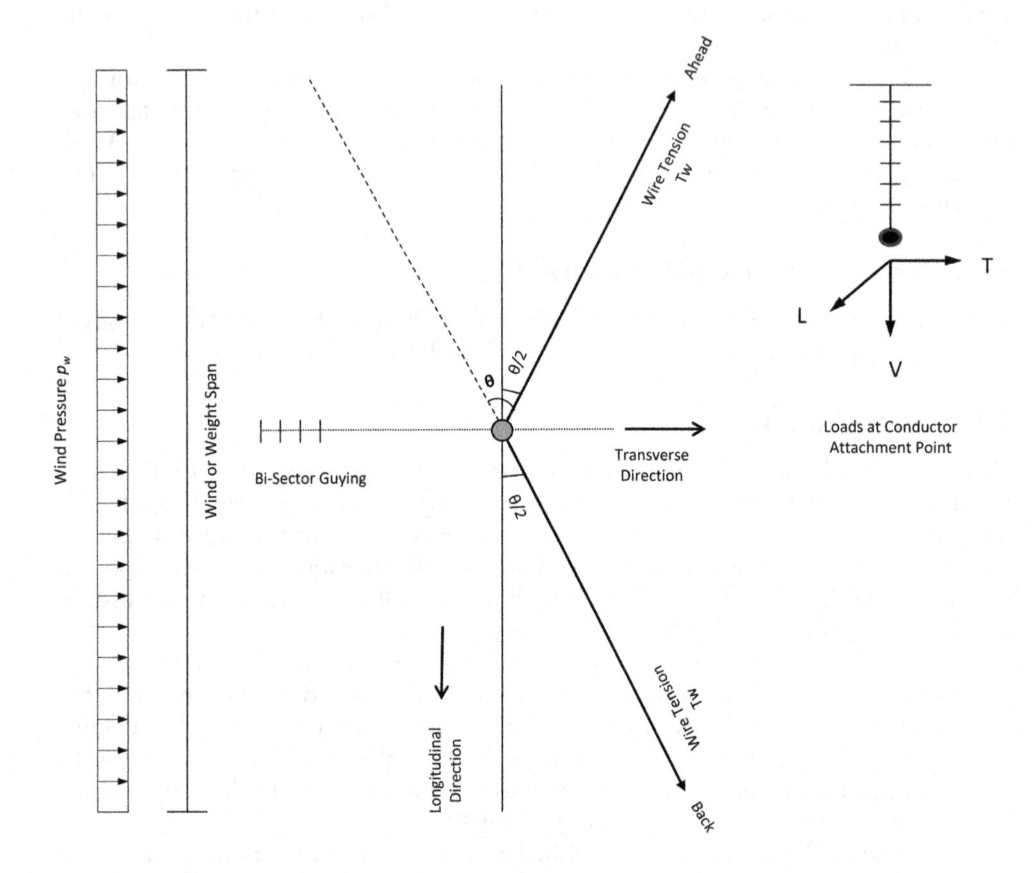

Figure 3.15 Calculation of Wire Loads.

w_{bw} = weight of bare wire (no ice) (plf.)
LF_v = load factor for vertical load
LF_t = load factor for transverse load (wind)
LF_{wt} = load factor for wire tension
S_{wt} = weight or vertical span (ft.)
S_{wd} = wind or horizontal span (ft.)
T_w = wire Tension for the load case (lbs.)
θ = line angle (deg.)
p_w = wind pressure on wire (psf.)

Note that the equation for vertical loads includes weight of glazed ice at 57 pcf (8.95 kN/m^3). For a case where line angle is zero, the equations reduce to the situation of a perfect tangent structure. The values obtained from these formulae depend on the parameters related to that specific load case. For example, for extreme wind case, all load factors are usually 1.0. Also, some utilities add the weight of workers to the vertical load component to account for construction-related loads.

The following Examples 3.2 and 3.2a illustrate the process of calculating point loads and ground line moments for an unguyed deadend pole.

Example 3.2

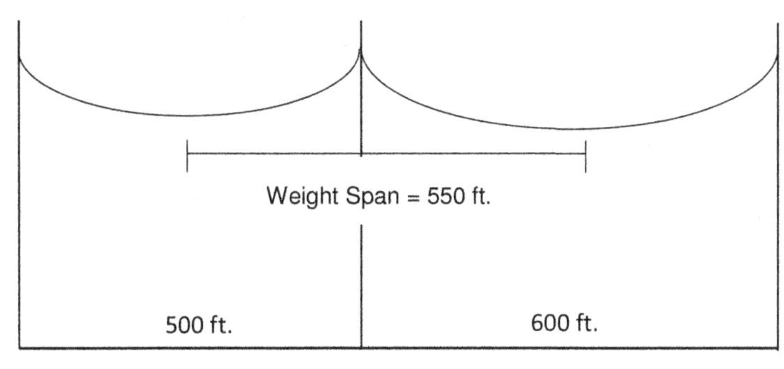

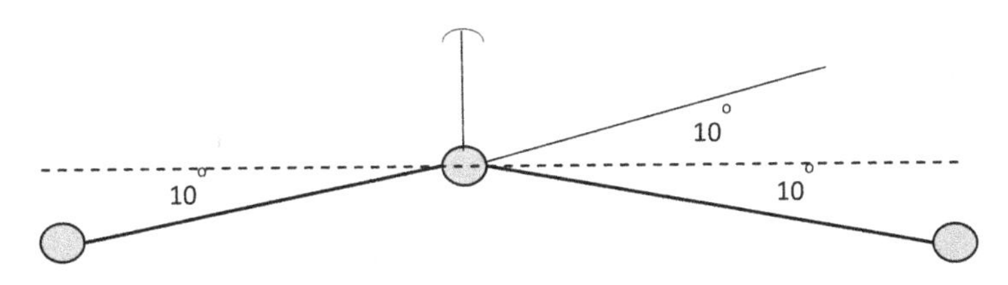

For the 2-span system, determine the point loads due to conductor for a line angle of 20° for the following situations:

(a) All Load Factors = 1.00
(b) Vertical LF = 1.50, Wind LF = 2.50 and Wire Tension LF = 1.65
(c) Determine the force in the guy wire supporting the transverse load for Case 'b' for a 45° guy angle.

Assume level ground and the following data:

Conductor tension T_W = 5,000 lbs. (22.25 kN)
Conductor diameter = 1 in. (2.54 cm)
Conductor weight = 1.0 plf. (14.59 N/m)
Wind pressure = 21 psf. (1.01 kPa)
No ice on wires. Guying is bi-sector type.

Solution:

For level ground span, with attachment points at the same elevation:

Wind span = Weight span = Average span
Average span = (500 + 600)/2 = 550 ft. (167.6 m).
Use Equations 3.2a and b.

(a) Vertical load V = (550) (1) (1) = 550 lbs. (2.45 kN)
Transverse load T = component due to wind + component due to wire tension
= (21) (550) (1/12) + (2) (Sin (10°)) (5000)
= 2,699 lbs. (12.01 kN)

Since the conductor tension is same in both spans, net longitudinal load L is zero.

(b) For the given load factors:
Vertical load V = (550) (1.50) = 825 lbs. (3.67 kN)
Transverse load T = component due to wind + component due to wire tension
= (2.5) (21) (550) (1/12) + (1.65) (2) (Sin (10°)) (5000)
= 5,271 lbs. (23.46 kN)

Since the conductor tension is same in both spans, net longitudinal load L is zero.

(c) Guy force = (5271) [1/Cos (45°)] = 5271/0.7071 = 7,455 lbs. (33.17 kN)

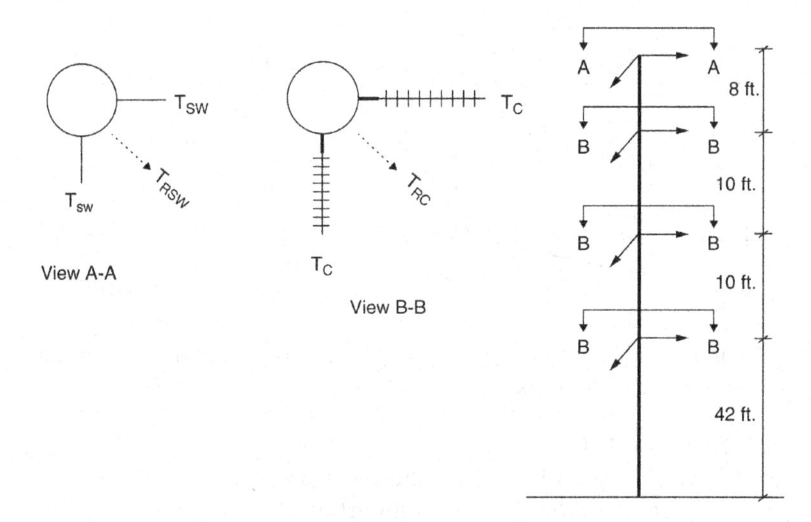

Example 3.2a Consider the 90°, unguyed, single circuit, deadend transmission pole in the figure. Wire tensions for various weather cases and their load factors have been determined as follows (consider only cable tensions; neglect wind forces on wires and pole):

Weather/ Load Case	*Conductor Tension T_C (lbs.)*	*Shield Wire T_{SW} (lbs.)*	*Load Factor LF_V*	*Load Factor LF_T*	*Load Factor LF_{WT}*
NESC Heavy	4000	3000	1.50	2.50	1.65
NESC Extreme Wind	3420	2170	1.10	1.10	1.10
Extreme Ice	7160	5170	1.10	1.10	1.10
NESC Ext. Ice w/Conc. Wind	5450	3960	1.10	1.10	1.10
Broken Wires	4000	3000	1.10	1.10	1.10

Determine the maximum ground line moment for the pole assuming 10% moment due to P – Δ effects. The span in each direction is 330 ft. (100.6 m).

Solution:

The design ground line moment (GLM) will be determined using the factored tensions of the wires for each load case.

Weather/Load Case	*Factored Conductor Tension T_C (lbs.)*	*Factored Shield Wire Design Tension T_{SW} (lbs.)*
NESC Heavy	4000 × 1.65 = 6600	3000 × 1.65 = 4950
NESC Extreme Wind	3420 × 1.10 = 3760	2170 × 1.10 = 2390
Extreme Ice	7160 × 1.10 = 7880	5170 × 1.10 = 5690
NESC Ext. Ice w/ Conc. Wind	5450 × 1.10 = 6000	3960 × 1.10 = 4360
Broken Wires	4000 × 1.10 = 4400	3000 × 1.10 = 3300

Resultant tensions for various load cases:
The first four (4) are cases where all wires are intact. The last one has all wires cut on one side of the pole.

NESC Heavy $T_{RC} = \sqrt{(6600^2 + 6600^2)} = 9{,}333$ lbs. (41.5 kN) (see Note 1)
$T_{RSW} = \sqrt{(4950^2 + 4950^2)} = 7{,}000$ lbs. (31.2 kN)
NESC Ext Wind $T_{RC} = \sqrt{(3760^2 + 3760^2)} = 5{,}317$ lbs. (23.7 kN)
$T_{RSW} = \sqrt{(2390^2 + 2390^2)} = 3{,}380$ lbs. (15.0 kN)
Extreme ice $T_{RC} = \sqrt{(7880^2 + 7880^2)} = 11{,}145$ lbs. (49.6 kN) ← CONTROLS
$T_{RSW} = \sqrt{(5690^2 + 5690^2)} = 8{,}047$ lbs. (35.8 kN) ← CONTROLS
NESC Ext. Ice $T_{RC} = \sqrt{(6000^2 + 6000^2)} = 8{,}485$ lbs. (37.8 kN)
w/Conc. Wind $T_{RSW} = \sqrt{(4360^2 + 4360^2)} = 6{,}166$ lbs. (27.4 kN)
Broken Wires $T_{RC} = \sqrt{(4400^2 + 0^2)} = 4{,}400$ lbs. (19.6 kN)
$T_{RSW} = \sqrt{(3300^2 + 0^2)} = 3{,}300$ lbs. (31.2 kN)

Extreme Ice load case governs. This case is associated with no wind, either on wires or poles (see Table 2.17a).

Bending Moment at ground line due to resultant controlling design wire tensions is:

$M = [(8{,}047)(70) + (11{,}145)(42 + 52 + 62)]/1000 = 2301.9$ kip-ft. (3121.4 kN-m)
Wind on wires = 0
Wind on pole = 0
$P - \Delta$ Moment = (0.10) (2301.9) = 230.2 kip-ft. (312.1 kN-m)
Total GLM = 2301.9 + 230.2 = 2532 kip-ft. (3433.5 kN-m) (see Note 2)

Note 1: In Equation (3.2b), if we neglect the first component due to wind pressure, the equation simplifies to 2 sin (90/2)(4000)(1.65) = 9,333 lbs. (41.5 kN).
Note 2: The wire tensions contribute a major component of the load on the deadend poles. However, in real life, the reader must consider moment due to wind on wires and wind on pole in addition to the moments associated with wire tensions.

3.3.4 Loading schedules

The various loads applied on transmission structures are specified to the manufacturers/vendors by means of Loading Schedules (also known as Loading Trees). Here the wire loads – vertical, transverse and longitudinal, associated with different weather cases and determined by various methods – are listed in a tabular form. An outline of the structure and the insulator attachment points are also shown.

Loading Schedules (also called Loading Trees) can be determined by several means:

1. Spreadsheets (usually for tangent and angle structures on level terrains).
2. PLS-CADD™ or PLS-CADD/LITE™ (for deadend and large angle structures on uneven, rolling terrains).

These Loading Trees constitute the most important communication between the engineer and the steel fabricator. Figures 3.16, 3.17 show commonly-used format of loading schedules for tangent lattice towers and H-Frames. Figure 3.18 refers to an angle/deadend steel pole; the loading table is left blank as an exercise for students. Only critical load cases are generally shown on these loading schedules; other additional cases (example: uplift case at sub-zero temperature) can be added at the discretion of the engineer. For angles and deadends, the loads generally refer to the bi-sector orientation (see below).

For strain and deadend structures, loading schedules often contain separate loads for the 'back' and 'ahead' spans if requested by the fabricator. If a full PLS-CADD™ model of a transmission line is available with all structures spotted and all wires strung, then the load trees for each structure can be extracted in print form from the line model in the required coordinate system. For greater control over design data, it is often preferred to use PLS-CADD/LITE™ to model each individual structure using a **bi-sector** model (by importing the structure file from the project's structure family library). Briefly, in the bi-sector format, the transverse axis is aligned with the line bisecting the line angle; the longitudinal axis is orthogonal to this transverse axis. These loads thus obtained are later converted to tabular form to be inserted onto the load schedule sheet.

With reference to Figures 3.16, 3.17 and 3.18, the loading trees include:

(a) NESC loading: Heavy (or Medium or Light), Extreme Wind, Extreme Ice with Concurrent Wind
(b) Others: Extreme Ice, Construction
(c) Broken Wire loading: wires cut, deadends

Broken Wire cases for angle/deadend structures usually consist of broken conductor(s), broken ground wire(s) or both, simulating a condition of unbalanced tension. For pure deadends or failure containment structures, this implies ALL wires cut on one side. Some utilities use a NESC Heavy weather condition for broken wires but with all load factors equal to 1.0.

Loads due to snapped conductors

ASCE Manual 74 (2010) recommends a procedure based on Residual Static Load (RSL) which is a final effective static tension in a wire after all the dynamic effects of a wire breakage have subsided. This RSL is a function of Span/Sag ratio and Span/Insulator ratio and is an unbalanced longitudinal load that acts on a support structure in a direction away from the initiating failure event. It is applied in one direction only and for lattice towers is given by $RSL = Wire\ Tension * Longitudinal\ Load\ Factor$.

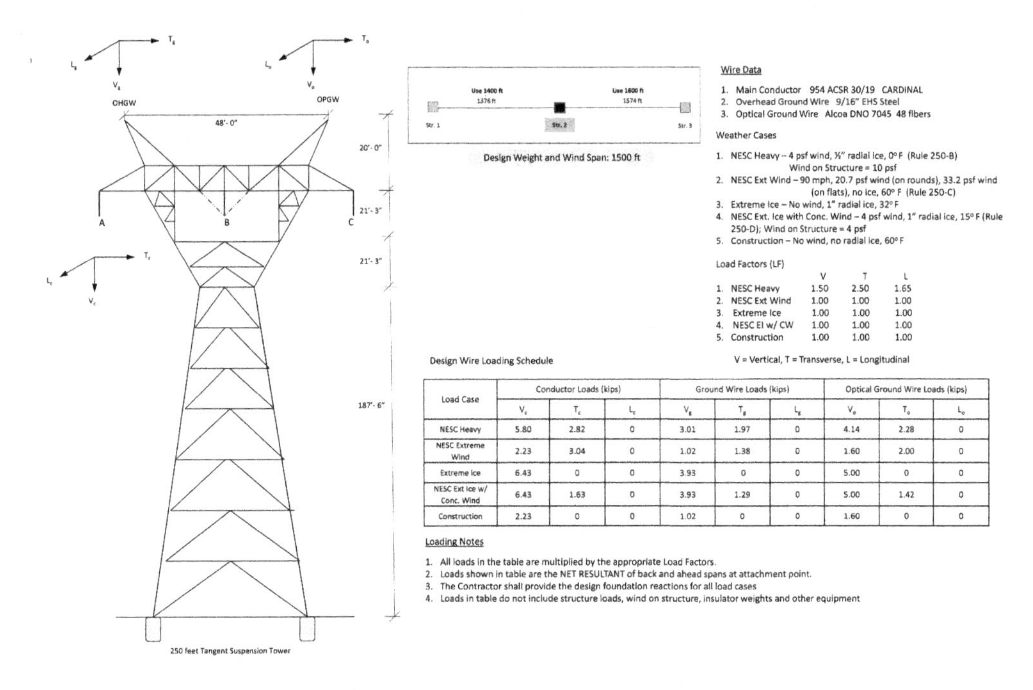

Load Case	Conductor Loads (kips)			Ground Wire Loads (kips)			Optical Ground Wire Loads (kips)		
	V_c	T_c	L_c	V_g	T_g	L_g	V_o	T_o	L_o
NESC Heavy	5.80	2.82	0	3.01	1.97	0	4.14	2.28	0
NESC Extreme Wind	2.23	3.04	0	1.02	1.38	0	1.60	2.00	0
Extreme Ice	6.43	0	0	3.93	0	0	5.00	0	0
NESC Ext Ice w/ Conc. Wind	6.43	1.63	0	3.93	1.29	0	5.00	1.42	0
Construction	2.23	0	0	1.02	0	0	1.60	0	0

Figure 3.16 Loading Schedule – Lattice Tower.

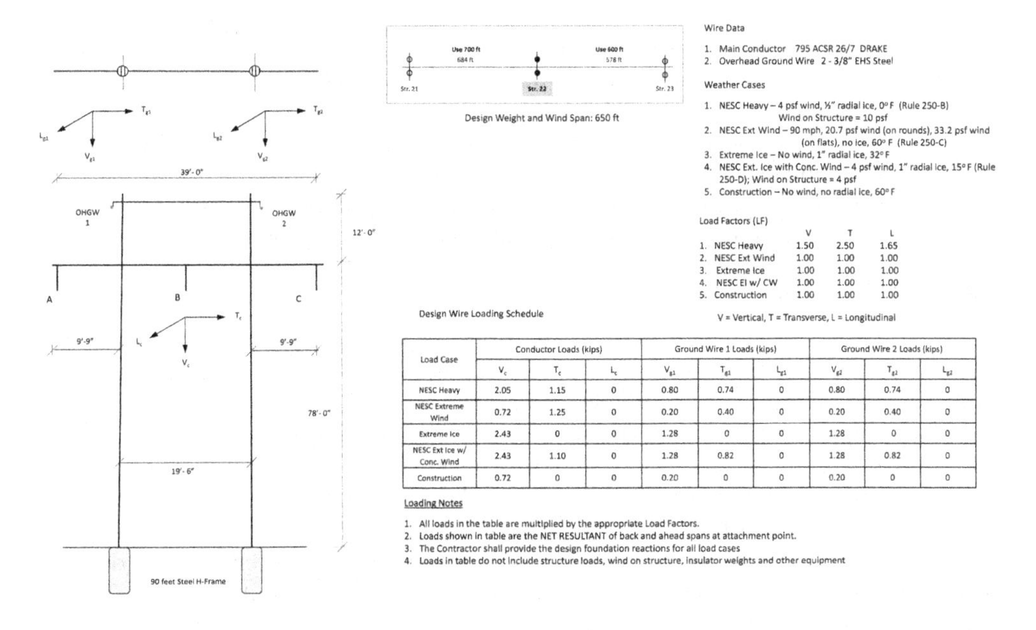

Load Case	Conductor Loads (kips)			Ground Wire 1 Loads (kips)			Ground Wire 2 Loads (kips)		
	V_c	T_c	L_c	V_{g1}	T_{g1}	L_{g1}	V_{g2}	T_{g2}	L_{g2}
NESC Heavy	2.05	1.15	0	0.80	0.74	0	0.80	0.74	0
NESC Extreme Wind	0.72	1.25	0	0.20	0.40	0	0.20	0.40	0
Extreme Ice	2.43	0	0	1.28	0	0	1.28	0	0
NESC Ext Ice w/ Conc. Wind	2.43	1.10	0	1.28	0.82	0	1.28	0.82	0
Construction	0.72	0	0	0.20	0	0	0.20	0	0

Figure 3.17 Loading Schedule – H-Frame.

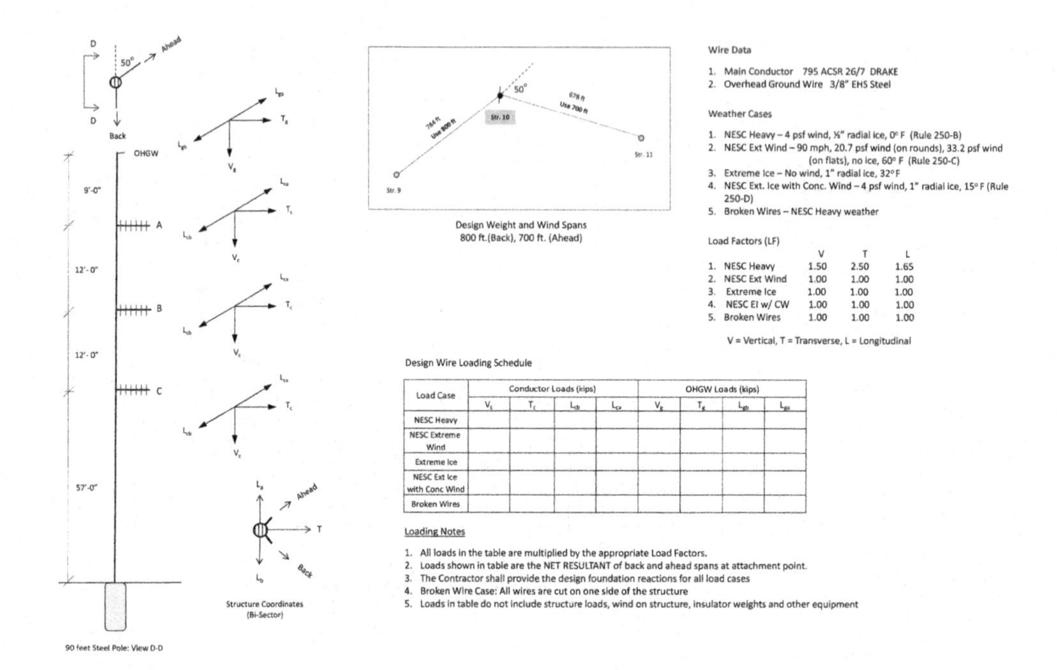

Figure 3.18 Loading Schedule – Angle Deadend.

RSL is generally calculated for bare wire (i.e.) no ice, no wind loading condition at an average temperature. Effects of wind are not considered when using RSL. Some utilities specify NESC Heavy weather condition for tensions; however, the longitudinal load factors are generally taken as 1.0.

For a single circuit structures, unbalanced longitudinal loads can be applied at any single conductor phase or any one ground wire support. For double circuit structures, unbalanced longitudinal loads could be applied to any two conductor phases or one or two ground wire supports or one conductor and one ground wire support.

3.3.5 Deflection limits

Limits on free-standing pole/structure deflections are often prescribed for various reasons which include aesthetics, reducing $P-\Delta$ effects (see below), maintaining the require phase clearances, maintaining conductor separation from structure surface and other objects. Some utilities also take into account the impact of structure deflection on vertical clearance to ground, typically in case of angle structures. These limits are defined by RUS for steel and concrete poles. Appendix 5 discusses these limitations in detail. For routine steel pole designs, engineers often limit the pole top movement to 1% to 2% of the pole heights for normal operating conditions. Specifying low deflection limits will result in a large, stiffer and more expensive structure. In a majority of the cases, the issue is left to the decision of the utilities and conveyed to structure designers at the fabricator. Some utilities request the manufacturer to camber the steel poles at angle locations so that the poles become straight and plumb after installation.

For concrete poles, the effects of deflections are more critical than steel; but owing to the difference in material behavior, the limits are different. ASCE Manual 123 (2012) requires appropriate concrete modulus be used in determining elastic (pre-cracking) and inelastic (post-crack) deflections.

For composite poles, tip deflection is a function of the materials used and the pole geometry. FRP (fiber-reinforced polymer) materials have a very high strength-to-stiffness ratio and often transmission poles made from such materials can be designed to be flexible. ASCE Manual 104 (2003) provides guidance with reference to deflections permitted in composite poles.

3.3.6 P-Delta analysis

This refers to the secondary bending effects on a transmission pole due to lateral deflection 'Δ' of the pole due to wind or other load actions. The movement of the vertical load application points produces small additional bending moments on the pole to add to those produced by wire and wind loads. Computer programs for transmission structure analysis include these effects. The exact magnitude of the secondary moments varies from structure to structure and can be determined only as a function of the geometry of the system. For quick manual calculations for preliminary pole sizing, it is sometimes assumed that P-Delta (Δ) moments are approximately 10% of the total moment on the pole.

ASCE Manual 111 (2006) uses the Gere-Carter Method for estimating these $P - \Delta$ loads for tapered poles. This method, in general, is conservative than other sophisticated methods such as the finite element analysis approach.

3.4 STRUCTURAL ANALYSIS

This section deals with the theoretical basis for analysis procedures for various transmission structure forms and types. Computer programs such as PLS-POLE™ and TOWER™ are generally employed by most utilities; some utilities also have their own in-house programs for structural analysis, in addition to general purpose finite element programs. However, it is important to understand the underlying basis and limitations of these programs. The basics of analysis of various systems will be illustrated below. Equations governing design checks will be covered in Section 3.5.

3.4.1 Single tangent poles

The analysis of single, unguyed poles is governed by bending. The controlling flexural stress in the tangent pole is a result of the effects of factored transverse and vertical loading on the structure.

Selection of the appropriate wood, concrete, steel or composite pole for a given design bending situation is also possible in other ways:

(a) Specifying the ultimate lateral load that can be applied 2 ft. below the pole tip
(b) Specifying the loading tree and load cases from which the manufacturer can deduce the size of pole needed (concrete, composite and steel only)
(c) Specifying the maximum horizontal wire span the pole can resist for a given conductor/ground wire/loading configuration (Allowable Spans)

3.4.2 H-Frames

Section 13.5 of the RUS Bulletin 200 (2015) gives the governing design formulae for various wood H-Frame configurations. The equations are based on critical points of flexure at specified pole locations. Structure strength, expressed in allowable spans, refers to these points. The most general H-Frame assembly often encountered in practice is the braced frame shown below. The labels are self-explanatory; C is at the location of point of contraflexure (point where bending moment changes its sign) and G is the ground line.

Governing Equations for Analysis.

The point of contraflexure is calculated using the following equation:

$$x_0/x = C_G\,(2\,C_G + C_D)/2\,(C_G^2 + C_G C_D + C_D^2) \tag{3.3a}$$

where:
x_0 = distance of C to ground line
x = distance from ground line to D
C_G = circumference of pole at ground line
C_D = circumference of pole at brace location

For general H-Frame assemblies, the ratio of x_0/x varies from 0.55 to 0.70.

Equations for limiting horizontal spans (HS) are as shown below:

Horizontal Span: Limited by Pole Strength at 'B'

$$HS_B = \{[\varphi M_B] - [LF * q * y_1^2 * (2d_t + d_B)/6]\}/(LF * p_g * y_1) \tag{3.3b}$$

Horizontal Span: Limited by Pole Strength at 'E'

$$HS_E = \{[\varphi M_E] - [LF * q * y^2 * (2d_t + d_E)/6]\}/½(LF * p_t * y_0) \tag{3.3c}$$

Horizontal Span: Limited by Pole Strength at 'D'

$$HS_D = \{[\varphi M_D] - [LF * q * (h - x_0) * (x_1) * (d_t + d_C)/2]\}/½(LF * p_t * x_1) \tag{3.3d}$$

Horizontal Span: Limited by Pole Strength at 'G'

$$HS_G = \{[\varphi M_G] - [LF * q * (h - x_0) * (x_0) * (d_t + d_C)/2]\}/½(LF * p_t * x_0) \tag{3.3e}$$

Horizontal Span: Limited by X-Brace Strength

$$HS_X = \{[\varphi * XBS * b] - [2 * LF * q * (h - x_0)^2 * (2d_t + d_C)/6]\}/(LF * p_t * h_2) \tag{3.3f}$$

Horizontal Span: Limited by Uplift

$$HS\,p_t\,h_2 - VS\,w_g b - 1.5\,VS\,w_c b = W_1 b + W_p b + X - Y \tag{3.3g}$$

Horizontal Span: Limited by Bearing

$$HS\,p_t\,h_2 + VS\,w_g b + 1.5\,VS\,w_c b = W_2 b - W_p b + X - Y + W_1 b \tag{3.3h}$$

where:
φ = Strength Factor
M_n = Bending Moment at Location "n", n = B,E,D,G (lb-ft.)
$W_1 = \pi\, F_{sf}\, D_e\, d_{ave}/FS$ (lbs.)

$W_2 = (1/4)\pi d_{bt}^2 (Q_u)/FS$ (lbs.)
W_p = Weight of Pole (lbs.)
Q_u = Ultimate Bearing Resistance of Soil (psf.)
b = pole spacing (ft.)
p_t = Total Horizontal Force per unit length due to Wind on Conductors and Ground Wire (plf.) (plf.)
$p_t = 3p_c + 2p_g$ or $3p_c + p_g + p_f$ (plf.)
$P_t = (HS)\ (3p_c + 2p_g)$ or $(HS)\ (3p_c + p_g + p_f)$ (lbs.)
$X = q(h - x_0)(d_t + d_C)x_0$
$Y = 2qh^2(2d_t + d_G)/6$
F_{sf} = Ultimate Skin Friction Resistance of Soil (psf.)
D_e = Depth of Embedment of Pole (ft.)
d_{ave} = Average Pole Diameter below ground level (ft.)
p_c = Wind Load per foot of Conductors (plf.)
p_g = Wind Load per foot of Overhead Ground wire (plf.)
w_c = Weight per foot of Conductor (plf.)
W_g = Weight per foot of Overhead Ground Wire (plf.)
p_f = Wind Load per foot of Fiber Optic Wires (plf.)
q = Wind Pressure (psf.)
h = Height of Structure above ground (ft.)
h_1 = Height to load P_t from ground (ft.)
h_2 = Height to load P_t from point C (ft.)
d_t = Pole Diameter at Top (ft.)
d_{bt} = Pole Diameter at Butt (ft.)
d_G = Pole Diameter at Ground Line (ft.)
d_n = Pole Diameter at Location "n", n = B,E,D,C,G (ft.)
FS = Factor of Safety
LF = Load Factor
x_1 = distance between C and D (ft.)
y_0 = Distance from E to P_t (ft.)
y_1 = Distance from pole top to cross arm (ft.)
y = Distance of point E from pole top (ft.)
XBS = Strength of X-Brace in compression (lbs.)
HS = Horizontal Span (ft.)
VS = Vertical Span (ft.)

For definitions of various items, refer to Figure 3.19.

3.4.3 Angle structures

The behavior of single-pole and three-pole guyed running angle structures is basically governed by interaction of guys and anchors and the design tensions transferred to the guy wires due to the line angles. For systems with multiple guy wires, it is also required to check buckling of the poles since the vertical component of the guy forces act on the poles axially. A typical guying guide is shown in later in Chapter 5 (see Figure 5.7); spreadsheets can be developed to calculate the allowable spans of a given pole-wire-guy-anchor system.

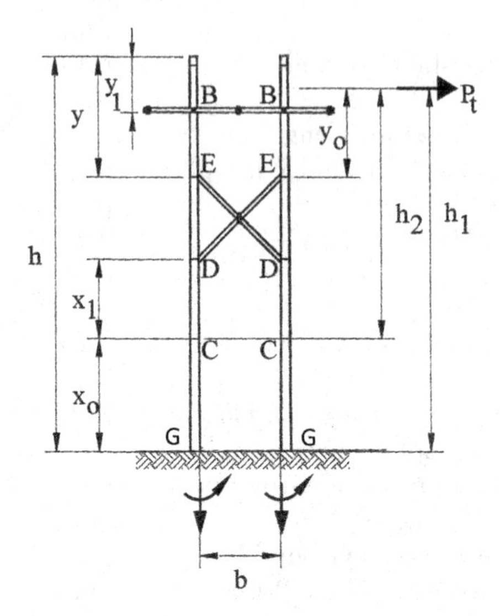

Figure 3.19 H-Frame Parameters (Source: RUS/USDA).

3.4.4 Deadends

The behavior of single-pole guyed angle/deadend structures is also governed by interaction of guys and anchors and the line design tensions transferred to the guy wires due to the line angles. One additional load is the component of the guy force transferred to the pole axially. Therefore, for systems with multiple guy wires, it is also required to check buckling of the poles.

If the structure has more wire (conductor + shieldwire) attachments than guy wires, and the highest guy wire attachment is below the lower-most wire, then the pole needs to be analyzed and checked for both bending and buckling. However, if the structure has the same number of wires and guy wires, and the wire and guy wire attachments are more or less at the same elevation, then the pole just need to be checked for buckling.

3.4.4.1 *Wood pole buckling*

Guyed angle and deadend poles, in addition to conductor and equipment weights, also carry vertical (axial) components of guy tensions. This axial force is directly dependent on the number of guy wires supporting the pole. Unlike concrete poles, wood poles are not strong in compression and are susceptible for buckling. Therefore, the buckling strength of guyed wood poles must be checked.

The RUS Bulletin 153 (2001) provides an equation for the critical vertical column load P_{cr} (in pounds) that can be imposed on a wood pole. The equation is derived from the theory of tapered columns.

$$P_{cr} = \pi EA^2/F_v K_a(K_u * H_{gb})^2 \quad (3.4)$$

where:
E = modulus of elasticity of wood (psi)
A = cross sectional area of the pole at 2/3 of the distance from the ground line to the bottom-most guy attachment (in^2)
$= \left(\frac{0.25}{\pi}\right) * [\{(C_b - C_t)(H_p - 0.667 * H_{gb})/(L_p - L_b)\} + C_t]^2$
K_a = conversion constant (576/ft^2)
H_{gb} = Height of bottom-most guy attachment from ground (ft.)
K_u = un-braced length coefficient (0.7 for bisector guying and 2.0 for deadend guying)
F_v = Factor of Safety = 1.50 (minimum)
C_b = Circumference of Pole at 6 ft. from the butt (in.)
C_t = Circumference of Pole at top (in.)
H_p = Pole height above ground (ft.)
L_b = ANSI point to Bottom of Pole (6 ft.)
L_p = Total Pole Length (ft.)

The most common wood pole buckling equation is however the one proposed by Gere and Carter (1962).

$$P_{cr} = P_A\, P^* \tag{3.5a}$$

$$P^* = (d_g/d_a)^{\alpha} \tag{3.5b}$$

where:
P_{cr} = critical load for a tapered column with circular cross section (lbs)
$P_A = \pi^2 EI_A/4L^2$ for Fixed-Free Column with $\alpha = 2.7$
$= 2\pi^2 EI_A/L^2$ for Fixed-Pinned Column with $\alpha = 2.0$
$= \pi^2 EI_A/L^2$ for Pinned-Pinned Column with $\alpha = 2.0$
P^* = a multiplier depending on end conditions
E = modulus of elasticity (psi)
I_A = moment of Inertia at guy attachment (in^4)
d_g = diameter at ground line (in.)
d_a = diameter at point of guy attachment (in.)
L = distance from ground line to point of guy attachment (in.)

When using the Gere-Carter formula for NESC District loads with load factors, strength factors between 0.50 and 0.65 are recommended. This gives a net safety factor of 3.3 to 2.5. For extreme wind loads, strength factors between 0.50 and 0.65 are suggested giving a net safety factor of 2.0 to 1.50. For full deadends, a much lower strength factor is recommended. Programs such as PLS-POLE™ conduct buckling checks automatically while performing a non-linear structural analysis.

3.4.5 Lattice towers

Utilities impose rigorous design requirements on lattice towers for bigger safety margins under severe structural and environmental loadings. The goal is to see that no tower member suffers permanent, inelastic deformation and that foundations are capable of sustaining imposed compressive and uplift forces. While suspension (tangent) towers are relatively easier to design, towers used at angles and deadends demand higher strength. At deadends, one of the design conditions is the broken wire case (i.e.) all wires cut on one side of the tower, creating a highly imbalanced loading situation.

Lattice towers are generally analyzed as 3-dimensional space trusses made up of steel angle members connected by bolts. The members carry compressive or tensile loads; therefore effective slenderness ratio kL/r governs design in most cases. The following slenderness limits are usually adopted for towers:

Legs ≤ 150
Others ≤ 200
Redundant ≤ 250
Tension only ≤ 500

where:
L = length of the member
k = slenderness factor depending on restraint at the ends
r = radius of gyration of angle member about one of the axes

3.4.5.1 *End restraints*

The primary consideration in the design of individual members in a lattice tower is the amount and type of restraint offered by the bolted connections at the member ends. Increasing the number of bolts at an end will always increase the amount of rotational restraint; and increasing restraint will lessen the effects of load eccentricity. The relationship between end restraint and eccentricity is qualitatively understood; but it is very difficult to mathematically quantify the actual joint stiffness in 3-dimensional space.

Some simple assumptions may be used in routine designs without sacrificing accuracy. A single bolt is considered as a hinge not offering any restraint; 2 bolts offer partial restraint and any joint with 3 bolts and above approaches full fixity. This idealization can help determine the appropriate slenderness ratios of members. The other consideration is whether the angle member is connected by one leg or both legs. Obviously, an angle bolted on both the legs is doubly restrained against rotation and thereby negate the effects of eccentricity of load. Main leg members of a tower are a good example of this situation.

To ensure integrity of connections and prevent tear-outs, minimum end and edge distance – along and perpendicular to the line of force in an angle – must be maintained. Figure 3.20 shows the definitions of end, edge and gauge distances (e, f, g_1 and g_2, respectively) and bolt spacing 's' associated with typical lattice tower angles.

3.4.5.2 *Crossing diagonals*

Almost all transmission towers contain crossing diagonals – X-type members – with a bolt in the middle. Structurally, these diagonals are part of a tension-compression system where a crossing tension member provides out-of-plane bracing support for the compression member. This support helps in reducing the effective buckling length for the compression member but is dependent on the load level in the tension diagonal. Programs such as TOWER™ contain various provisions to define such diagonals.

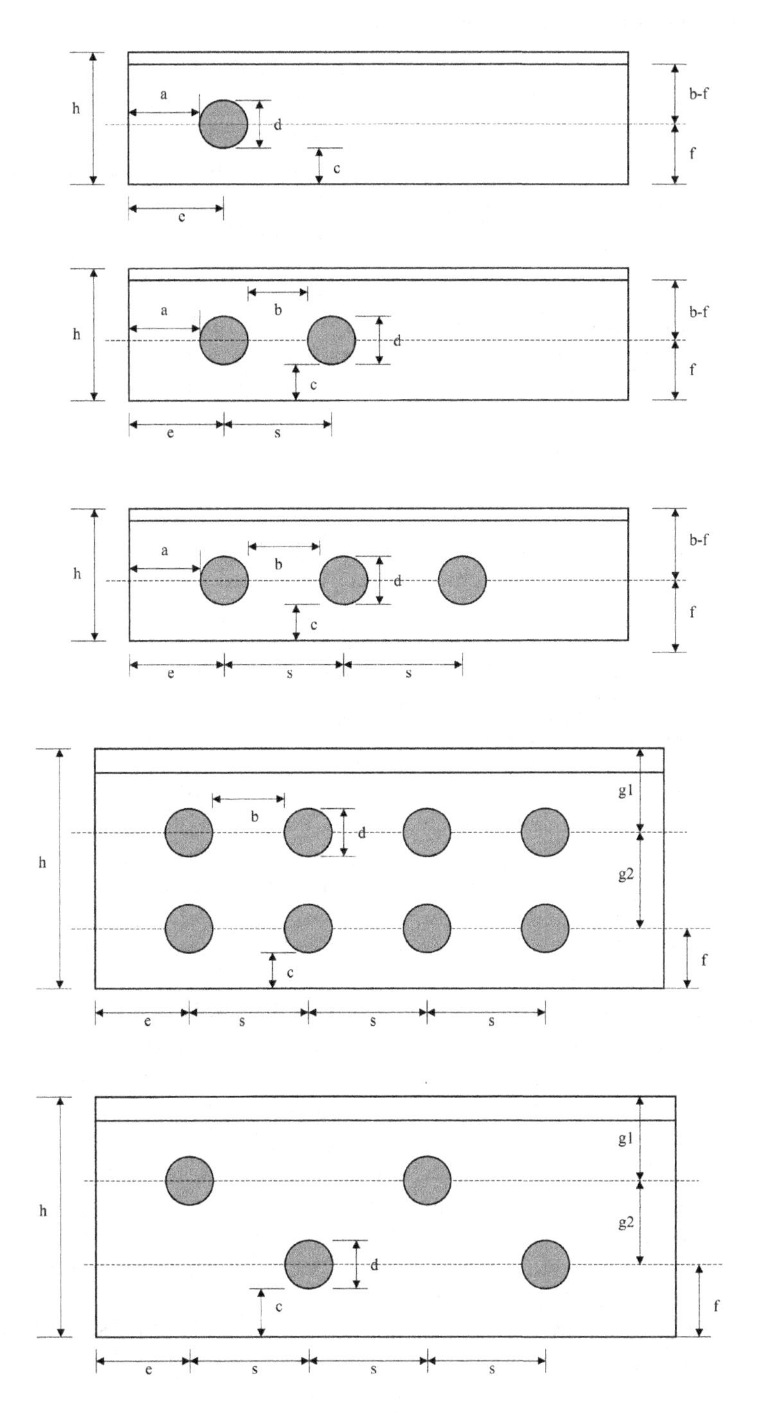

Figure 3.20 End, Edge, Gage Distance and Bolt Spacing.

3.4.6 Substation structures

Outdoor electrical substations and switchyards are a collection of various equipment, structures and components where electrical energy (typically high voltage) is modified. These substation structures support above grade items such as switches, circuit breakers, insulators, arresters, rigid bus and transformers. In the USA, analysis and design of substation systems is governed by ASCE Substation Structure Design Guide 113 (2008), RUS Bulletin 300 (2001) for rural systems and IEEE Standard 693 (2005).

Substation structures are essentially classified into three types, based on function:

Line Support Structures (LSS) – These are also called take-off or strain structures, deadend or line termination structures and internal strain bus. These can be single or multi-bay, truss or steel pole type. Major forces sustained are conductor and ground wire tensions (full or slack) and wind. LSS are critical components of a substation and are designed to withstand large stresses although on a non-catastrophic failure basis.

Equipment Support Structures (ESS) – These are switch stands, bus support stands, lightning arrestor stands, and line trap stands etc. ESS are designed mainly as vertical cantilever beams for short circuit forces and wind. Stresses rarely control but deflections must be checked.

Distribution Structures – Intended for low voltage applications, these are mostly comprised of steel beams and columns (truss or tube), may have multiple bays but are usually one-bay wide. They support switches and other equipment and are designed for rigidity at equipment location.

Structural profile configurations in substations basically fall under three types: Lattice, Solid and Semi-Solid. Lattice-type systems consist of steel angles framing into a box truss, both vertically and horizontally. Solid-type systems are made of wide flange shapes, pipes, round or tapered poles or rectangular tubes. Connections are either bolted or welded. Solid- type configurations are widely used for LSS, in either an A-Frame setup or single pole. Cross arms may be square or rectangular tubes or round/tapered polygonal members. Semi-solid types are made from wide-flange shapes or pipes as main members and use steel angles as braces in-between.

Design loads depend on whether they are intended for LSS or ESS. All LSS and their components, however, must withstand stresses induced by *factored* loads. For LSS, design loading is similar to that of a transmission line structures. Load Factors (LF) are specified for vertical (V), wind (W) and wire tension (T) forces. For ESS, design loading includes all applicable wind, ice, short circuit and dead loads; wind plus short circuit loads, however, produce maximum stresses. Ice is not expected to control design. Since all loads contain load factors, Ultimate Strength Design (or USD) is appropriate for LSS. Stiffness is an important requirement for ESS and the goal is to limit deflection under wind. Sections a bit larger than necessary are usually specified. Allowable Stress Design (ASD) is appropriate for ESS, while conforming to NEMA-SG6 (2006) rules, per RUS Bulletin 300.

3.4.6.1 Seismic considerations

Seismic analysis may be necessary at places with high earthquake risk. Seismic loads are generally considered as environmental load situations and are not combined with ice or wind loads, but may be combined with short circuit loading or operational loading. Substation structures are divided into four types with references to seismic loading:

ST1 Single- or Multi-bay Rack (not supporting equipment)
ST2 Single- or Multi-bay Rack (supporting equipment and conductors)
ST3 Rigid Isolated Support (supporting equipment)
ST4 Flexible Isolated Support (supporting equipment)

ST3 and ST4, in voltage classes higher than 121 kV, and within seismic zones 3 and 4, should be designed as per IEEE 693. For other types of structures and situations, design procedures are outlined in ASCE Substation Guide 113.

3.4.6.2 Deflection considerations

In addition to stresses, deflection limitations are imposed on ESS. Excessive movement or rotation of substation structures and components can affect mechanical/electrical operation of the equipment, reduce clearances, induce stresses in insulators/connectors etc. Disconnect switches are highly sensitive to deflections but overhead line dead-ends are not. Lattice-type systems, and A-Frames with solid sections, do not present any deflection problems. ASCE Guide 113 defines the following classes of substation structures for deflection purposes:

Class A – support equipment with mechanical mechanisms where structure deflection could impair or prevent proper operation.
Examples: group switches, vertical switches, ground switches, circuit breaker supports and circuit interrupters.

Class B – support equipment without mechanical components but where excessive structure deflection could result in compromised phase-to-phase or phase-to-ground clearances, stresses in equipment, fittings or bus.
Examples: support structures for rigid bus, surge arresters, metering devices, power transformers, hot-stick switches and fuses.

Class C – support equipment relatively insensitive to deflection or stand-alone structures that do not carry any equipment.
Examples: support structures for flexible bus, masts for lightning shielding, dead-end structures for incoming transmission lines. SG-6 does not give any specific limits but deflections here refer to limiting $P - \Delta$ stresses.

Multiple Class – Combination of any above classes

Tables 3.2, 3.3 and 3.4 show the basic load conditions, load cases for ultimate strength design (USD) and deflection limits for substation structures. Figure 3.21 from ASCE Guide 113 defines various structure classes and spans graphically for determining deflections.

Substation structures have a wide range of ground line reactions due to applied forces and therefore a wide variety of foundation types. These include drilled shafts, spread footings and slabs on grade, among others. Drilled shafts are typically used for LSS and steel poles while spread footings are used for circuit breakers and small

Table 3.2 Basic Loading Conditions – Substation Structures.

Loading Condition	*Wire Loaded Substation Structures*	*Switch and Interruption Supports*	*Rigid Bus Supports*	*Other Equipment Supports*
NESC*	YES	NO	NO	NO
Extreme Wind	YES	YES	YES	YES
Combined Ice and Wind	YES	YES	YES	YES
Earthquake	YES	YES	YES	YES
Short Circuit	NO	YES	YES	NO**
Construction and Maintenance	YES	YES	YES	YES
Equipment Operation	NO	YES	NO	YES
Deflection	YES	YES	YES	YES

(with permission from ASCE).
**Local Codes, such as GO-95, may also be applicable.*
***Engineer shall determine if this load effect is significant.*

Table 3.3 Ultimate Strength Design Cases and Load Factors – Substation Structure.

Load Case	*Load Factors and Combinations*
1	$1.1\ D + 1.2\,W\ I_{FW} + 0.75\ SC + 1.1\ T_W$
2	$1.1\ D + 1.2\ I_W\ I_{FI}{}^{*} + 1.2\,W_I\ I_{FIW}{}^{**} + 0.75\ SC + 1.1\ T_W$
3	$1.1\ D + 1.0\ SC + 1.1\ T_W$
4	$1.1\ D + 1.25\ E\ (\text{or}\ E_{FS})\ I_{FE} + 0.75\ SC + 1.1\ T_W$

(with permission from ASCE).
**The importance factor for ice is applied to the thickness.*
***The importance factor for wind with ice I_{FIW} is 1.0.*

D = structure and wire dead load
W = extreme wind load
W_I = wind load in combination with ice
I_W = ice load in combination with wind
E = earthquake load, F_E (without I_{FW})
E_{FS} = earthquake load reactions from first support imposed on the rest of the structure (without I_{FW})
T_W = horizontal wire tension
SC = short circuit load
I_F = importance factors
I_{FW}, I_{FI} = importance factors for wind and ice loads
Earthquake Load $F_E = (S_a/R)\ W_d\ (I_{FE})\ (I_{MV})$ applied at center of gravity of structure
R = structure response modification factor, a function of structural system
(e.g.: cantilever = 2.0)
I_{FE} = importance factor for earthquake loads
W_d = dead load
S_a = spectral response acceleration
I_{MV} = 1.0 for single mode behavior
= 1.5 for multiple vibration modes

transformers. Heavy oil-filled transformers need slab-on-grade footings which are generally designed not to exceed the allowable bearing pressure at site as determined by the Geotechnical engineer.

If frost is present at substation location, then spread footings must be seated below frost depth or a minimum of 12 in. (30.5 cm), whichever is maximum. Design loads on most footings include axial compression, uplift, bending and shear. Most utilities

Table 3.4 Deflection Limitations.

		Maximum Structure Deflection as Ratio of Span Length		
		Structure Class		
Member	*Direction of Deflection*	*Class A*	*Class B*	*Class C*
Horizontal*	Vertical	1/200	1/200	1/100
Horizontal*	Horizontal	1/200	1/100	1/100
Vertical**	Horizontal	1/100	1/100	1/50

(with permission from ASCE).
**Spans for horizontal members should be clear span between vertical supports; for cantilever beams, the distance to the nearest vertical support.*
***Spans for vertical members should be vertical distance from foundation to point under consideration.*

adopt the Ultimate Strength Design philosophy for structural design of foundations, often nominally increasing the ultimate reactions by 10% or so for additional margin of safety.

RUS recommends that soil borings be taken at critical locations (i.e.) deadend structures and heavy transformers. Bearing capacity, ground water level and other soil parameters must be determined. Possibility of differential settlement in silts and silty sands must be checked.

3.4.7 Special structures

Special situations in transmission lines arise at locations such as river crossings, storm structures and air break switches. These are discussed briefly below.

3.4.7.1 *Anti-cascade structures*

Overhead Transmission Lines often face extreme events such as severe ice or wind loads, which damage line sections and affect power supply to customers. Even when the best design criteria are employed, there is always a risk to overhead lines when extreme wind or ice storms *exceed* the design criteria. This damage can occur in poles or supporting structures, insulators or guy wires, depending on the weakest point. Utilities must therefore consider the possibilities of severe wind storms and icing while planning for High Voltage Lines.

Some approaches to limiting the impact of ice and wind loads on overhead lines include:

a) Better forecasting of maximum wind and ice loads
b) Careful design approaches to minimize the risk of failures, while simultaneously reducing the potential consequences of such events

One preferred way of reducing the chance of several miles of cascading line collapse (domino effect) is to install an in-line or tangent deadend or storm structure at chosen locations. Mitigation approaches also include strengthening existing deadends at critical locations by adding or using stronger guy wires and strain insulators along with better hardware.

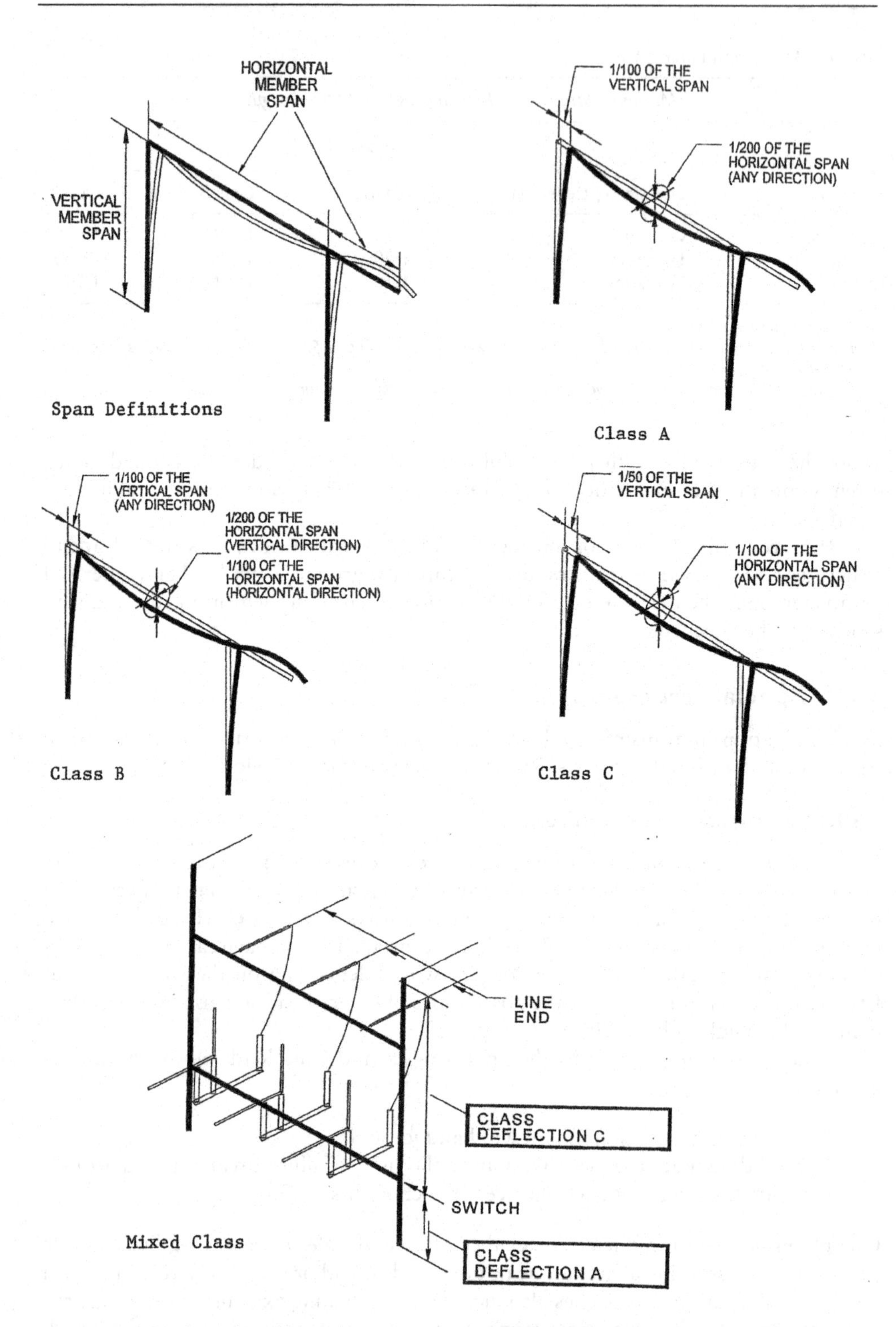

Figure 3.21 Substation Structures – Classes for Deflections (with permission from ASCE).

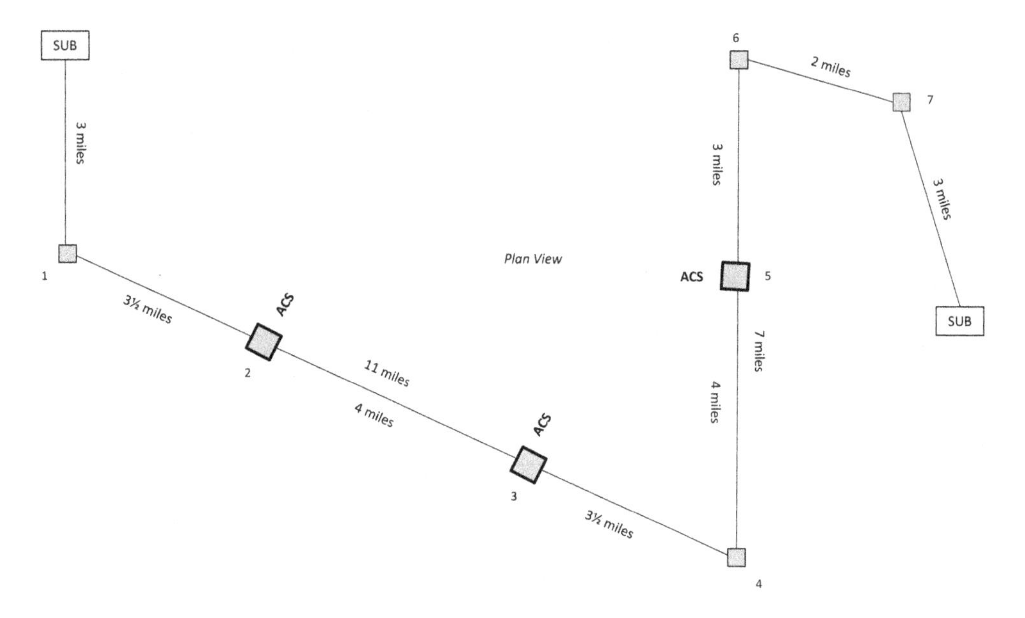

Figure 3.22 Anti-Cascade Structures.

Storm or damage mitigation structures are typically installed every 4 to 5 miles of a transmission line. Theoretical structure configuration is that of a guyed in-line deadend. This means, even if one section of the line is damaged, the other remains intact. The structure introduced in the line to prevent cascading failures is known as "Anti-Cascade Structure" or ACS. Figure 3.22 shows typical anti-cascade structure locations on a transmission line.

3.4.7.2 Long span systems

Figure 3.23 shows typical river crossing structures in a transmission line. Long span designs are very rare and make complex demands on various design-related items such as loadings, wire strengths and foundations as well as regulatory and environmental impacts. The design criteria fall outside the normal scope used on other routine cases; long spans mean much larger loads, larger tensions, increased scope for Aeolian vibration and larger foundation loads, not to mention custom design and installation of dampers on all wires. Galloping checks are therefore an important means of accepting a particular conductor or ground wire.

Not all conductors are amenable to a long span situation and special wires may be needed in the span. Special wires in turn demand special attachment hardware and handling. It is also difficult to choose an optimum optical ground wire given the demand for large tensions. The structures themselves at each end of the river (long) span are much taller than the others, requiring transition structures to gradually reduce the height to normal levels. It is common to design these transition structures as a full deadend capable of resisting either large tensions from the special wires or unbalanced tensions due to wire changes at the transition points.

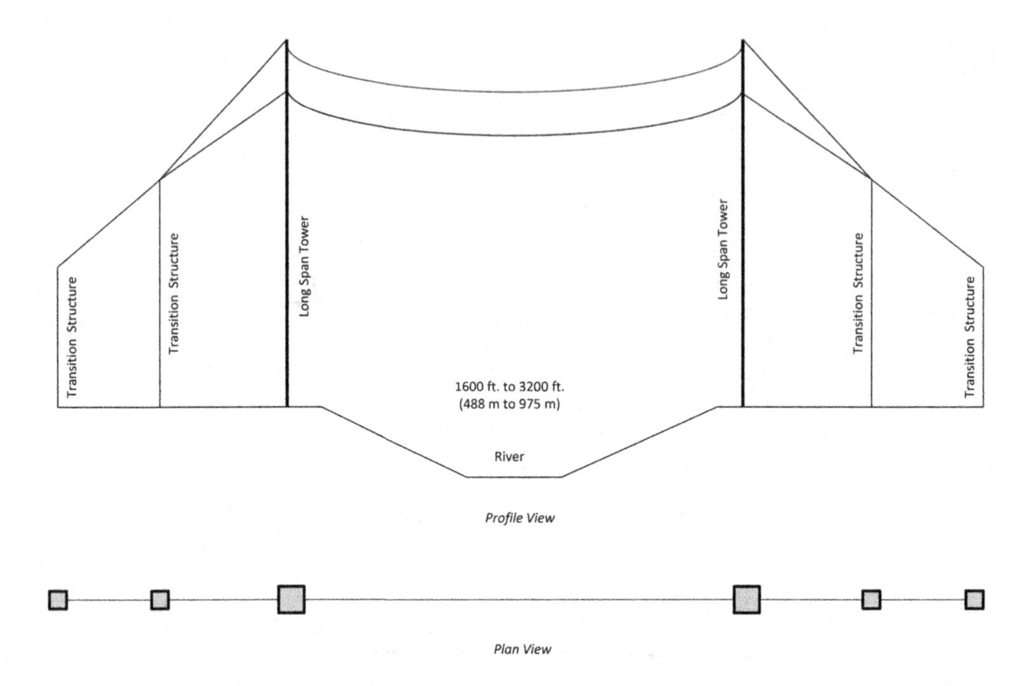

Figure 3.23 Long Span River Crossing Structures.

Constructability, including access to large construction vehicles, is a very important issue after design. Communities living within the vicinity of such large structures often are known to express public opposition, mostly based on aesthetics. Other impediments include special markers, lighting beacons and height constraints if located within the proximity of an airport.

3.4.7.3 *Air-break switches*

Two-way and three-way phase-over-phase (vertical) and low-profile horizontal phase configuration switches are designed specifically for switching applications on transmission lines. They provide economical sectionalizing, and tap and tie switching points for circuit control. These phase-over-phase switches can be mounted on a single pole, minimizing ROW requirements. Installation on a single pole significantly reduces costs of land and equipment that a conventional switching substation requires. Switches rated 69 kV and below can be mounted on any suitable structure; for 115 kV and above, they need laminated wood, steel or concrete poles. For side-break style switches, the operating effort to open and close the switch is minimal even at high voltages.

Figure 3.24 shows a steel pole- mounted three-way air break switch (the framing drawing of a 161 kV switch structure is shown later in Section 3.5.3.5). Analysis and design of switch structures is a specialized process. Briefly, the structures are designed as 3-way deadends with the third wire usually a slack span into another line or substation. Deflection is one of the design criteria and guying is often used for laminated wood poles, especially in the slack span plane.

Figure 3.24 Air Break Switch.

3.4.7.4 *Line crossings*

Figure 3.25 shows a situation where one line crosses another. Crossing clearances are defined on the assumption that the upper circuit is of higher voltage. In most cases, the lower line is a distribution (or another transmission) circuit. Depending on the voltages

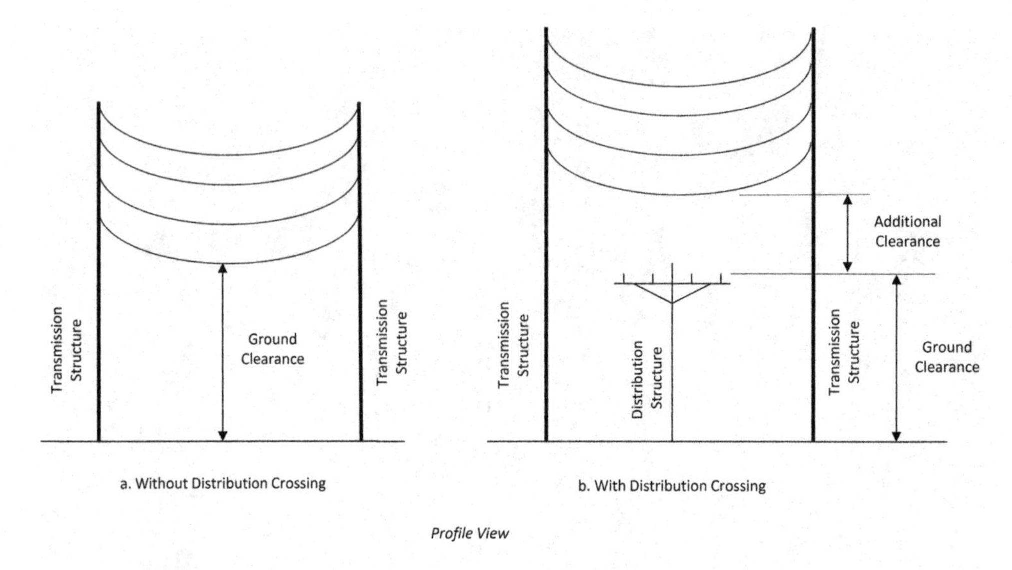

Figure 3.25 Line Crossings.

involved, this situation demands taller structures due to extra clearance required. A special case arises when the two lines are owned by different utilities. Both the RUS Bulletin 200 and NESC provide guidance for wire and structure clearances for all these cases.

3.5 STRUCTURE DESIGN

The actual design of a transmission structure not only involves checking the strength and suitability of various components, but also ensuring that all materials and components meet the requirements of the governing standards or codes. Important items such as grounding are also a part of final structure design configuration and are briefly reviewed.

3.5.1 Strength checks

The following sections cover code-mandated design checks for wood poles, steel poles, lattice towers, concrete and composite poles. Formulae given in this section refer mostly to U.S. codes. Comparable equations from other codes are listed in Appendix 15.

3.5.1.1 Wood poles

The behavior of wood poles used as transmission structures is more complex than steel poles. The basic difference is in the material: wood is orthotropic with low flexural strength. Wood poles are generally sized for normal stresses due to bending and axial loads. A variety of other factors including moisture content, effect of bolt holes, defects and environmental deterioration etc. are also often considered.

The equations for computing the pole section moment capacity at a given elevation are:

Ultimate Moment Capacity	$M_u = S * MOR$	(3.6a)
Allowable Moment Capacity	$M_a = \varphi * S * MOR$	(3.6b)
Structure Strength Usage	$(f_a + f_b)/(MOR * \varphi)$	(3.6c)

where:
φ = Strength Factor
S = Section Modulus = $\pi d^3/32$
d = diameter of pole
f_a = axial stress in pole
f_b = bending stress in pole
MOR = Designated Fiber Stress or Modulus of Rupture (psi)

If the pole is checked per ANSI O5.1, then pole usage must also be calculated at each segment along the pole length with the MOR adjusted for height.

Wood pole grounding

The grounding of a wood pole begins at the pole top where the copper grounding wire is attached to the overhead ground or shield wire with a clamp. From that location, the grounding wire continues down the length of the pole, with down lead clamps every 12 in. (30.5 cm) intervals, and to a ground rod that is installed at a given distance from the pole. The grounding wire is run 18 in. (45.7 cm) below the ground and clamped to the rod. RUS Specifications 810 and 811 (1998) provides another way of grounding wood poles by using a butt wrap where the grounding wire from the pole top travels all the way down the pole and is wrapped in 3 or 4 rounds near the butt of the pole. For H-Frame type wood structures, the same grounding process is applied at both the poles. For additional information on several other methods of H-Frame grounding, the reader is referred to the above mentioned RUS Bulletins.

The grounding of a guyed wood pole depends on whether the shield wire is guyed or unguyed. If it is unguyed, then the procedure for a regular wood pole is adopted. If the shield wire is guyed, the grounding wire is clamped to the shield wire, and then bonded to the guy wire. The guy at the shield wire location then becomes the ground wire and the anchor acts as a ground rod. For additional information on other methods of guyed pole grounding, the reader is referred to the above mentioned RUS Bulletins.

The detail drawings given at the end of this Chapter shows several grounding techniques for wood poles.

Examples 3.3, 3.4, 3.5, 3.6, 3.7 and 3.8 given below illustrate various concepts associated with wood pole design, namely, selection of pole class, buckling, selection of H-Frames and allowable spans for H-Frames and single poles.

Example 3.3 A 55 ft. (16.76 m) Class 1 Douglas Fir (DF) wood pole shown below is subject to transverse loads from a 4-wire distribution circuit (3 phases and 1 neutral). Each load is 500 lbs. (2.23 kN). Assume 18 psf. (Extreme Wind) and neglect moments due to vertical loads. Assume P-Delta effects of 10% and that center of gravity of pole at half the height above ground. RUS Standards apply. Is Class 1 adequate?

Solution:

The pole stands 47.5 ft. (14.48 m) above the ground and embedded 7.5 ft. (2.29 m) into the ground.

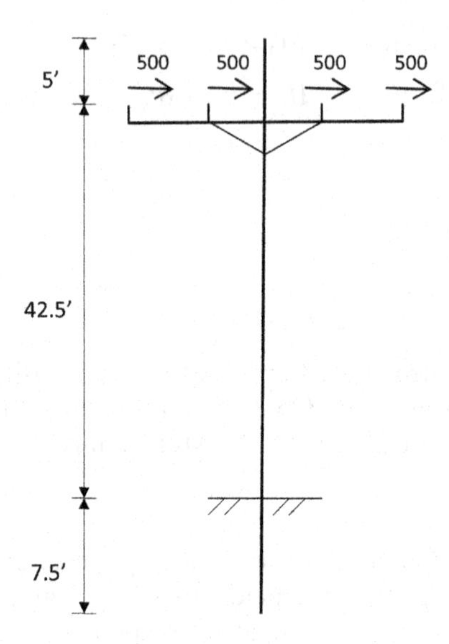

Properties of Class 1 DF pole from pole tables:

d_t = 8.6 in. (21.6 cm) (Table A2.4)
d_{GL} = 14.6 in. (37.1 cm)
Moment due to wire loads = (4) (500)*(47.5 – 5)/1000 = 85 kip-ft. (115.26 kN-m)
Moment due to wind on pole = (18) (47.5) [(8.6 + 14.6)/(2)(12)](47.5/2)/1000 = 19.63 kip-ft. (26.6 kN-m)
(with pole CG assumed approximately at half pole height above ground. Exact value can be computed using a trapezium shape for the pole).
Total Moment applied at GL = 85 + 19.63 = 104.63 kip-ft. (141.88 kN-m)
Add 10% second order effects: 104.63 + 10.46 = 115.1 kip-ft. (156.1 kN-m)
Ultimate capacity of a Class 1 DF pole can be calculated using pole classes, Appendix 2, Table A2.1 as: (47.5–2) (4500)/1000 = 204.75 kip-ft. (277.6 kN-m)
Strength Reduction factor = 0.75 (for Extreme Wind; Table 2.15a and Load Factor of 1.0)
Available Capacity = (0.75) (204.75) = 153.6 kip-ft. (208.2 kN-m) > 115.1 kip-ft.
Therefore Pole is adequate.

Note: Extreme Wind load case is not required for poles shorter than 60 ft. (18.3 m) above ground per NESC. However, RUS requires all poles be checked for this load case regardless of height.

Example 3.4 Consider the following pole situation. Assume an Extreme Wind pressure of 21 psf. (1.0 kPa) and 8% second order effects. Determine the most suitable wood pole class. Note the incline (gain base) at the horizontal post insulators. Use Strength Factor for High Wind = 0.75 and Load Factor of 1.0.

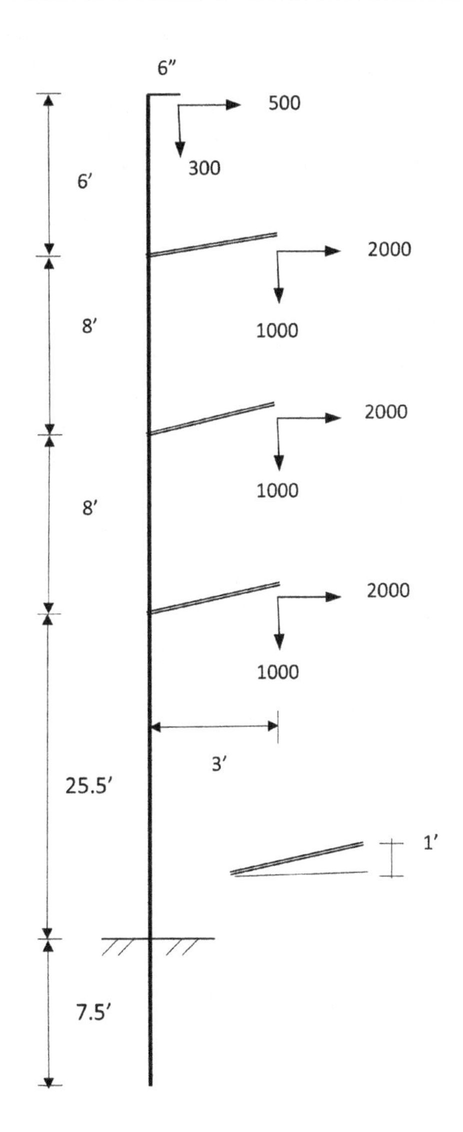

Solution:

The pole is 47.5 ft. above ground.

Moment due to vertical loads M_{VL} = [(300) (½) + (3)(1000) (3)]/1000 = 9.15 kip-ft. (12.41 kN-m)

Moment due to horizontal loads M_{HL} = (500) (47.5/1000) + (2000)([(41.5 + 1) + (33.5 + 1) + (25.5 + 1)])/1000 = 230.75 kip-ft. (312.9 kN-m)

Pole size not given; so assume Class H2 Douglas Fir (DF) pole

Pole diameter at top = 9.9 in. (25.1 cm)

Pole diameter at GL = 16.35 in. (41.5 cm)

Average diameter from GL to top = (9.9 + 16.35)/2 = 13.1 in. (33.3 cm)

Moment due to wind on pole M_{WP} = (21) (47.5) (13.1/12) (47.5/2) (1/1000) = 25.9 kip-ft. (35.1 kN-m)

(with pole CG assumed approximately at half pole height above ground. Exact value can be computed using a trapezium shape for the pole).

Total moment due to wire and wind loads = 9.15 + 230.75 + 25.9 = 265.8 kip-ft. (360.4 kN-m)

P-Delta secondary effects = 8% of total moment = (0.08) (265.8) = 21.3 kip-ft. (28.8 kN-m)

Total applied moment = 265.8 + 21.3 = 287.1 kip-ft. (389.3 kN-m)

Strength Factor = 0.75

Required lateral load capacity for determining pole class = (287.1) (1000/[(47.5-2) (0.75)] = 8,413 lbs. (37.4 kN)

Lateral Load rating for Class H4 = 8,700 lbs. (Appendix 2, Table A2.1)

USE 55 ft. Class H4.

(The student should re-check pole strength with the revised pole diameter of a H4 pole).

Note: Extreme Wind load case is not required for poles shorter than 60 ft. (18.3 m) above ground per NESC. However, RUS requires all poles be checked for this load case regardless of height.

Example 3.5 Determine the ultimate buckling capacity of the multi-guyed wood pole system for the 90 deg. DDE (double deadend) guying configuration. Use the Gere-Carter formula. The pole is guyed in both planes and all guys are inclined to the pole at 45° angle. Other data is as follows:

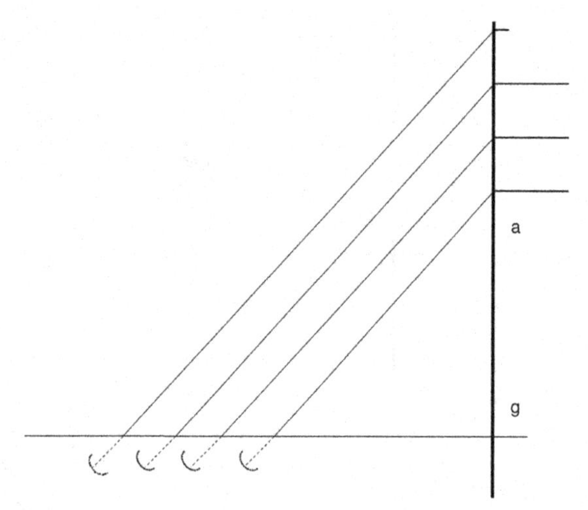

E = 1800 ksi (12.4 GPa)

Spacing between phase wires = 10 ft. (3.05 m)

Distance from lowest phase to ground L = 46 ft. (14 m)

Pole top diameter = 8.6 in. (21.84 cm)

Pole diameter at ground line = 16.8 in. (42.9 cm)

Solution:

The largest unsupported column is the 46 ft. segment between the ground and the lowest guy. Therefore, using the Gere-Carter formula (Equation 3.5) for this segment and assuming fixed-pinned end conditions:

$P_{cr} = 2\pi^2 EI_a/L^2 * (d_g/d_a)^2$
Pole taper = (16.8 − 8.6)/(46) = 8.2/46 = 0.187 in./ft.
Pole diameter at lowest guy = 8.6 + (30) (0.187) = 14.21 in. (36.1 cm) = d_a
I_A = moment of inertia at the location of the lowest guy = (3.1416) (14.21^4/64) = 2001.5 in^4 (83307.1 cm^4)
P_{cr} = [(2) (3.1416^2) (1800) (2001.5)/[(46) (12)]^2] $(16.8/14.21)^2$ = 326.2 kips (1451.7 kN)
Use a factor of safety of 3.0 since this is a DDE.
Design buckling capacity = 326.2/3 = 108.7 kips (483.9 kN)

Example 3.6 Discuss the various situations where you would recommend the below H-Frame systems.

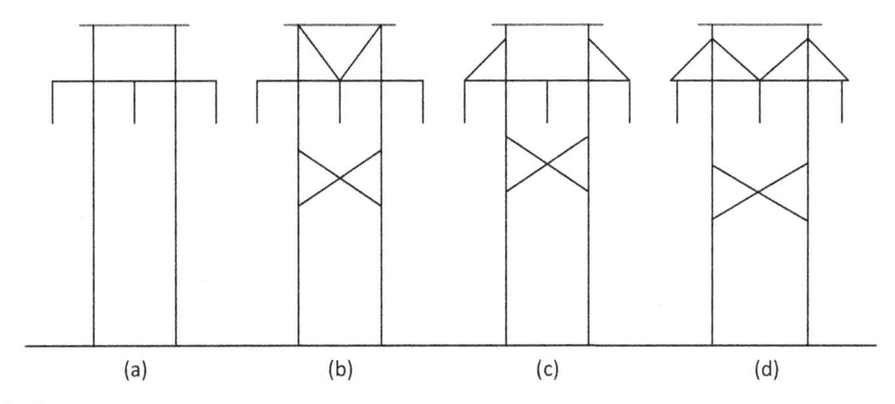

Solution:

(a) Preferred at small to moderate spans on flat terrains. Lateral soil pressure more important for foundation checks given absence of X-bracing.
(b) Used at moderate to large horizontal (wind) spans where transverse loads due to wind control design. Bearing and uplift are more dominant than lateral soil pressure.
(c) This design is used for larger vertical (weight) spans on uneven and hilly terrains.
(d) Preferred choice for large wind and weight spans and higher voltages. A double cross arm can be adopted for large weight spans; an extra X-Brace can help reduce frame bending.

Example 3.7 For the 69 kV braced wood H-Frame shown below, with three (3) phase conductors and two (2) overhead ground wires, determine the maximum allowable wind and weight spans, for Extreme Wind load case, based on the strength of X-Braces. Use 10% reduction in span to account for P-Delta effects and a vertical span to horizontal span ratio of 1.15 to account for uneven terrain.

The following data applies to the problem:

Poles 55 ft. Class 1 DF; Embedment = 7.5 ft. (2.29 m)
Height of pole above ground = h = 55 − 7.5 = 47.5 ft. (14.48 m)
Pole spacing = b = 10.5 ft. (3.20 m)
Pole diameter at top = d_t = 8.6 in. (21.84 cm) or 0.716 ft. (See Table A2.4)
Pole diameter at ground line G = d_G = 14.6 in. (37.1 cm) or 1.216 ft.

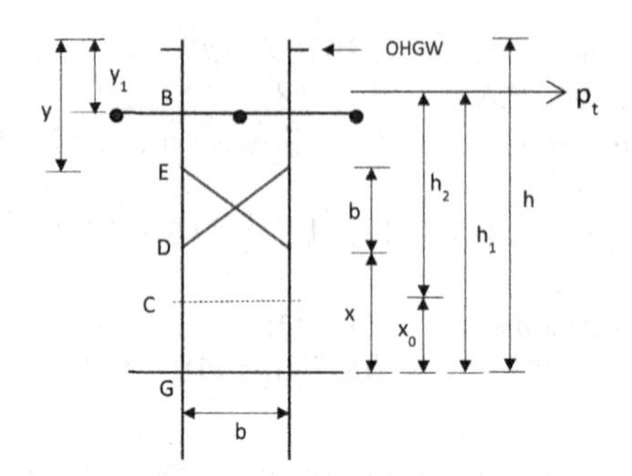

Pole diameter at B = $d_B = 9.48$ in. (24.1 cm) or 0.790 ft.
Pole diameter at E = $d_E = 10.37$ in. (26.3 cm) or 0.864 ft.
Pole diameter at D = $d_D = 11.7$ in. (29.7 cm) or 0.975 ft.
Pole diameter at C = $d_C = 12.93$ in. (32.8 cm) or 1.078 ft.
Moment capacity at ground line = $M_G = 204.75$ kip-ft. (277.6 kN-m) (See Example E3.3)
Distance of OHGW from pole top = $g = 6$ in. (0.15 m)
Distance of Cross Arm from pole top = $y_1 = 7$ ft. (2.13 m)
Distance of X-Brace from pole top = $y = 14$ ft. (4.27 m)
$x = h - y - b = 47.5 - 14 - 10.5 = 23$ ft. (7.01 m)
Load Factor LF = 1.00 (Extreme Wind)
Wind pressure $q = 21$ psf. (1 kPa) (Extreme Wind)
Weight of bare conductor per unit length = $w_c = 1.10$ plf. (16.06 N/m)
Weight of bare ground wire per unit length = $w_g = 0.40$ plf. (5.84 N/m)
Wind load per unit length on conductor = $p_c = 1.75$ plf. (25.55 N/m)
Wind load per unit length on ground wire = $p_g = 0.70$ plf. (10.22 N/m)
(Wind loads refer to Extreme Wind case, bare wire with no ice).
XBS = Strength of X-Braces = 28, 300 lbs. (125.94 kN)

Solution:

Determine location of p_t:
p_t = total horizontal force per unit length = $(2)(p_g) + (3)(p_c) = (2)(0.70) + (3)\ (1.75) =$ 6.65 plf. (97.1 N/m)
To determine location of p_t, use moment equilibrium about pole top.
(p_t)(distance of p_t from pole top, z) = $(2)(p_g)(g) + (3)(p_c)(y_1)$
From which, distance of p_t from pole top, $z = [(2)(p_g)(6/12) + (3)(p_c)(7)]/p_t$
$= [(2)\ (0.70)(0.5) + (3)(1.75)(7)]/6.65 = 5.63$ ft. (1.72 m)
$h_1 = h - z = 47.5 - 5.63 = 41.9$ ft. (12.76 m)

Determine x_o:
$d_D = 11.7$ in. $\rightarrow C_D = \pi d_D = 36.76$ in. (93.36 cm) or 3.063 ft.
$d_G = 14.6$ in. $\rightarrow C_G = \pi d_G = 45.87$ in. (116.51 cm) or 3.822 ft.

From Equation 3.3a:
$x_0/x = C_G(2C_G + C_D)/2(C_G^2 + C_G C_D + C_D^2) = 0.573$
$x_o = (0.573)(x) = (0.573)(23) = 13.18$ ft. (4.02 m)

$h_2 = h_1 - x_o = 41.9 - 13.18 = 28.72$ ft. (8.75 m)

Strength Factor $\varphi = 0.75$ (Extreme Wind)
Maximum Horizontal Span (HS) based on X-Brace strength is given by Equation 3.3f.
$HS_X = \{[\varphi * XBS * b] - [2 * LF * q * (h - x_0)^2 * (2d_t + d_C)/6\}/(LF * p_t * h_2)$
$= \{[(0.75)(28300)(10.5)] - [(2)(1.0)(21)(47.5 - 13.18)^2((2)(0.716) + 1.078)]\}/6$
$/ [(1.0)(6.65)(28.72)]$
$= (222,862.5 - 20964.2)/190.988 = 1,058.5$ ft. (322.64 m)
Reduce 10% due to P-Delta effects.
$HS_X = (0.90)\ (1058.5) = 952.65$ ft. (290.4 m)
Maximum Vertical Span (VS) for VS/HS ratio 1.15:
$VS = (1.15)\ (952.65) = 1{,}095.55$ ft. (333.9 m)

Example 3.8

(a) Determine the allowable horizontal span for the tangent pole with post insulators shown below for the case of Extreme Wind at 21 psf (1005 Pa). The pole is a 75 ft (22.86 m) Class 1 Douglas-Fir wood pole with a ground line bending moment capacity of 283.5 kip-ft (384.4 kN-m). Other data is as follows:

Load Factor for Vertical Load, $LF_v = 1.0$
Load Factor for Wind Load, $LF_t = 1.0$

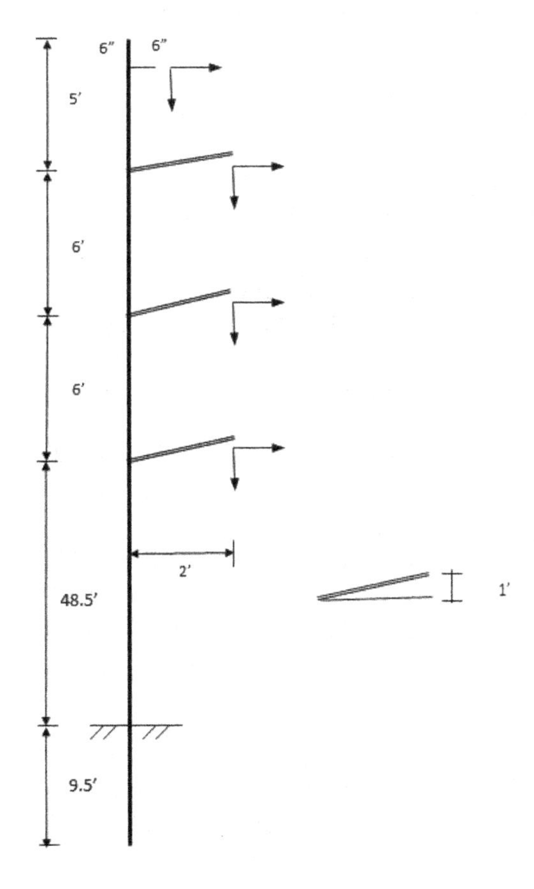

Strength Factor for Wood Pole SF = 0.75 for Extreme Wind
The overhead ground wire is 3/8" EHS and conductors are 477 ACSR 18/1, Pelican. Insulators are horizontal posts as shown. Assume level terrain and ignore deflection effects. Also assume wind span = weight span.

(b) Determine the effects of using $LF_v = LF_t = 1.10$
(c) Determine the allowable span if bundled conductors are used (2 wires per insulator).

Solution:

(a) Let the allowable horizontal span be H.
From wire tables in RUS Bulletin 200, the following data on the ground wire and conductor are obtained.

Diameter ground wire = 0.360 in. (9.1 mm)
conductor = 0.814 in. (20.7mm)
Vertical $w_{gw} = 0.273$ plf (3.98 N/m)
$w_{con} = 0.518$ plf (7.56 N/m)

Compute unit loads due to wind on wires.

Wind $p_{gw} = (21)(0.36)(1/12) = 0.630$ plf (9.19 N/m)
$p_{con} = (21)(0.814)(1/12) = 1.425$ plf (20.8 N/m)

Compute vertical and transverse (wind) loads on the pole for a span of H.

Vertical Loads
Due to ground wire weight = $(w_{gw})(H)(LF_v)$
Due to conductor weight = $(w_{con})(H)(LF_v)$
Wind Loads
Due to wind on ground wire = $(p_{gw})(H)(LF_t)$
Due to wind on conductor = $(p_{con})(H)(LF_t)$

Compute bending moment at ground line due to above loads for $LF_v = LF_t = 1.0$.
Moment due to vertical load at ground wire = w_{gw} H LF_v (6/12) = 0.5 w_{gw} H
Moment due to vertical load at conductor = w_{con} H LF_v(2)(3 wires) = 6 w_{con} H
Moment due to wind load at ground wire = p_{gw} H LT_t (75 − 9.5 − 0.5) = 65 p_{gw} H
Moment due to wind load at conductor = p_{con} H W_{LF} (49.5 + 55.5 + 61.5) = 166.5 p_{con} H

Pole Data pole top diameter = 8.6 in. 21.8 cm)
ground line diameter = 16.3 in. (41.4 cm)
average diameter = (8.6 + 16.3)/2 = 12.45 in. (31.6 cm)
pole height above ground = 65.5 ft. (19.96 m)

Moment due to wind on pole = (21) (12.45/12)(0.5)(65.5^2)/1000 = 46.74 kip-ft. (63.4 kN-m)

Total Applied Moment =
$M_A = 0.5\ w_{gw}\ H + 6\ w_{con}\ H + 65\ p_{gw}\ H + 166.5\ p_{con}\ H + (46.74)\ (1000)$
$= [0.5\ w_{gw} + 6\ w_{con} + 65\ p_{gw} + 166.5\ p_{con}]\ H + 46{,}740$

$= [(0.5)(0.273) + (6)(0.518) + (65)(0.63) + (166.5)(1.425)](H) + 46,740$
$= (281.46)(H) + 46,740$ lb-ft
$M_A \leq (SF)(M_{cap}) \leq (0.75)(283.5)(1000)$ lb-ft
$281.46\ H \leq 212{,}625 - 46{,}740 = 165{,}885$ lb-ft
$H = 589.4$ ft. (179.6 m)

(b) For load factors of 1.10, the allowable horizontal span will be $H = 589.4/1.10 = 535.8$ ft. (163.3 m)
That is, increasing the load factors to 1.10 resulted in a 9% reduction in allowable span.

(c) For bundled conductors, the parameters affected will be the weight and wind load at conductor points.
$w_{con} = (2)(0.518) = 1.036$ plf (15.12 N/m)
$p_{con} = (2)(1.425) = 2.850$ plf (41.6 N/m)
$M_A = [(0.5)(0.273) + (6)(1.036) + (65)(0.63) + (166.5)(2.85)](H) + 46{,}470 \leq (0.75)(283{,}500)$
$521.83\ H \leq 165{,}885$
$H = 317.9$ ft. (96.9 m)
That is, bundling the conductors resulted in almost half the allowable span.

(Note: In a 2-wire horizontal bundle, one conductor shields the other; but this shielding is ignored for conservative purposes).

3.5.1.2 *Steel poles*

Steel poles are sized for bending, axial and shear stresses, supplemented by local buckling checks given the width-to-thickness ratios and stress interaction. The equations for computing the pole section moment capacity at a given elevation are the same as wood except that geometrical properties are a function of shaft thickness and diameter and are given in Appendix A3.

The overall usage of a tubular steel pole is determined with reference to the most highly stressed quadrant of the cross section. This is given by the ASCE interaction equation shown in PLS-POLE™:

$$\text{Structure Strength Usage} \qquad \sqrt{[(f_a + f_b)^2 + 3(f_v + f_t)^2]}/(F_a * \varphi) \tag{3.7}$$

where:
φ = Strength Factor = 1.00
f_a = normal stress due to axial load
f_b = normal stress due to bending
f_v = shear stress due to shear force
f_t = shear stress due to torsion
F_a = Allowable or permissible combined stress per ASCE 48-11

The allowable combined stress F_a is based on D/t (diameter-to-thickness ratio) for round poles and w/t (flat width-to-thickness ratio) for multi-sided polygonal poles.

Section properties for various steel pole cross sections are given in Appendix 3. Other design checks per ASCE 48-11 are as follows:

Permissible Compressive Stress: Rectangular, Hexagonal and Octagonal Members

$$F_a = F_y \qquad \text{when } \frac{w}{t} \leq \frac{260\Omega}{\sqrt{F_y}} \tag{3.8a}$$

$$F_a = 1.42F_y\left[1 - \left(\frac{0.00114}{\Omega}\right)\sqrt{F_y}\left(\frac{w}{t}\right)\right] \qquad \text{when } \frac{260\Omega}{\sqrt{F_y}} < \frac{w}{t} \leq \frac{351\Omega}{\sqrt{F_y}} \tag{3.8b}$$

$$F_a = \frac{104{,}980\Phi}{\left(\frac{w}{t}\right)^2} \qquad \text{when} \frac{w}{t} > \frac{351\Omega}{\sqrt{F_y}} \tag{3.8c}$$

Note: If the axial stress is greater than 1 ksi (6.9 MPa), Equations 3.9 of Dodecagonal members shall be used for rectangular members.

Permissible Compressive Stress: Dodecagonal Members

$$F_a = F_y \qquad \text{when} \frac{w}{t} \leq \frac{240\Omega}{\sqrt{F_y}} \tag{3.9a}$$

$$F_a = 1.45F_y\left[1 - \left(\frac{0.00129}{\Omega}\right)\sqrt{F_y}\left(\frac{w}{t}\right)\right] \qquad \text{when} \frac{240\Omega}{\sqrt{F_y}} < \frac{w}{t} \leq \frac{374\Omega}{\sqrt{F_y}} \tag{3.9b}$$

$$F_a = \frac{104{,}980\Phi}{\left(\frac{w}{t}\right)^2} \qquad \text{when} \frac{w}{t} > \frac{374\Omega}{\sqrt{F_y}} \tag{3.9c}$$

Permissible Compressive Stress: Hexdecagonal Members

$$F_a = F_y \qquad \text{when } \frac{w}{t} \leq \frac{215\Omega}{\sqrt{F_y}} \tag{3.10a}$$

$$F_a = 1.42F_y\left[1 - \left(\frac{0.00137}{\Omega}\right)\sqrt{F_y}\left(\frac{w}{t}\right)\right] \qquad \text{when } \frac{215\,\Omega}{\sqrt{F_y}} < \frac{w}{t} \leq \frac{412\Omega}{\sqrt{F_y}} \tag{3.10b}$$

$$F_a = \frac{104{,}980\Phi}{\left(\frac{w}{t}\right)^2} \qquad \text{when} \frac{w}{t} > \frac{412\Omega}{\sqrt{F_y}} \tag{3.10c}$$

where:

F_a = compressive stress permitted
F_y = specified minimum yield stress
w = flat width of a side
t = wall thickness
Ω = 1.0 for stress in ksi; 2.62 for stress in MPa
Φ = 1.0 for stress in ksi; 6.90 for stress in MPa

Stress Interaction – Round Members

$$\frac{f_a}{F_a} + \frac{f_b}{F_b} \leq 1.0 \tag{3.11}$$

where:
f_a = compressive stress due to axial loads
F_a = compressive stress permitted
f_b = compressive stress due to bending moment
F_b = bending stress permitted

$$F_a = F_y \quad \text{when } D_o/t \leq 3800\ \Phi/F_y \tag{3.12a}$$

$$F_a = 0.75\,F_y + 950\Phi t/D_o \quad \text{when } 3800\Phi/F_y < D_o/t \leq 12000\ \Phi/F_y \tag{3.12b}$$

$$F_b = F_y \quad \text{when } D_o/t \leq 6000\ \Phi/F_y \tag{3.12c}$$

$$F_b = 0.70\,F_y + 1800\Phi t/D_o \quad \text{when } 6000\Phi/F_y < D_o/t \leq 12000\ \Phi/F_y \tag{3.12d}$$

D_o = outside diameter of the tubular section
t = wall thickness
Φ = 1.0 for stress in ksi; 6.90 for stress in MPa

Shear Stress Interaction

$$\frac{VQ}{Ib} + \frac{Tc}{J} \leq F_v \quad \text{where } F_v = 0.58\,F_y \tag{3.13}$$

where:
F_v = shear stress permitted
V = shear force
Q = moment of section about neutral axis (NA)
I = moment of inertia
T = torsional moment
J = torsional constant of cross section
c = distance of NA to extreme fiber
b = 2*wall thickness, t

Bending Stresses

$$Mc/I \leq F_t \quad \text{or} \quad Mc/I \leq F_a \tag{3.14}$$

where:
M = bending moment
F_t = tensile stress permitted
F_a = compressive stress permitted

Stress Interaction – Combined Stresses

Polygonal Members

$$\left\{[(P/A) + (M_x c_y/I_x) + (M_y c_x/I_y)]^2 + 3\left[\frac{VQ}{It} + \frac{Tc}{J}\right]^2\right\}^{0.5} \leq F_t \text{ or } F_a \tag{3.15a}$$

Round Members

$$\left\{[(P/A)+(M_x c_y/I_x)+(M_y c_x/I_y)]^2+3\left[\frac{VQ}{It}+\frac{Tc}{J}\right]^2\right\}^{0.5}\leq F_t \text{ or } F_b \quad (3.15b)$$

where:
F_t = tensile stress permitted = P/A_g or P/A_n
F_a = compressive stress permitted
F_b = bending stress permitted
M_x = bending moment about X-axis
M_y = bending moment about Y-axis
I_x = moment of inertia about X-axis
I_y = moment of inertia about Y-axis
c_x = distance of point from Y-axis
c_y = distance of point from X-axis
t = wall thickness
P = axial force on member
A = cross-sectional area
V = total resultant shear force
Q = moment of section about neutral axis
T = torsional moment
J = torsional constant of cross section

Tensile Stresses

$$P/A_g \leq F_t \quad \text{when } F_t = F_y \quad (3.15c)$$

$$P/A_n \leq F_t \quad \text{when } F_t = 0.83F_u \quad (3.15d)$$

where:
F_u = specified minimum tensile stress
A_g = gross cross sectional area
A_n = net cross sectional area

Anchor Bolts (per ASCE-48-11)

Minimum development length in concrete $L_d = l_d\alpha\beta\gamma$ (3.16a)

l_d = basic development length defined for various bolt sizes

For bars up to and including # 11 (i.e.) $1\frac{3}{8}$ in. diameter: $l_d = 1.27\Gamma A_g F_y/\sqrt{f'_c}$ (3.16b)

or $l_d = 0.400\Phi d F_y$ (use larger)

For bars # 14 (i.e.) 1¾ in. diameter: $l_d = 2.69\Theta F_y/\sqrt{f'_c}$ (3.16c)

For bars # 18J (i.e.) 2¼ in. diameter: $l_d = 3.52\Theta F_y/\sqrt{f'_c}$ (3.16d)

where:
A_g = gross area of anchor bolt
A_{sr} = required tensile stress area of anchor bolt

F_y = specified minimum yield stress of bolt material
f'_c = compressive strength of concrete
d = anchor bolt diameter
$\Gamma = 1.00$ for stress in *ksi* and area in in^2; 0.015 for stress in MPa and area in mm^2
$\Phi = 1.00$ for stress in *ksi* and diameter in inches; 0.145 for stress in MPa and diameter in mm
$\Theta = 1.00$ for stress in ksi; 9.67 for stress in MPa
$\alpha = 1.0$ if F_y is 60 ksi and 1.2 if F_y is 75 ksi
$\beta = 0.8$ if bolt spacing $\geq$ 6 in. (15.2 cm); 1.0 if spacing is less than 6 in. (spacing measured center-to-center of anchor bolt)
$\gamma = A_{sr}/A_g$

ASCE 48-11 (2011) recommends that the absolute minimum length of an anchor bolt using deformed rebars shall not be less than 25 times the bar diameter. This requirement is to prevent the usage of unusually short bars.

Steel davit arms

The most common structures with davit arms are tangent (suspension) and light angle poles. Longer davit arms facilitate greater phase separation without increase in structure height. Upswept davit arms are often used for aesthetic reasons with the upsweep ranging from 6 in. (15.2 cm) to 18 in. (45.7 cm). The length and vertical spacing of davit arms depend on required phase separation for that particular voltage, clearances for insulator swing, galloping conditions etc. Arms are also employed for shield wires to reduce the shielding angle and improve lightning protection. Such arm lengths are determined solely by the required shielding angle.

Other than some nominal guidance offered in ASCE 48-11, there is no standardized design method for steel davit arms. Designs are generally based on empirical approaches, finite element analysis as well as inferences from full-scale testing. Davit arms with suspension or light angle insulators are usually designed as a cantilever subject to vertical and transverse loads. Stress checks for steel davit arms are similar to those of steel pole shafts described in the previous section. If PLS-POLE™ program can be used for design, then axial, shear, bending and torsional stresses produced by each load case are checked.

Connections to the steel pole usually involve brackets with pin-type bolts transferring flexure and shear effects. For double circuit structures, where davit arms are needed on both sides of the pole, the connection involves brackets with through plates. In a majority of cases, the pole maker or fabricator designs, fabricates and details the davit arms. Wind-induced vibration of unloaded davit arms is a big concern during construction; weights are often suspended from the arm tips to provide some damping. Another means of reducing vibration effects is to tie the arm tips with a cable to the pole.

Items such as vangs welded to steel poles for installing running angle or strain insulators are designed for full tension effects. Vangs are fabricated and welded to the pole shaft prior to galvanizing.

Base plates

Currently only one design guide is available – AISC (2006) – to provide guidance for analysis and design of base plates for tubular steel poles supported by concrete piers.

Pole fabricators also have in-house and proprietary design processes which vary widely from one manufacturer to another. However, ASCE 48-11 outlines a procedure based on effective bend lines and a 45° bend-line limitation. The manual also gives equations to calculate effective anchor bolt loads, base plate stresses and plate thickness.

Steel pole grounding

Steel transmission poles are usually provided with a metal grounding pad welded to the side of the pole about 12 in. (30.5 cm) to 18 in. (45.7 cm) above the ground or base plate. The shield wire is bonded to the steel pole near the pole top with a copper (grounding) wire and a stainless steel nut. Grounding rods, if used, are usually installed about 3 ft. (0.9 m) from the pole connected to the ground pad.

Examples 3.9, 3.10 and 3.11 given below illustrate some concepts associated with steel pole design, namely, thickness requirement and determination of number of anchor bolts for a pole transferring moment to the base plate and concrete pier.

Example 3.9 An 80 ft. (24.4 m) steel pole has been rated for a GL moment capacity of 280 kip-ft. (379.7 kN-m). The pole diameters are 8.7 in. (22.1 cm) at the top and 20.4 in. (51.8 cm) at the butt. These are mean diameters measured to the midpoint of the thickness across the flats. Assume 12-sided cross section, uni-axial bending about X-axis, a material yield stress of 65 ksi (448.2 MPa) and standard embedment of 10% + 2 ft. Neglect axial, shear and torsional stresses. Determine the approximate pole shaft thickness required. Verify if thickness satisfies local buckling criteria.

Solution:

Assume w/t criterion is satisfied so that $F_a = F_t = F_y$
$M_{GL} = 280$ kip-ft. = 3,360 kip-in = M_x and $M_y = 0$; $P = V = T = 0$
80 ft. pole implies 70 ft. above ground and 10 ft. embedment.
Pole taper = (20.4 – 8.7) /80 = 0.146 in. /ft.
Pole diameter at GL = 8.7 + (0.146) (70) = 18.9 in. (48.1 cm)
From Equation 3.15a:
$(M_x\ C_y)/I_x \leq F_t$ where I_x is the Moment of Inertia.
Use Table A3.4 for cross sectional geometric properties.
Assume maximum moment occurs at a location defined by $\alpha = 90$ deg.
$C_y = 0.518\ (D + t)\ \text{Sin}\ (\alpha) = 0.518\ (D + t)\ (1.0) = 0.518\ (D + t)$
Moment of Inertia of a 12-sided pole = $I_x = 0.411\ D^3\ t$
$I_x/C_y = 0.411\ D^3\ t/[0.518\ (D + t)]$
Therefore $3360 = 0.411\ D^3 t/[0.518\ (D + t)]\ (65)$
Solving for "t", we have t = 0.184 in or 3/16 in. (4.76 mm)

Local Buckling Checks:
For a 12-sided pole, the flat width is given by $w = 0.268\ (D - t - 2\ BR)$ per Appendix A3.4
Assuming bend radius BR is approximately equal to 4 times t, we have:
w = 0.268 (18.9 – (3/16) – (8)(3/16)) = 4.61 in. (11.72 cm)
Width-to-thickness ratio = w/t = 4.61/0.1875 = 24.58
Limiting w/t ratio for 12-sided sections = $240/(65)^{0.5} = 29.76$ (Equation 3.9a with $\Omega = 1.0$)
24.58 < 29.76 OK

Therefore our assumption about width-to-thickness ratio is valid.
(See Table A3.8 to verify this analysis).

Example 3.10 For the dual use pole system shown below, determine the required pole height. The transmission circuit is 3-phase 161 kV and the under-build distribution is 34.5 kV. Assume sag of the transmission and distribution conductors as 8.0 ft. (2.44 m) and 6.0 ft. (1.83 m), respectively. Use Tables 2.6a-1, 2.6b-1 and 2.7. Assume a buffer of 2.0 ft. (0.61 m) for ground clearance.

Solution:

The process basically involves determining the various wire spacing associated with the system. From Table 2.6a-1:
Required ground clearance for 34.5 kV = 18.7 ft. (5.7 m)
From Table 2.7:
S = shield wire to phase separation = 4.3 ft. (1.31 m)
P = phase to phase clearance = 6.7 ft. (2.04 m)
Sag of transmission conductor = 8.0 ft. (2.44 m)
C1 = clearance from sagged 161 kV wire to 34.5 kV wire = 7.0 ft. (2.13 m) (Table 2.6b-1)

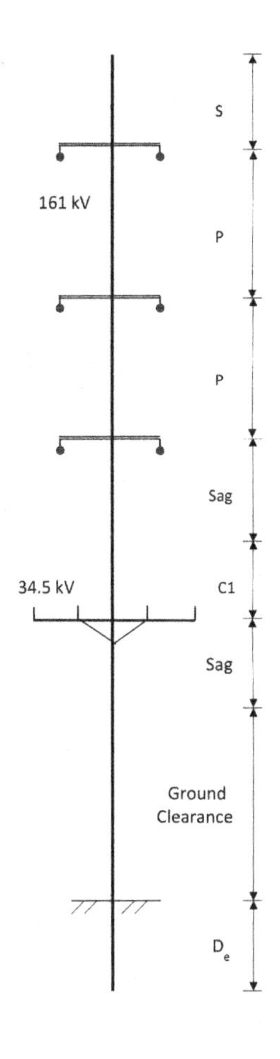

Sag of distribution conductor = 6.0 ft. (1.83 m)
Therefore required height of the pole above ground is:
H = 4.3 + 6.7 + 6.7 + 8.0 + 7.0 + 6.0 + 18.7 + 2.0 = 59.4 ft. (18.11 m)
Consider a 70 ft. (21.34 m) pole with 9 ft. (2.74 m) embedment D_e giving 61 ft. (18.6 m) above ground.
70 ft. Pole is adequate.

Example 3.11 Determine the approximate number of anchor bolts required for the following situation:

Pole GL diameter = 48 in. (121.9 cm)
Moment transmitted from pole loads = 4800 kip-ft. (6509 kN-m)
Use ASTM A615 # 14 anchor bolts (1.75 in. diameter) with a ultimate tensile strength of 100 ksi. (689 MPa). Assume 5 threads per inch. Consider bending effects only.

Solution:

(The aim of this problem is to illustrate a quick approximate method for determining the *preliminary* number of anchor bolts needed for a steel pole with a base plate. The exact number must be determined either by a detailed analysis per ASCE 48-11 or via the PLS-POLE™ program.)

The relationship between pole shaft, base plate, anchor bolt circle and pier is shown below. For the 48 in. diameter pole shaft, the approximate size of the anchor bolt circle is 54 in (see Figure 5.9 in Chapter 5). Assume all anchor bolts arranged in a circular fashion.

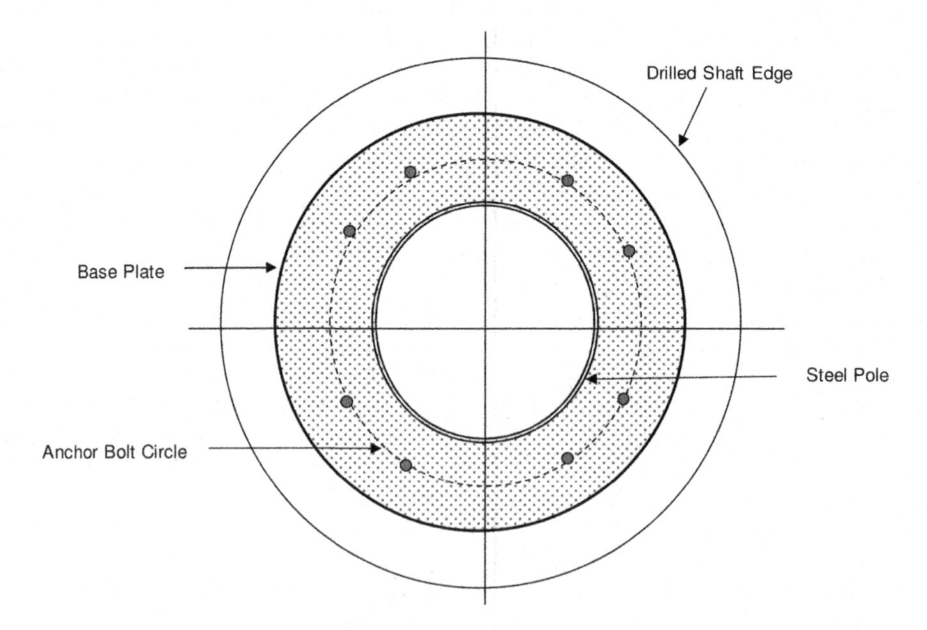

Distance to the nearest anchor bolt from shaft center = 54/2 = 27 in. (76.2 cm)
From ASCE 48-11, cross sectional stress area of an anchor bolt $A = (\pi/4)[d - (0.9743/n)]^2$

Where 'n' is the number of threads per unit length.
A = (3.1416/4) [1.75 – (0.9743/5)]^2 = 1.90 in^2 (1225.5 mm^2)
(Area reduction due to bolt threads is often considered during such calculations).
Permitted tensile stress in anchor bolt = 0.75 F_u
Capacity of a single anchor bolt = (1.90) (100)(0.75) = 142.5 kips (634.1 kN)
Number of anchor bolts required = Moment/(bolt capacity)(lever arm)
= (4800)(12)/(142.5)(27) = 14.9 or 15 per half circle
Use 30 anchor bolts for the whole pole, arranged in a circular fashion.

3.5.1.3 *Lattice towers*

Design checks given below for angle members in steel lattice towers refer to ASCE 10-15 (2015) and include allowable slenderness ratios and associated tensile and compressive stresses in steel angles. TOWER™ program also determines the usage level of each member group based on ASCE. Connection checks include bolt shear, bearing, block shear capacity and tensile rupture.

Compression Capacity is the minimum of:
(a) Member compressive strength based on slenderness ratio, kL/r
(b) Connection shear capacity
(c) Connection bearing capacity

Tension Capacity is the minimum of:
(a) Member tensile strength based on net section
(b) Connection rupture
(c) Connection shear capacity
(d) Connection bearing capacity

Angles in Compression:

Design compressive stress $$F_a = [1 - 0.5(\eta^2/C_c{}^2)]F_y \quad \text{when } \eta \le C_c \tag{3.17a}$$

$$= \pi^2 E/\eta^2 \quad \text{when } \eta > C_c \tag{3.17b}$$

where:
$\eta = kL/r$ = slenderness ratio
k = effective length coefficient
r = radius of gyration
L = unbraced length
F_y = minimum guaranteed yield stress of steel
$C_c = \pi\sqrt{2E/F_y}$
E = modulus of elasticity

Maximum permitted angle flat width-to-thickness ratio (w/t) = 25
If the w/t ratio exceeds the limit given by

$$\left(\frac{w}{t}\right)\lim = \frac{80\psi}{\sqrt{F_y}} \tag{3.18}$$

Then the revised design compressive stress (F_a) is obtained by replacing F_y in Equation 3.17a, and in C_c, with F_{cr} given below.

$$F_{cr} = \left[1.677 - \left(\frac{0.677\left(\frac{w}{t}\right)}{\left(\frac{w}{t}\right)_{\text{lim}}}\right)\right] F_y \quad \text{when } \left(\frac{w}{t}\right)_{\text{lim}} \leq \left(\frac{w}{t}\right) \leq \frac{144\psi}{\sqrt{F_y}} \tag{3.19a}$$

$$= \frac{0.0332\pi^2 E}{\left(\frac{w}{t}\right)^2} \quad \text{when } \left(\frac{w}{t}\right) > \frac{144\psi}{\sqrt{F_y}} \tag{3.19b}$$

In all equations above, $\psi = 1$ for stress in ksi and 2.62 if stress is in MPa.

Compressive capacity of the angle member is given by the design compressive stress from above equations times the gross cross sectional area.

Effective Lengths of Angle Members

For leg members bolted in both faces:

$$kL/r = L/r \quad \text{for } 0 \leq L/r \leq 150 \tag{3.20a}$$

Unsupported panels

For other compression members with concentric load at both ends:

$$kL/r = L/r \quad \text{for } 0 \leq L/r \leq 120 \tag{3.20b}$$

For other compression members with concentric load at one end and normal framing eccentricity (NFE) at the other:

$$kL/r = 30 + 0.75L/r \quad \text{for } 0 \leq L/r \leq 120 \tag{3.20c}$$

For other compression members with NFE at both ends:

$$kL/r = 60 + 0.50L/r \quad \text{for } 0 \leq L/r \leq 120 \tag{3.20d}$$

For other compression members unrestrained against rotation at both ends:

$$kL/r = L/r \quad \text{for } 120 \leq L/r \leq 200 \tag{3.20e}$$

For other compression members partially restrained against rotation at one end:

$$kL/r = 28.6 + 0.762L/r \quad \text{for } 120 \leq L/r \leq 225 \tag{3.20f}$$

For other compression members partially restrained against rotation at both ends:

$$kL/r = 46.2 + 0.615L/r \quad \text{for } 120 \leq L/r \leq 250 \tag{3.20g}$$

NFE is the Normal Framing Eccentricity which is defined as the condition when the centroid of the bolt pattern is located between the heel of the angle and the centerline of the connected leg. NFE may cause a reduction of up to 20% in the axial capacity of short, stocky, single angle struts.

Angles in Tension:

For angle members connected by one leg:

$$\text{Design tensile stress on net cross-sectional area} = F_t = 0.90\, F_y \quad (3.21a)$$

$$\text{Net Section Capacity } N_{cap} = A_{net} F_t \quad (3.21b)$$

(For angle members connected in both legs, the 0.90 factor is replaced by 1.0 (i.e.) $F_t = F_y$).

N_{cap} is the strength based on tearing of a member across its net area A_{net} which is defined as:

$$A_{net} = A_g - (d)(t)(n_b) \quad (3.21c)$$

where:
A_g = gross cross-sectional area of the angle section
d = bolt hole diameter (generally 1/16 in. or 1.6 mm more than bolt diameter)
n_b = number of holes
t = thickness of angle

If there is a chain of holes in a zigzag fashion in an equal leg angle, then the net width of an element h_n and the net area A_{net} must be determined as follows:

$$h_n = [2h - (n_b)(d) + n_g(s^2/4g)] \quad (3.22a)$$

$$A_{net} = (h_n)(t) \quad (3.22b)$$

where:
h = width of angle leg
n_b = number of bolt holes in the chain
n_g = number of gauge spaces in the chain
s = bolt spacing or pitch along the line of force
g = gauge length or transverse spacing of the bolts

If the centroid of the bolt pattern on the connected leg is outside the center of gravity of the angle, then all connections must be checked for block shear or rupture using Equation 3.23:

$$R_{BSH} = 0.60 A_v F_u + A_t F_y \quad (3.23)$$

where:
A_v = minimum net area in shear along a line of transmitted force for a single angle = (t) {a + (n_b − 1)(b)}
t = angle thickness
a = effective end distance = e − d/2 (See Figure 3.20)
b = s − d (See Figure 3.20)
s = bolt spacing, center to center
F_u = specified minimum tensile strength of the angle steel

F_y = specified minimum yield stress of the angle steel
A_t = (t)(c)
c = effective edge distance = f – d/2 (See Figure 3.20)
d = bolt hole diameter
e, f = as shown in Figure 3.20
n_b = number of bolt holes

Connection rupture often occurs due to insufficient edge and end distances as well as bolt spacing. ASCE 10-15 therefore specifies the following minimum values for the parameters.

Minimum end and edge distances

The minimum end distance 'e' (inches) shall be the largest value of:

$$e = 1.2\,P/tF_u \quad \text{or} \tag{3.24a}$$

$$= 1.3\,d \quad \text{or} \tag{3.24b}$$

$$= t + (d/2) \tag{3.24c}$$

P = force transmitted by bolt
d = nominal bolt hole diameter

Minimum edge distance 'f' (inches) shall not be less than:

(a) 0.85 e_{min} for a rolled edge
(b) 0.85 e_{min} + 0.0625 in. for a sheared or mechanically-cut edge where e_{min} is the largest value determined from Equations 3.24a, 3.24b and 3.24c.

Bolt spacing

The center-to-center distance between bolt holes shall not be less than:

$$s_{min} = 1.2\,P/tF_u + 0.6\,d \tag{3.24d}$$

Tower grounding

The process discussed for steel poles is also applicable to steel towers, except that grounding must be facilitated at a minimum of 2 tower legs. In situations where ground resistance is not optimum, all 4 legs can be grounded.

Examples 3.12, 3.13, 3.14 and 3.15 given below illustrate various concepts associated with tower design, namely, crossing diagonals, allowable compressive stress, net section areas and block shear determination.

Example 3.12 Determine the effective buckling lengths of the crossing diagonals of the tower panel shown below.

b = 11.7 ft. (3.57 m)
v = 13.25 ft. (4.04 m)
k = 1.483 ft. (0.452 m)

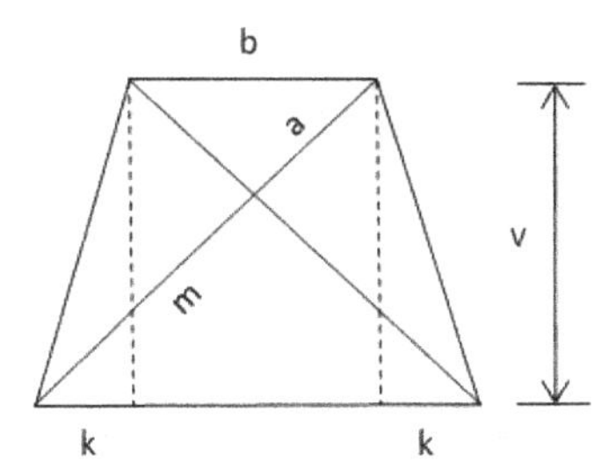

Solution:

The AISC Steel Manual's cross bracing equations were used to estimate the lengths of the main diagonals. The diagonals usually have a bolt at the meeting point which helps with reducing effective buckling lengths.

$$m = \sqrt{(b+k)^2 + v^2}$$
$$= \sqrt{(11.7 + 1.483)^2 + 13.25^2}$$
$= 18.69$ ft.(5.70 m) (full diagonal)
$a = 0.5mb/(b+k)$
$= (0.5)(18.69)(11.7)/(11.7+1.483)$
$= 8.294$ ft. (2.53 m) (shorter portion of diagonal)
$a' =$ longer portion of diagonal $= m - a = 18.69 - 8.294 = 10.40$ ft. (3.17 m)
The maximum unbraced length of the diagonal for buckling is 10.4 ft.
Length ratios:
Ratio $a/m = 8.294/18.69 = 0.444$
Ratio $a'/m = 10.40/18.69 = 0.556 \leftarrow$ controls for buckling

Example 3.13 For the 10 ft. (3.05 m) tower angle member shown, determine the design compressive stress F_a. Assume $F_y = 50$ ksi (344.8 MPa) and $E = 29{,}000$ ksi (200 GPa).

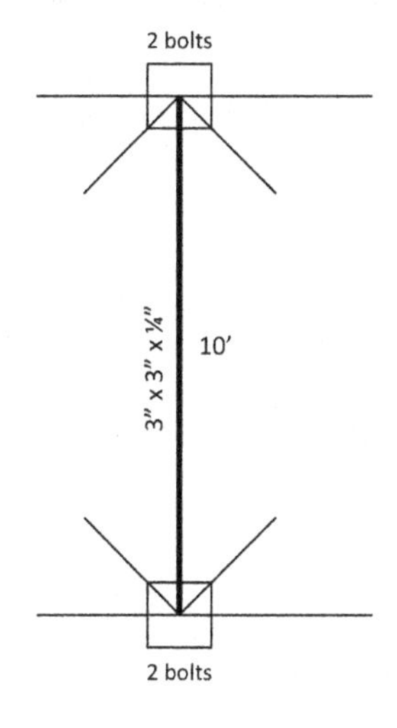

Solution:

For the 3" × 3" × ¼" angle:
Cross Sectional Area = 1.44 in^2 (929 mm^2)
Slenderness Ratios:
r_x = 0.93 in. (23.6 mm)
r_y = 0.93 in. (23.6 mm)
r_z = 0.59 in. (15 mm)

The presence of two bolts at each end and the member framing into other members at the joints provides for partial restraint. Assume normal framing eccentricity at both ends. Using Equation 3.20g:
$kL/r = 46.2 + 0.615\ L/r = 46.2 + (0.615)\ ((10)(12)/r) = 46.2 + 73.8/r$
About X- and Y-axes: $kL/r_x = kL/r_y = 46.2 + 73.8/0.93 = 125.6$
About Z-axis: $kL/r_z = 46.2 + 73.8/0.59 = 171.3$ – controls
The width-to-thickness ratio of the member is w/t = 3/0.25 = 12

Limiting w/t ratios:
$(w/t)_{lim1} = 80/\sqrt{F_y} = 80/7.071 = 11.3$
$(w/t)_{lim2} = 144/\sqrt{F_y} = 144/7.071 = 20.4$
$11.3 < 12 < 20.4$
$(w/t)_{lim1} < (w/t) < (w/t)_{lim2}$

Therefore, from Equation 3.19a:
$F_{cr} = [1.677 - (0.677)\ (12/11.3)]\ (50) = 47.93$ ksi (330.24 MPa)
$C_c = \pi\sqrt{(2)(29000)/47.93} = (3.1416)\ (34.78) = 109.3$
$kL/r > C_c$

Design compressive stress using Equation 3.17b:
$F_a = \pi^2 E/(kL/r)^2 = (3.1416^2)(29000)/171.3^2 = 9.8$ ksi (67.6 kPa)
Corresponding design compressive force = (1.44) (9.8) = 14.1 kips (62.8 kN)

Note: The above value is only from slenderness point of view. The connection's shear and bearing capacity must also be evaluated to determine which one controls.

Example 3.14 For the 5" × 5" × ½" steel angle shown, determine:

(a) Net area
(b) Net section capacity in tension
(c) If the end and edge distances shown are adequate. (assume force transmitted as 20 kips). Assume a rolled edge.

All bolts are ¾ in. (19 mm) diameter. F_y = 36 ksi (248 MPa) and F_u = 58 ksi (400 MPa).

Solution:

Angle properties: Area A_g = 4.75 in^2 (30.65 cm^2)
Bolt hole diameter = ¾ + 1/16 = 13/16 in. (2.06 cm)

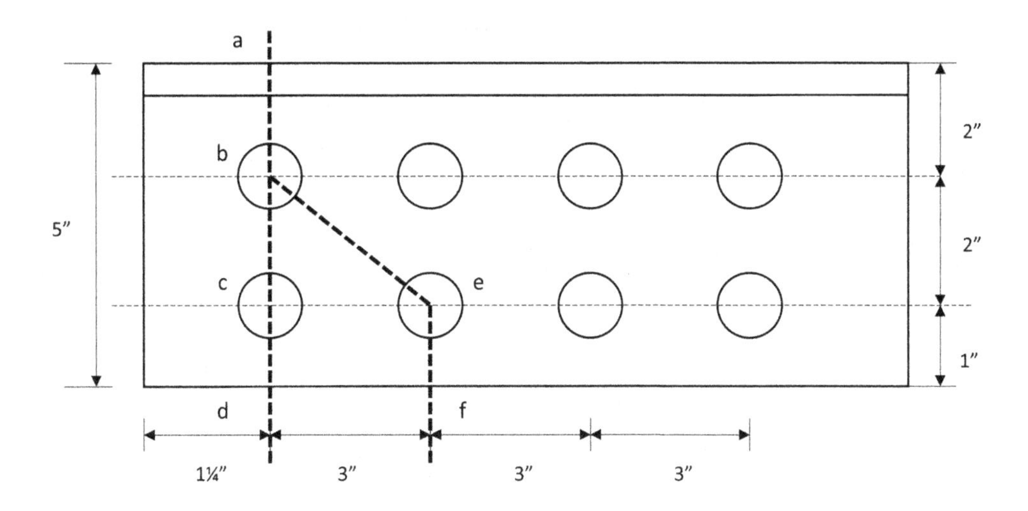

Tensile failure can occur along any of the two paths shown – *abcd* or *abef*.
s = pitch = 3 in. (76.2 mm)
g = gauge = 2 in. (50.8 mm)

(a) Net Area on *abcd* = (½) [(2)(5) – (2)(13/16)] = 4.1875 in^2 (2701.6 mm^2) – controls
Net Width on *abef* = [(2)(5) – (2)(13/16)] + [(3^2)/(4)(2)] = 9.5 in. (241.3 mm)
Net Area on *abef* = (½) (9.5) = 4.75 in^2 (3064.5 mm^2)

(b) Net Section capacity = (A_{net}) (0.9) (F_y) = (4.1875)(0.90)(36) = 135.68 kips (603.8 kN)

(c) Minimum end distance is the largest of:
t + d/2 = 0.50 + 0.75/2 = 0.875 in. (22.2 mm)
1.30d = (1.30) (0.75) = 0.975 in. (24.77 mm) – controls
(1.20)(20)/(58)(0.5) = 0.8275 in. (21 mm)
The end distance provided is 1.25 in. (31.75 mm) and is adequate.
Minimum edge distance is 0.85 e_{min} = (0.85) (0.975) = 0.83 in. (21 mm)
The edge distance provided in 1 in. (25.4 mm) and is adequate.

Example 3.15 Determine the block shear capacity of the above the 3" × 3" × ¼" steel angle. All bolts are ¾ in. (19 mm) diameter. F_y = 36 ksi (248 MPa) and F_u = 58 ksi (400 MPa).

Solution:

Bolt hole diameter = ¾ + 1/16 = 13/16 in. (2.06 cm)

From steel tables, the CG of the angle section is at a distance of 0.842 in. (21.4 mm) from the heel. The bolt pattern line is outside the CG. Equation 3.23 applies.

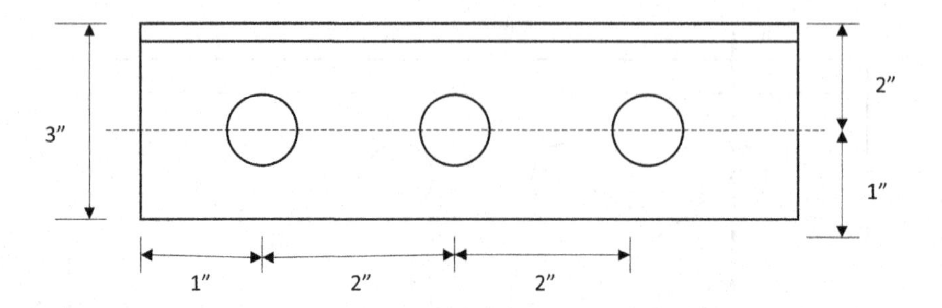

Force to cause rupture of the joint is given by $P = (0.60)(A_v)(F_u) + (A_t)(F_y)$
$A_v = (0.25)[(1 - (0.5)(13/16) + (2)(2 - 13/16)] = 0.742\ in^2\ (478.8\ mm^2)$
$A_t = (0.25)[(1 - (0.5)(13/16))] = 0.148\ in^2\ (95.76\ mm^2)$
$P = (0.60)(0.742)(58) + (0.148)(36) = 31.15$ kips (138.6 kN)

3.5.1.4 Concrete poles

The behavior of reinforced concrete poles used as transmission structures is more complex than steel given the basic difference in material – while steel is a uniform and isotropic material, concrete is anisotropic, non-linear with low tensile strength (i.e.) susceptible to cracking at even moderate bending. Concrete poles are therefore evaluated not only in terms of the ultimate moment (factored loads), but also the initial cracking moment at service loads and zero tension moment (no cracking). The initial cracking strength will be roughly 40% to 55% of the ultimate strength while zero tension strength is about 70% to 85% of the initial cracking strength. Element usage is determined only as a function of bending moment. Derivations of various equations are available in ASCE Manual 123 (2012).

The behavior of concrete poles subject to axial and flexural loads is an explicit function of f'_c, the concrete 28-day strength which in turn controls the Modulus of Elasticity, E_c, and thereby, the pole's bending resistance. Figures 3.26 and 3.27 show the assumed stress distribution used in deriving the relevant moment expressions. The equations for computing pole cross section moment capacity are based on section equilibrium and at a given elevation are as follows:

Ultimate moment capacity

$$\varphi M_u = \Sigma e_i A_{psi} f_{sei} + c\ C_c(1 - K) \quad i = 1 \text{ to } n \tag{3.25a}$$

where:
φ = Capacity reduction factor for axial loads and flexure
A_{psi} = Area of the '*i*'-th strand (in^2 or mm^2)
f_{sei} = Stress in the '*i*'-th strand (psi or kPa) = $\varepsilon_u\ E_{strand}$
$C_c = 0.85 f'_c A_a$
A_a = Area of concrete annulus in compressive stress block of depth $\beta_1 c = 0.85c$ (in^2 or mm^2) for $f'_c \leq 4000$ psi
(Value of β_1 is a function of concrete compressive strength; see ASCE Manual 123)

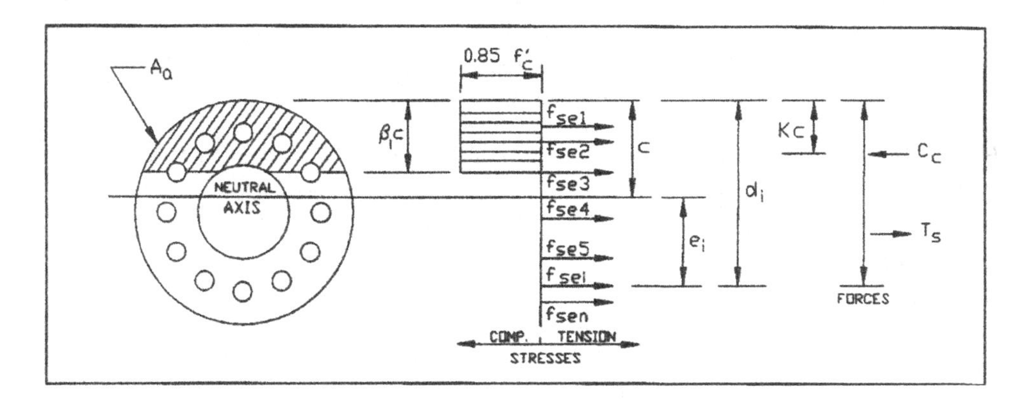

Figure 3.26 Assumed Concrete Stress Distribution (with permission from ASCE).

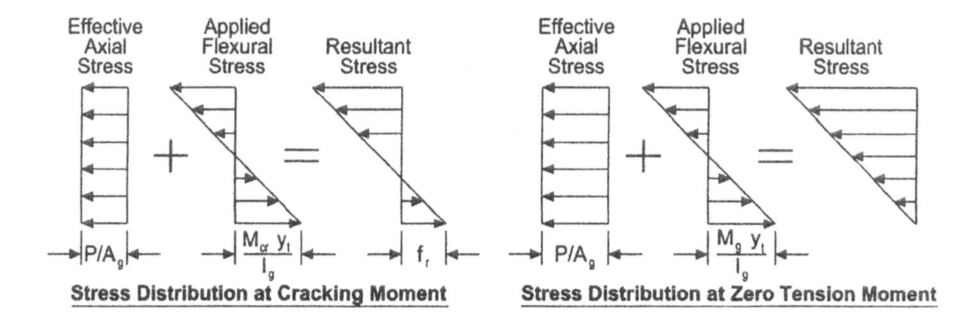

Figure 3.27 Stress Distribution – Cracking and Zero Tension (with permission from ASCE).

c = Depth of stress block (NA to extreme compressive fiber) (in. or mm)
$e_i = d_i - c$ (d_i is the distance of the 'i'-th strand from extreme compression fiber) (in. or mm)
K = Factor relating centroid of force C_c to Neutral Axis (NA)

Initial Cracking Strength $$M_{ic} = (f_r\ I_g/y_t) + (PI_g/A_g y_t) \quad (3.25b)$$

where:
f_r = Modulus of Rupture of concrete = $7.5\sqrt{f_c'}$ (psi) for normal weight concrete
A_g = Gross area of the cross section (in^2 or mm^2)
I_g = Gross moment of Inertia of the cross section (in^4 or mm^4)
y_t = Distance of extreme tensile fiber from centroidal axis (in or mm)
P = Effective Prestress Force (lbs. or N)
M_{ic} is approximately equal to 40% to 55% of M_u.

Zero Tension Capacity $$M_{zt} = (PI_g/A_g y_t) \quad (3.25c)$$

M_{zt} is approximately equal to 70% to 85% of M_{ic} or 28% to 47% of M_u.

The usage of concrete poles is determined for each of the above three definitions of bending moments with the appropriate strength factor, as specified for the design. For square concrete poles, the procedures are the same except that the bending moment M is replaced by the larger of the two moments about the principal axes, X and Y.

Example 3.16 Determine the cracking moment, zero tension moment and approximate ultimate moment capacity for the prestressed concrete pole with hollow circular section described below.

Outside diameter $d_o = 24$ in. (61 cm)
Inside diameter $d_i = 18$ in. (45.7 cm)
Concrete strength $= f_c' = 10{,}000$ psi. (68.9 MPa)
Modulus of rupture $= f_r = 7.5\sqrt{f_c'} = 750$ psi. (5.17 MPa)
Ultimate compressive strain = 0.003
Steel strands = 20 no's of ½ in. (12.7 mm) diameter 7-wire 270 ksi strands (1.86 GPa)
Initial prestress $= 0.50(f_{pu})$
f_{pu} = Specified tensile strength of prestressing tendons
Concrete cover = ¾ in. (19 mm)
Effective prestress loss = 25%

Solution:

(a) Cracking Moment
I_g = gross moment of inertia of cross section $= (\pi)(d_o^4 - d_i^4)/64 = 11{,}133$ in^4 (4.634 E5 cm^4)
A_g = gross area of cross section $= (\pi)(d_o^2 - d_i^2)/4 = 198$ in^2 (1.276 E3 cm^4)
y_t = distance to extreme fiber = 12 in. (30.5 cm)
Area of each tendon = 0.153 in^2 (98.7 mm^2)
P = total effective prestress force
$= (1 - 0.25)(20)(0.50)(270{,}000)(0.153) = 309,\ 825$ lbs. (1378.7 kN)
$M_{ic} = (f_r\ I_g/y_t) + (PI_g/A_g y_t)$ (Equation 3.25b)
$= (750)(11{,}133)/12 + (309{,}825)(11133)/(198*12)$
$= 2{,}147{,}530$ lb-in or 179 kip-ft. (242.7 kN-m)

(b) Zero Tension Moment
This is the second term of the above equation for cracking moment.
$M_{zt} = (309{,}825)(11133)/(198)(12) =$ 1,451,717.9 lb-in. or 120.97 kip-ft. (164 kN-m)

(c) Ultimate Moment

This involves using Equation 3.25a and the associated procedure.

Depth of the neutral axis 'c' must be determined from a trial-and-error approach using reinforced concrete design fundamentals. The procedure is too tedious to be reproduced here but is left to the student as an exercise.

For this problem, a computer analysis determined 'c' to be 8.5 in (21.6 cm). β_1 is 0.65 for 10,000 psi.
Corresponding C_c is found to be 329,953 lbs. (1468.3 kN)
Nominal ultimate moment capacity $M_n = 490$ kip-ft. (664.4 kN-m)

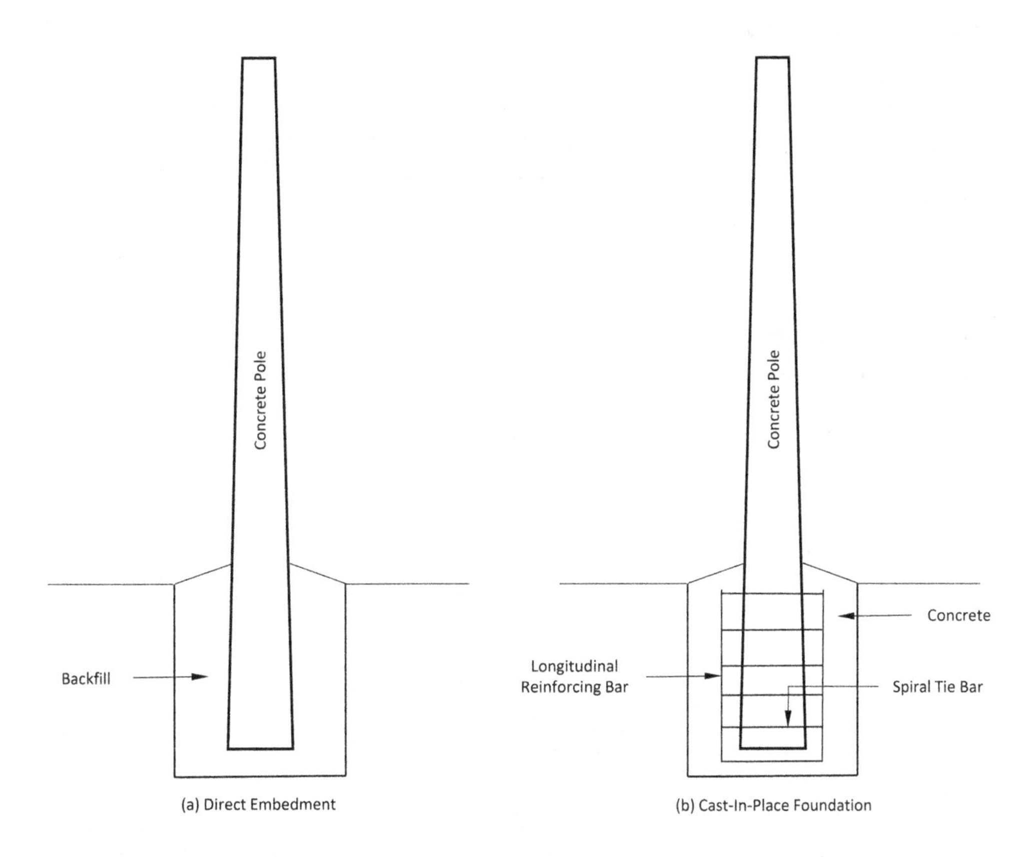

Figure 3.28 Concrete Pole Foundations (with permission from ASCE).

ACI Capacity Reduction Factor = 0.90
Ultimate moment capacity $M_u = (0.90)(490) = 441$ kip-ft. (598 kN-m)
$M_{ic}/M_u = 179/441 = 0.406$
The initial cracking moment is generally about 40% of the ultimate moment.

Embedment

Concrete poles can be embedded into the ground in a manner similar to wood poles. Cast-in-place foundations can also be used. Typical arrangements are shown in Figures 3.28a and 3.28b.

Concrete pole grounding

The grounding of concrete poles can be external, internal or both. For external ground, threaded inserts can be embedded in the pole for clamping the grounding wire to the pole's surface. Internal grounds can be embedded in the concrete or pulled through the central hollow section of the pole with grounding pads. In areas known for lightning strikes or high ground resistance (over 25 ohms), all hardware is bonded to the grounding system. Additionally, manufacturers provide an electrical bond between the pole steel reinforcing cage and the pole ground.

3.5.1.5 Composite poles

The design of composite poles is governed by both strength (flexural capacity) and stiffness (deflections). The most common way of selecting a composite pole is by using the design charts provided by the manufacturer – in terms of wood pole equivalency – and then check if the pole geometry is adequate for a given limiting deflection. RS Technologies (2012), for example, provides a design guide with various pole modules and lengths, and lateral load capacities (load applied 2 ft. from pole top) accompanied by tip deflections.

Composite pole grounding

Most composite poles can be grounded in the same manner as wood poles, with the ground wire affixed to the outer surface of the pole using wire clips and screws. Another option is to have the ground wire run internally through the pole, exiting to the ground rod through a hole in the base module.

3.5.2 Assemblies and parts

The size and quantity needed of each material/component of a transmission structure is tabulated on the primary assembly and sub-assembly drawings. This list can be linked to PLS-POLE™, TOWER™ and PLS-CADD™ modules. PLS-CADD™ contains a powerful material management system that allows import of parts and assemblies from existing databases, generate material lists and costs for both construction as well as structural and line optimization. The *Parts Editor* helps store data related to structure parts (stock number, description, unit price, manufacturer and supplier). These parts can be used to build 'assemblies' and/or 'sub-assemblies'.

Once the parts and assemblies library is developed in PLS-CADD™, the associated assemblies and/or parts can be linked to its structure model. If optimization capabilities of PLS-CADD™ are utilized, the total cost of each structure is automatically calculated. Finally, once the structures are spotted, PLS-CADD™ can provide a listing of material and parts (and labor, if needed) in several formats.

Figures 3.29 and 3.30 show the concept behind developing assembly drawings for various sub-assemblies and parts. These are typical schemes and can be adjusted based on the requirements of the utility, project, type of structure and material.

3.5.3 Framing drawings

The term "Structure Framing" refers to the general location and arrangement of various components in a transmission structure to meet specific loading, geographical, right-of-way and construction preferences of the project under consideration. All transmission line structure designs are accompanied by drawings showing the framing of the assemblies – cross arms, X-braces, guying attachments, OHGW attachments, end fittings on the cross arms, fastener holes, washers, grounding assemblies etc. The framing requirements vary by structure and material configuration; but they must all satisfy the design criteria established for that particular project.

In the case of wood structures, the framing drawings indicate the heights of the wood poles, cross arms or davit arms and x-braces (for H-Frames) and OHGW/OPGW

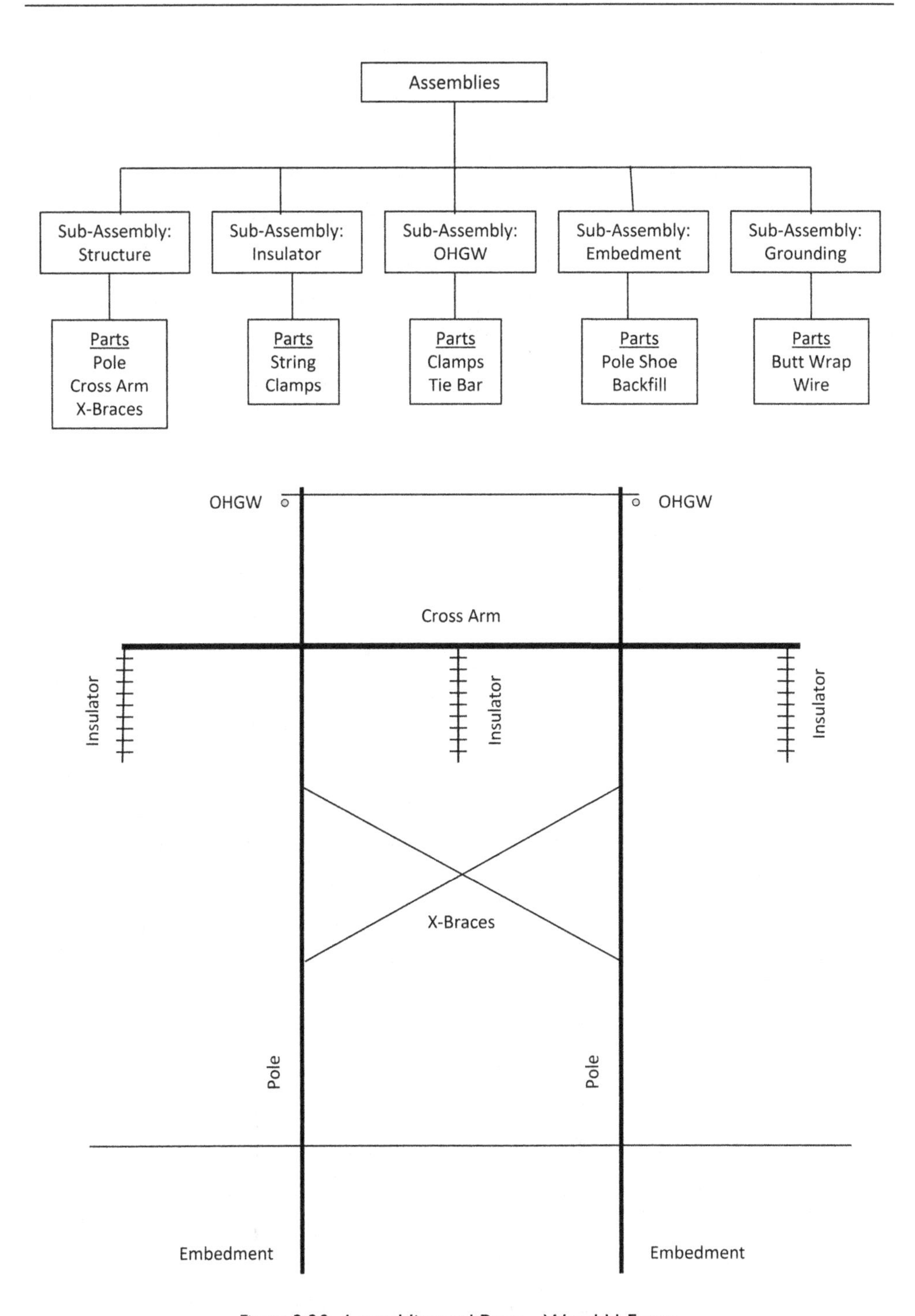

Figure 3.29 Assemblies and Parts – Wood H-Frame.

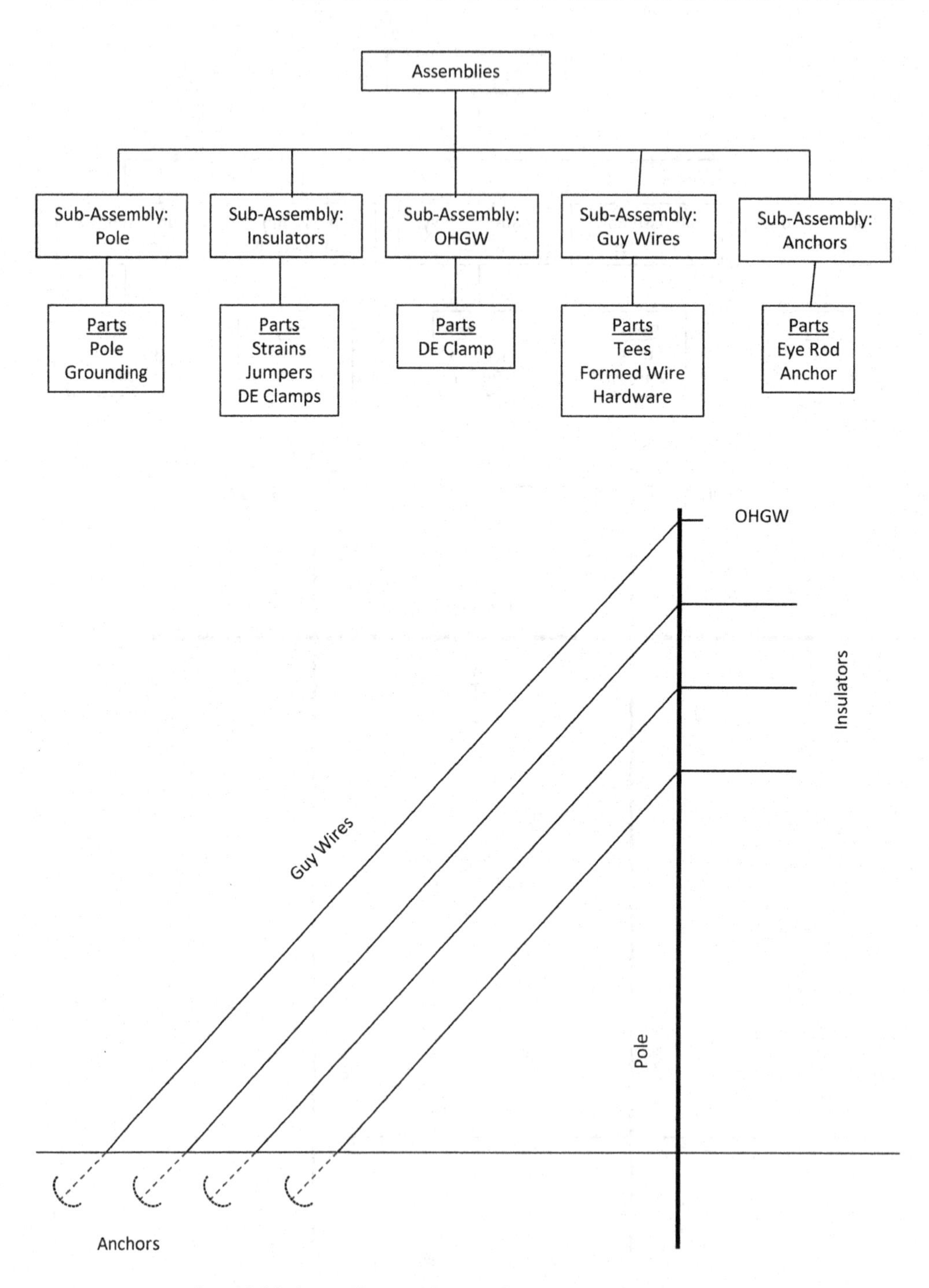

Figure 3.30 Assemblies and Parts – Guyed Wood Pole Deadend.

attachment details etc. The depth of embedment (for tangent poles) is also sometimes shown. Framing guidelines for most RUS-standard structures can be found in relevant RUS Bulletins. Hole drilling can be performed on-site for most wood poles.

In the case of steel structures, the framing drawings indicate the sizes of the tubular shafts, joints (splices or bolted flanges), insulator brackets, welded components, ladder clips, grounding nuts, davit arms, cross arms and x-braces (for H-Frames) and base plates/anchor bolt assemblies etc. The size of the concrete shaft foundation (or depth of embedment for tangent poles) is also often shown.

For concrete transmission structures, framing is more complex and time-consuming. This is because the prestressed concrete poles contain stressed tendons inside and hole-drilling must be performed carefully. In most cases, the engineer will specify the insulator and ground wire attachment locations, line angles and structure orientation to the manufacturer; the drilling will be performed during fabrication.

For lattice steel towers, structure framing is usually more complex and extensive. Assembly drawings show each angle member individually, their bolt patterns and location in the tower. Leg members are more carefully highlighted.

All data relevant to the structure, components and hardware will be listed on framing and assembly drawings. The high level of detail shown on these drawings depends upon various factors such as structure configuration, vendor/fabricator requirements, owner's inventory stock numbers and manufacturer catalog numbers etc. Most of this information is reflected in the PLS parts and assembly databases so that a utility can track these items as part of their asset management process.

In the following sections, the component assembly drawings related to several structure families at various voltages will be discussed. The figures accompanying the discussion serve to illustrate the wide variation in the content of the drawings as a function of structural material and usage. These are only examples of typical framing drawings and a format used by RUS/USDA for structural systems.

3.5.3.1 69 kV family

Figures 3.31a to 3.31 g shows the assembly drawings for a family of wood structures for 69 kV applications. The first two refer to tangent poles but with different insulators (post and suspension mounted on braced cross arms) while the third sketch (Figure 3.31c) shows a popular configuration of an H-Frame system. The angle structure of the fourth drawing (Figure 3.31d) is similar to the tangent system of Figure 3.31b but uses swinging brackets to facilitate small line angles.

Figures 3.31e and 3.31f show single pole angle system and deadend. Note the use of a horizontal post insulator with jumpers for Type-1 deadend. The last sketch Figure 3.31g is a common form of a 3-pole angle deadend used at substations at the beginning (and end) of a line; one side is strung at full-tension while the other, going into the substation, is strung at reduced tensions. The number of anchors required for these deadends is a function of down guy tensions; for slack or low-tension spans, two down guys can share an anchor (see Figure 3.6a).

3.5.3.2 161 kV structures

Figures 3.32a to 3.32i shows the assembly drawings for a family of structures for 161 kV applications. The first four sketches show the drawings for insulators supported

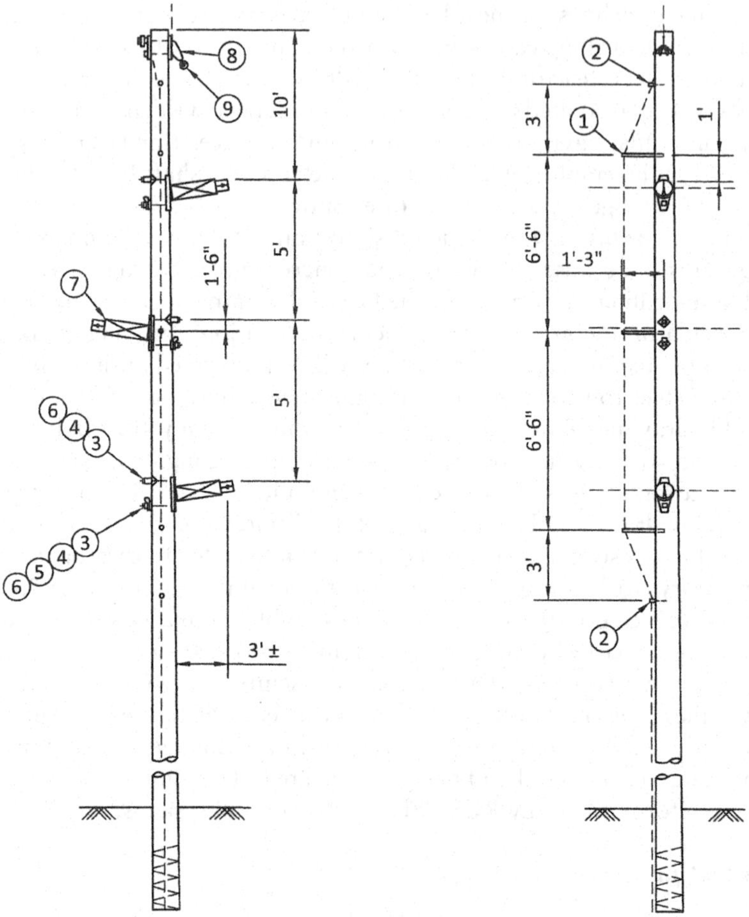

[SOURCE : RUS / USDA]

Item	Description	Quantity
1	Rod, Fiberglass	3
2	Wireholder	2
3	7/8" Bolt	6
4	Curved Washer	6
5	Spring Washer	3
6	7/8" Locknut	6
7	Insulator,Horizontal Post,W/Clamp	3
8	OHGW Support Assembly	1
9	OPGW Assembly, Tangent	-

1 ft = 30 cm 1 in = 25.4 mm

Figure 3.31a 69 kV Tangent Structure with Horizontal Posts.

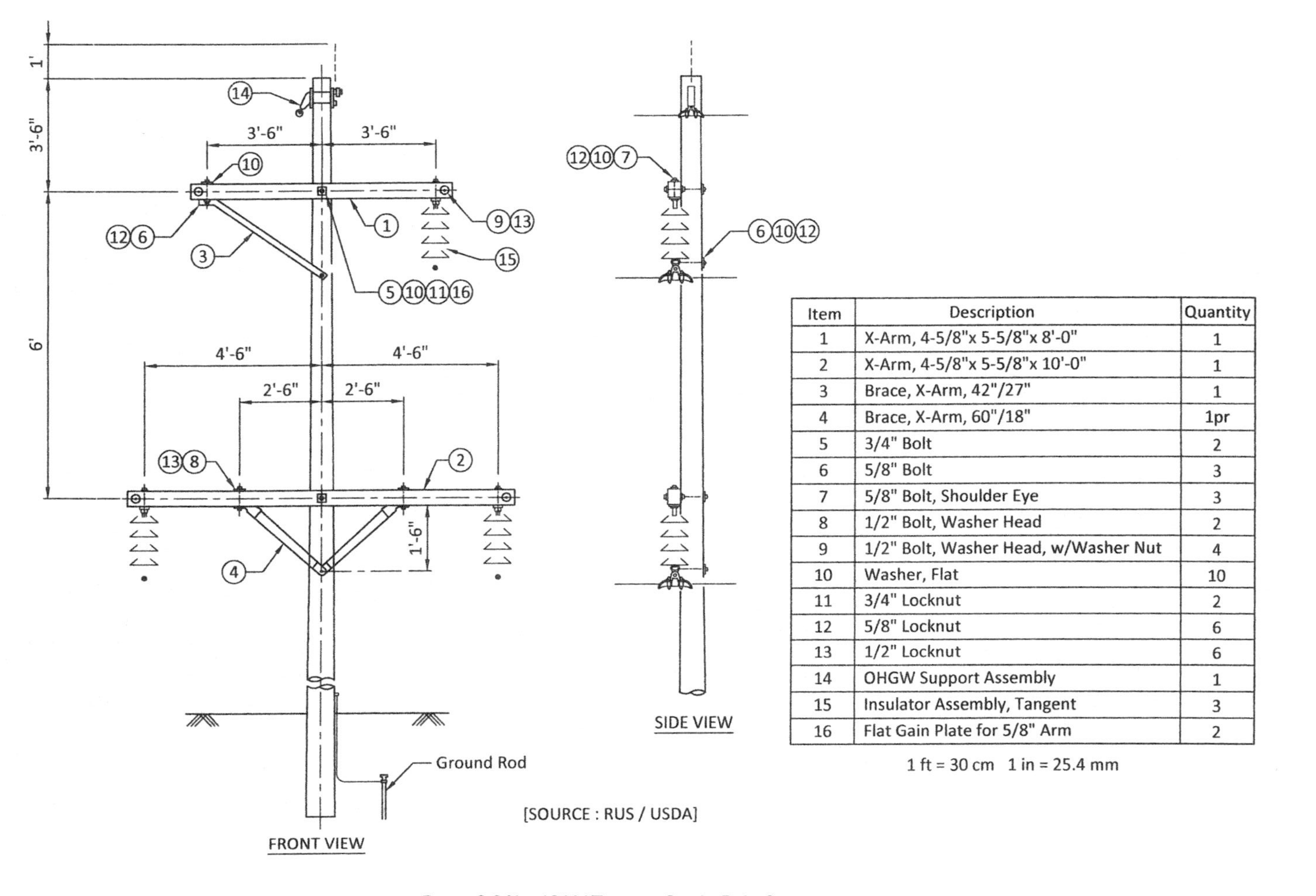

Item	Description	Quantity
1	X-Arm, 4-5/8"x 5-5/8"x 8'-0"	1
2	X-Arm, 4-5/8"x 5-5/8"x 10'-0"	1
3	Brace, X-Arm, 42"/27"	1
4	Brace, X-Arm, 60"/18"	1pr
5	3/4" Bolt	2
6	5/8" Bolt	3
7	5/8" Bolt, Shoulder Eye	3
8	1/2" Bolt, Washer Head	2
9	1/2" Bolt, Washer Head, w/Washer Nut	4
10	Washer, Flat	10
11	3/4" Locknut	2
12	5/8" Locknut	6
13	1/2" Locknut	6
14	OHGW Support Assembly	1
15	Insulator Assembly, Tangent	3
16	Flat Gain Plate for 5/8" Arm	2

1 ft = 30 cm 1 in = 25.4 mm

[SOURCE : RUS / USDA]

Figure 3.31b 69 kV Tangent Single Pole Suspension.

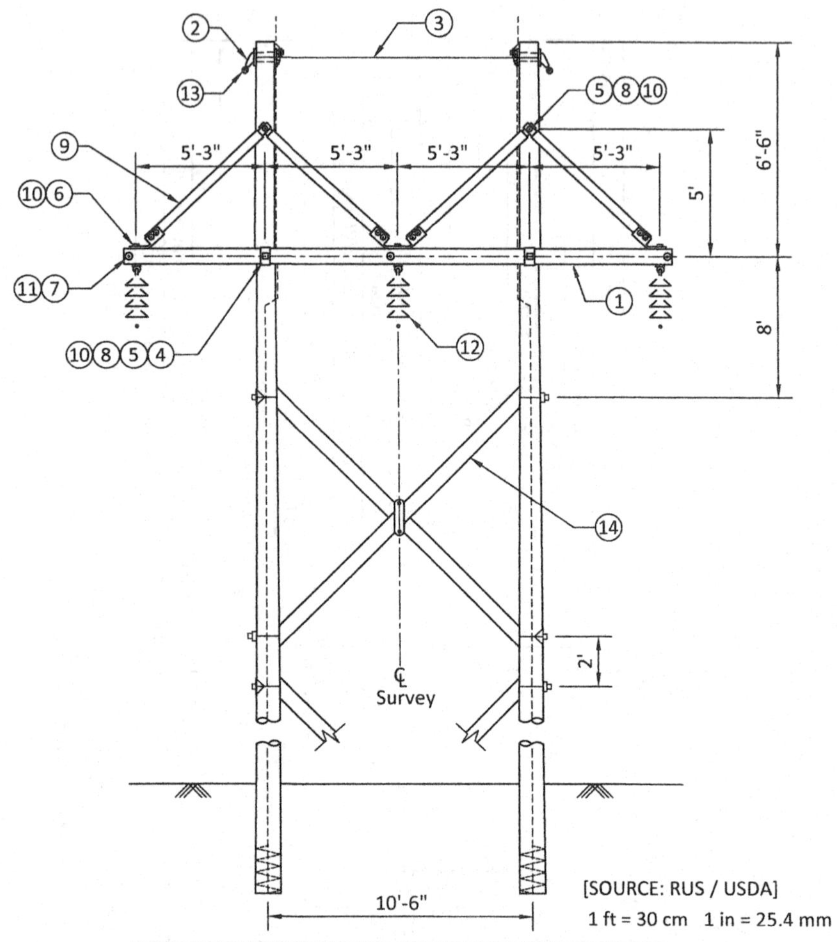

Item	Description	Quantity
1	X-Arm, 5-5/8"x7-3/8"x22'-0"	1
2	OHGW Support Assembly, double bolt	2
3	Grounding Assembly Cross Tie	1
4	Plate, X-Arm Reinforcing	2
5	3/4" Bolt	4
6	3/4" Bolt, Shoulder Eye	3
7	1/2" Bolt, Washer Head, w/Washer Nut	3
8	Curved Washer	4
9	Pair, X-Arm Braces	2
10	3/4" Locknut	7
11	1/2" Locknut	3
12	Insulator Assembly, Tangent	3
13	OHGW Assembly, Tangent	2
14	X-Brace Assembly	1

Figure 3.31c 69 kV Tangent H-Frame.

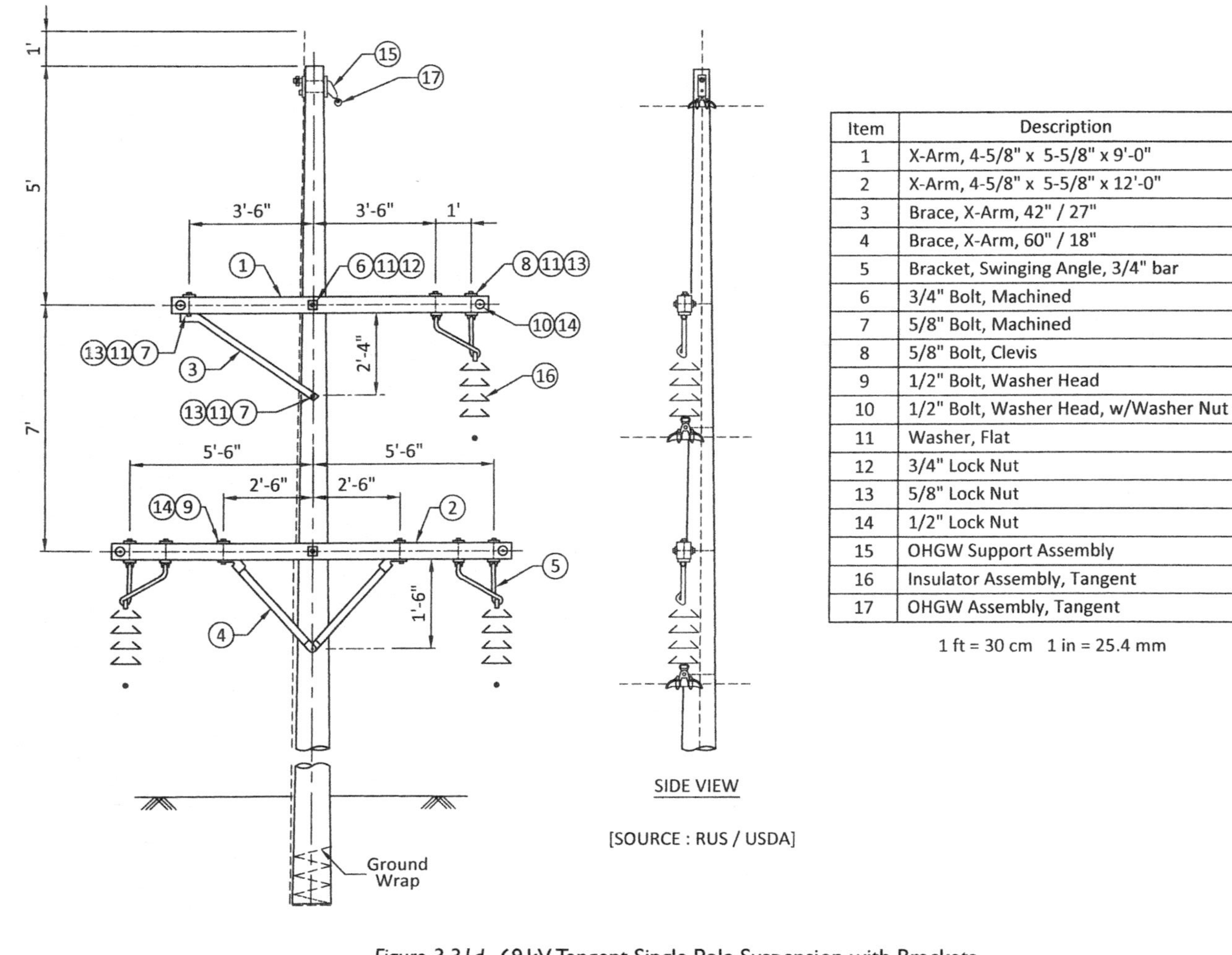

Item	Description	Quantity
1	X-Arm, 4-5/8" x 5-5/8" x 9'-0"	1
2	X-Arm, 4-5/8" x 5-5/8" x 12'-0"	1
3	Brace, X-Arm, 42" / 27"	1
4	Brace, X-Arm, 60" / 18"	1pr
5	Bracket, Swinging Angle, 3/4" bar	3
6	3/4" Bolt, Machined	2
7	5/8" Bolt, Machined	3
8	5/8" Bolt, Clevis	6
9	1/2" Bolt, Washer Head	2
10	1/2" Bolt, Washer Head, w/Washer Nut	4
11	Washer, Flat	13
12	3/4" Lock Nut	2
13	5/8" Lock Nut	9
14	1/2" Lock Nut	6
15	OHGW Support Assembly	1
16	Insulator Assembly, Tangent	3
17	OHGW Assembly, Tangent	1

1 ft = 30 cm 1 in = 25.4 mm

[SOURCE : RUS / USDA]

Figure 3.31d 69 kV Tangent Single Pole Suspension with Brackets.

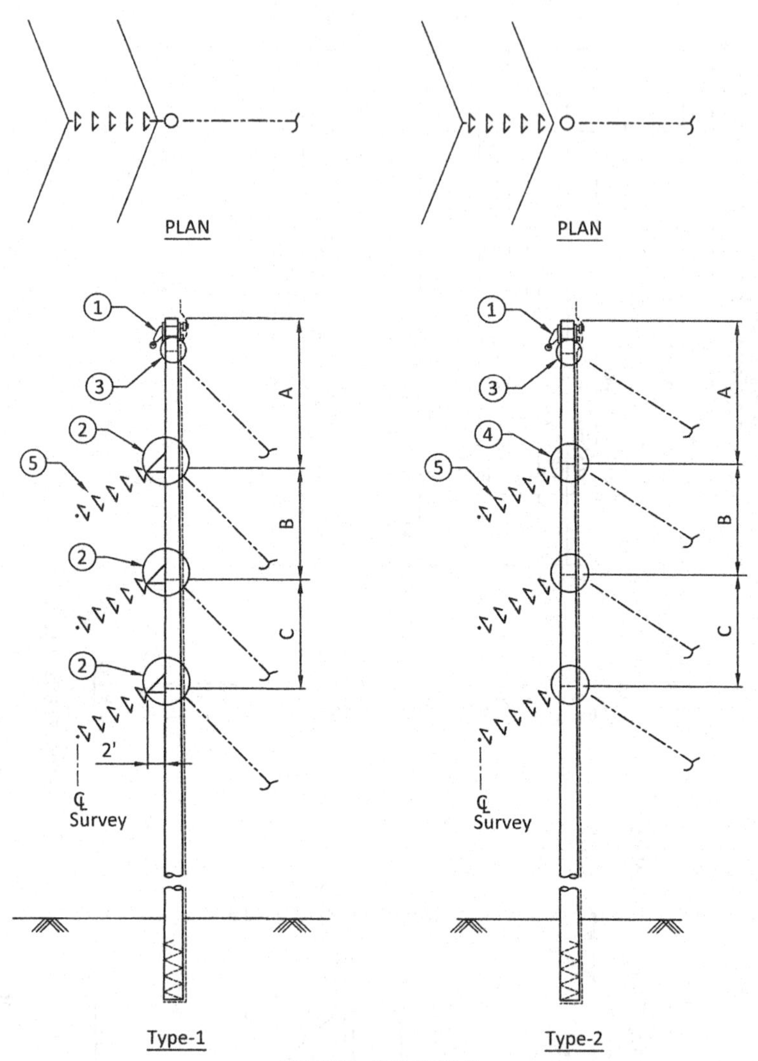

[SOURCE : RUS / USDA]

Item	Description	Quantity	
		Type 1	Type 2
1	OHGW Support Assembly, double bolt	1	1
2	Bracket Assembly	3	-
3	Guy Attachment	1	1
4	Guy Attachment	-	3
5	Insulator Assembly, Angle	3	3

	Voltage	
	34 kV, 46 kV	69 kV
A	6'-0"	7'-0"
B	6'-0"	7'-0"
C	6'-0"	7'-0"

1 ft = 30 cm 1 in = 25.4 mm

Figure 3.31e 69 kV Medium and Large Vertical Angles.

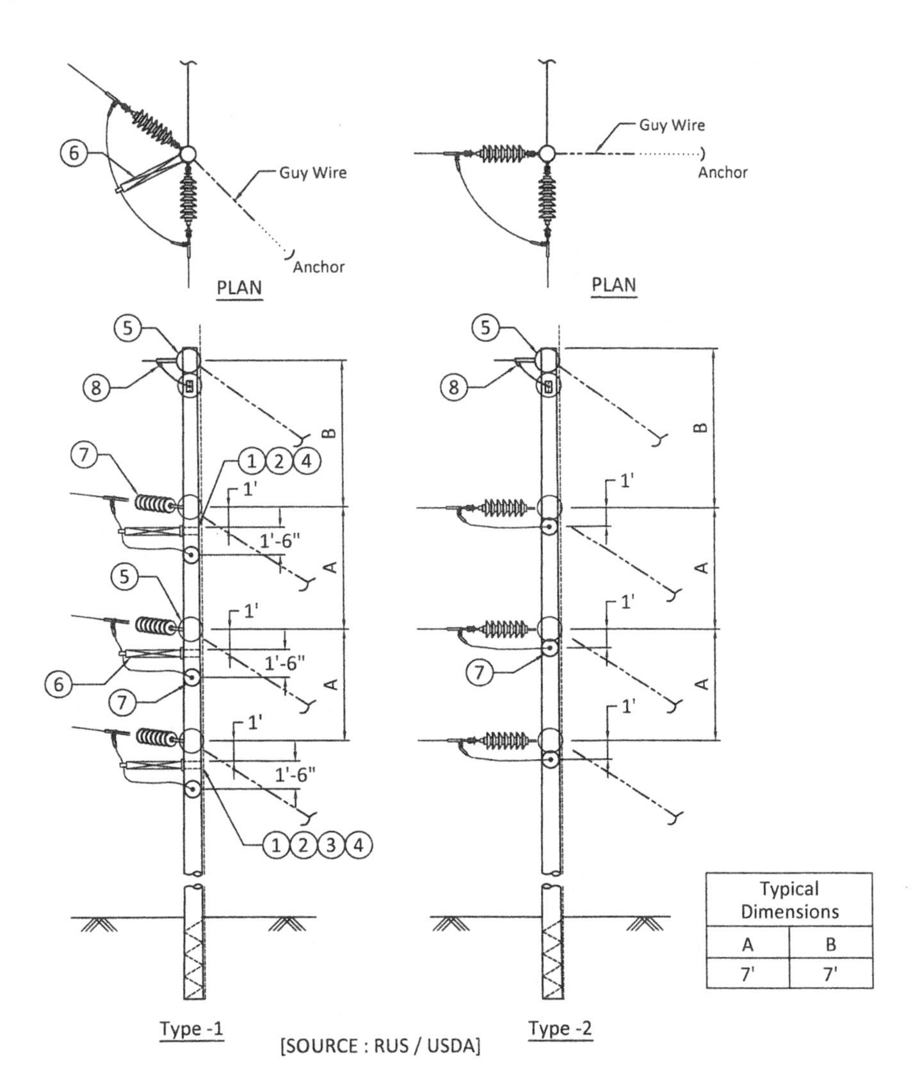

Typical Dimensions	
A	B
7'	7'

Item	Description	Quantity	
		Type 1	Type 2
1	3/4" Bolt	6	-
2	Curved Washer	6	-
3	Spring Washer	3	-
4	3/4" Locknut	6	-
5	Guy Attachment	8	8
6	Insulator, Horizontal Post, w/Clamp	3	-
7	Insulator Assembly, Deadend	6	6
8	OHGW Assembly, Deadend	1	1

1 ft = 30 cm 1 in = 25.4 mm

Figure 3.31f 69 kV Vertical Double Deadend.

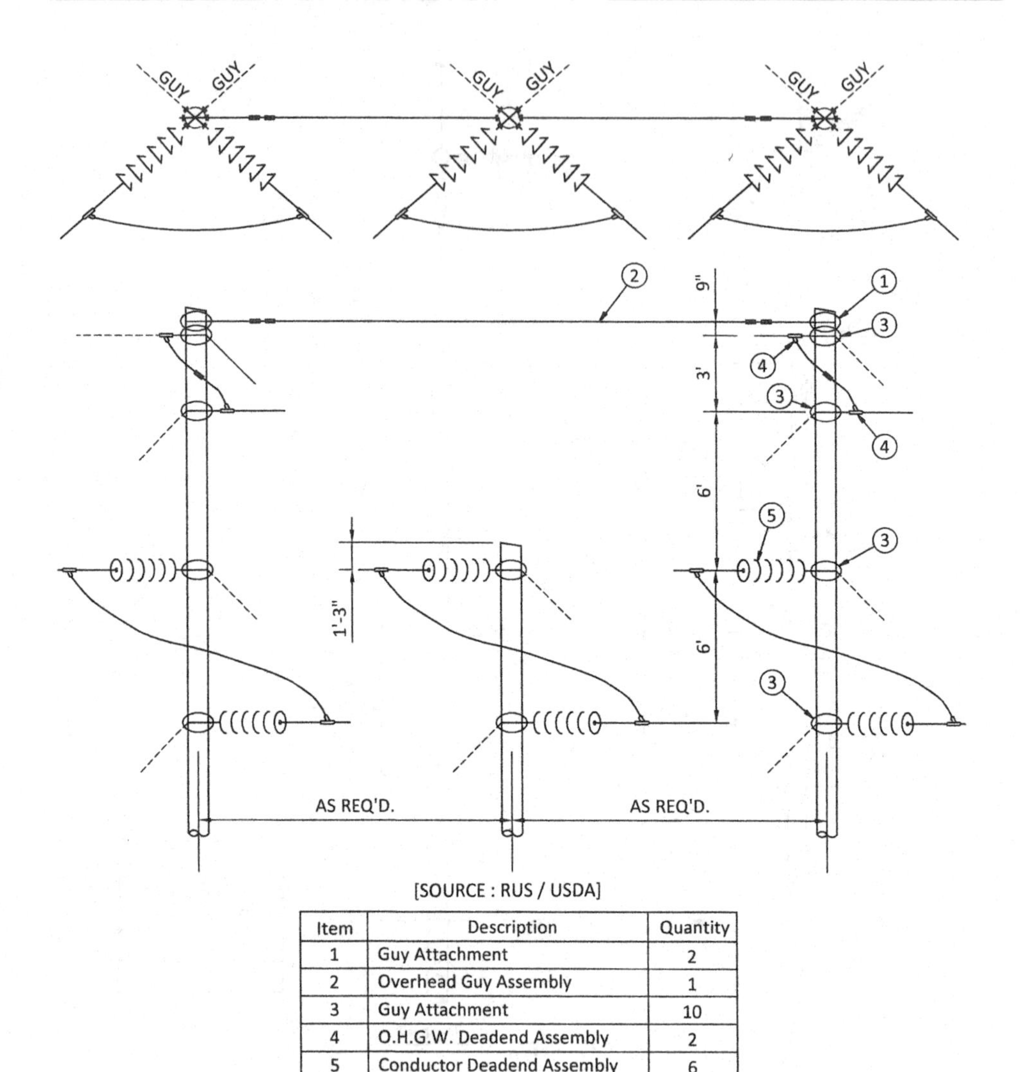

Item	Description	Quantity
1	Guy Attachment	2
2	Overhead Guy Assembly	1
3	Guy Attachment	10
4	O.H.G.W. Deadend Assembly	2
5	Conductor Deadend Assembly	6

1 ft = 30 cm 1 in = 25.4 mm

Figure 3.31g **69 kV Large Angle Deadend Structure.**

on steel upswept davit arms in various forms, single and double circuit applications including unequal voltages.

Figures 3.32e and 3.32f show structures with braced line post insulators, for standard use as well as long span use (with greater phase separation). A medium angle 3-pole configuration with swing brackets is shown in Figure 3.32g. This system can also be used for 138 kV applications. Note that as the voltage is increased from 138 kV to 161 kV, the pole spacing increased by 1.5 ft. (0.46 m).

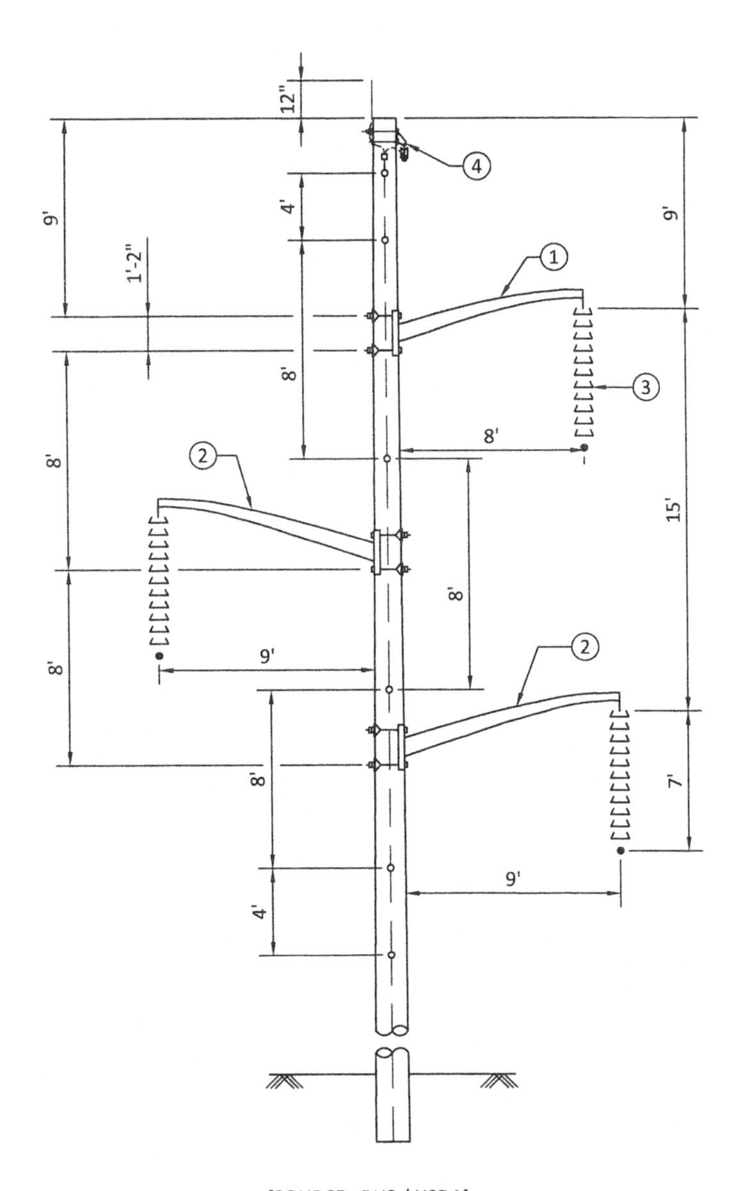

[SOURCE : RUS / USDA]

Item	Description	Quantity
1	Galv. Steel Suspension Arm Assembly	1
2	Galv. Steel Suspension Arm Assembly	2
3	Polymer Insulator Assembly-Tangent	3
4	OHGW Support Assembly	1

1 ft = 30 cm 1 in = 25.4 mm

Figure 3.32a 161 kV Single Circuit Tangent Structure with Steel Upswept Arms.

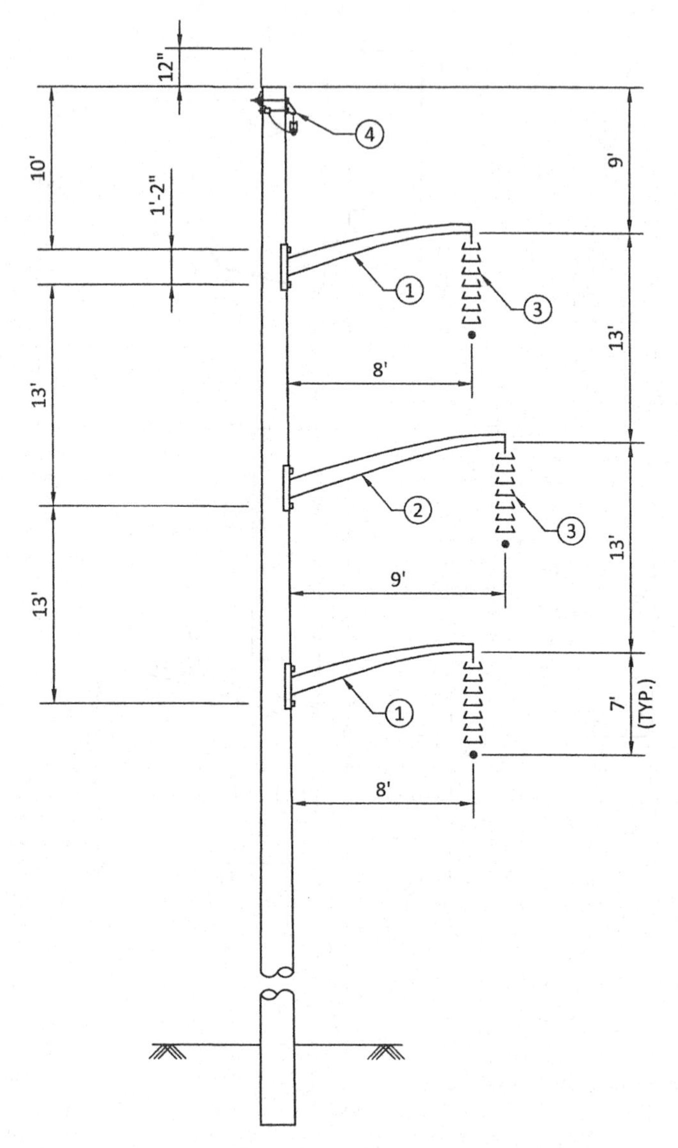

[SOURCE : RUS / USDA]

Item	Description	Quantity
1	Steel Suspension Arm Assembly	2
2	Steel Suspension Arm Assembly	1
3	Polymer Insulator Assembly	3
4	OHGW Support Assembly	1

1 ft = 30 cm 1 in = 25.4 mm

Figure 3.32b 161 kV Single Circuit Tangent Structure with Steel Upswept Arms.

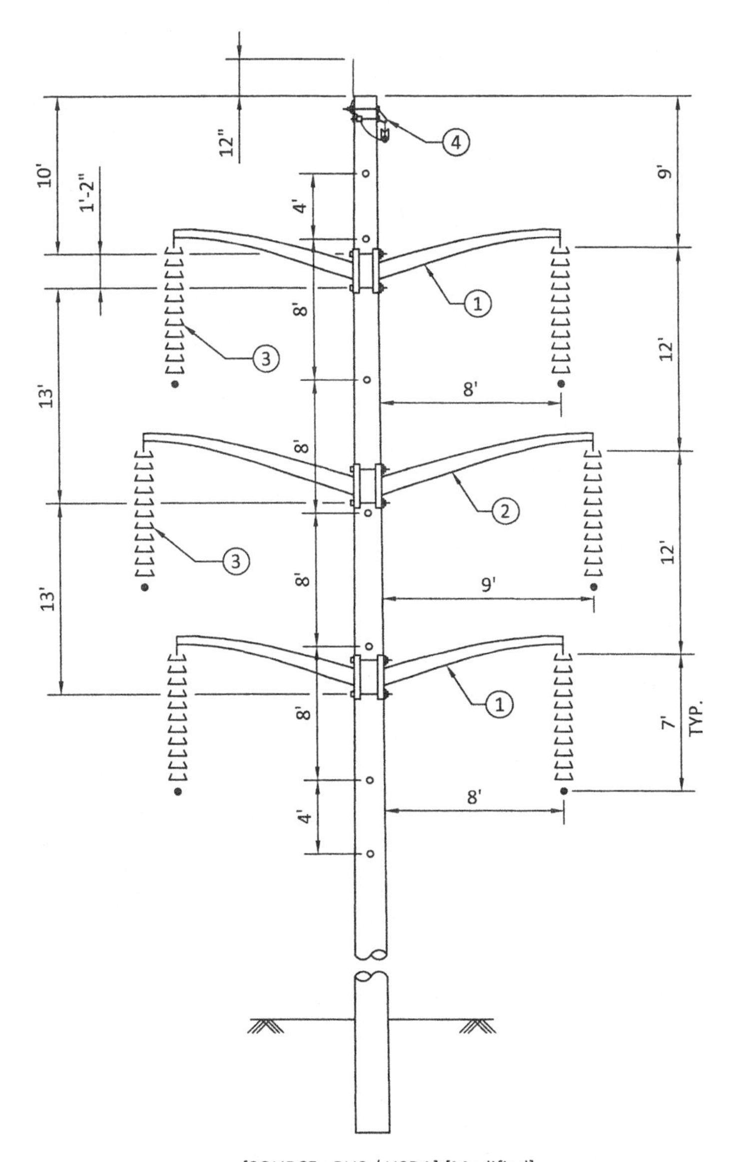

[SOURCE : RUS / USDA] [Modified]

Item	Description	Quantity
1	Galv. Steel Suspension Arm Assembly	2
2	Galv. Steel Suspension Arm Assembly	1
3	Polymer Insulator Assembly-Tangent	6
4	OHGW Support Assembly	1

1 ft = 30 cm 1 in = 25.4 mm

Figure 3.32c 161 kV Double Circuit Tangent Structure with Steel Upswept Arms.

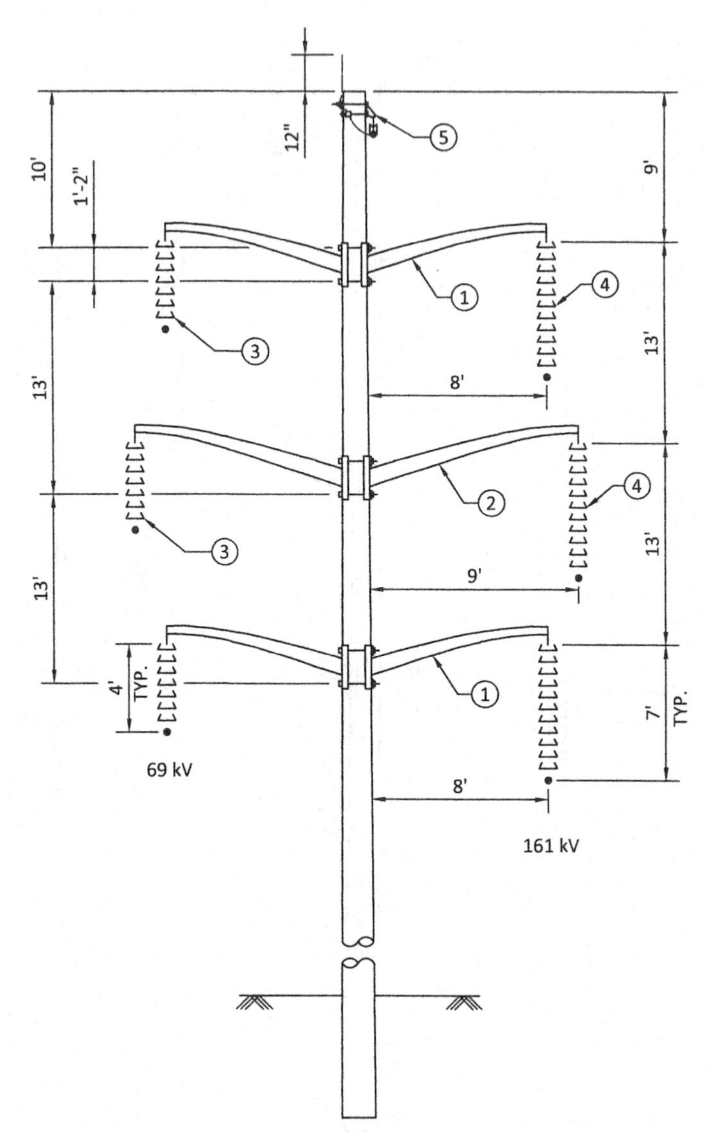

[SOURCE : RUS / USDA] [Modified]

Item	Description	Quantity
1	Steel Suspension Arm Assembly	2
2	Steel Suspension Arm Assembly	1
3	Polymer Insulator Assembly-Tangent 69 kV	3
4	Polymer Insulator Assembly-Tangent 161 kV	3
5	OHGW Support Assembly	1

1 ft = 30 cm 1 in = 25.4 mm

Figure 3.32d 69 kV–161 kV Double Circuit Tangent Structure with Steel Upswept Arms.

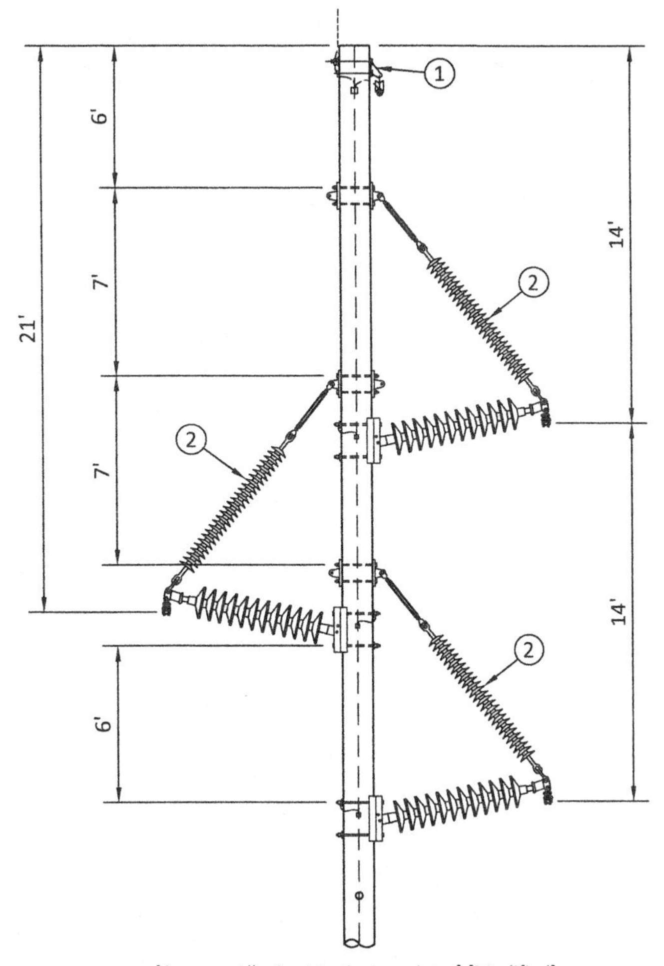

[Source : Allgeier Martin Associates] [Modified]

Item	Description	Quantity
1	OHGW Assembly, Tangent	1
2	Conductor Support Assembly	3

1 ft = 30 cm 1 in = 25.4 mm

Figure 3.32e 161 kV Tangent Structure with Braced Line Posts.

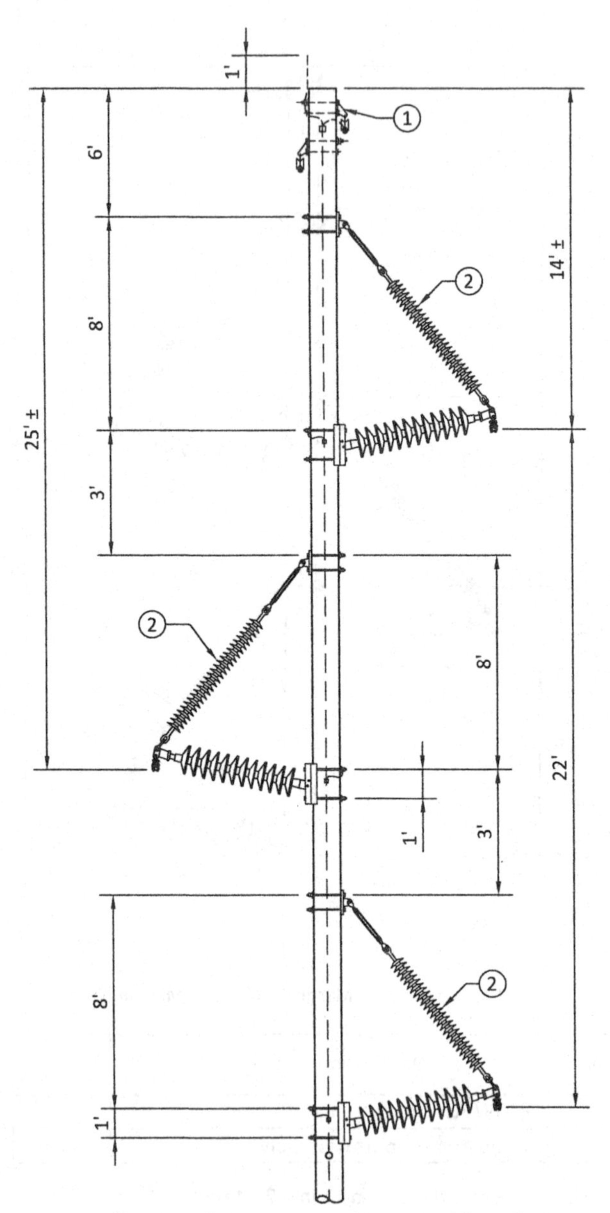

[Source : Allgeier Martin and Associates] [Modified]

Item	Description	Quantity
1	OHGW Assembly, Tangent	2
2	Conductor Support Assembly	3

1 ft = 30 cm   1 in = 25.4 mm

Figure 3.32f 161 kV Long Span Tangent Structure with Braced Line Posts.

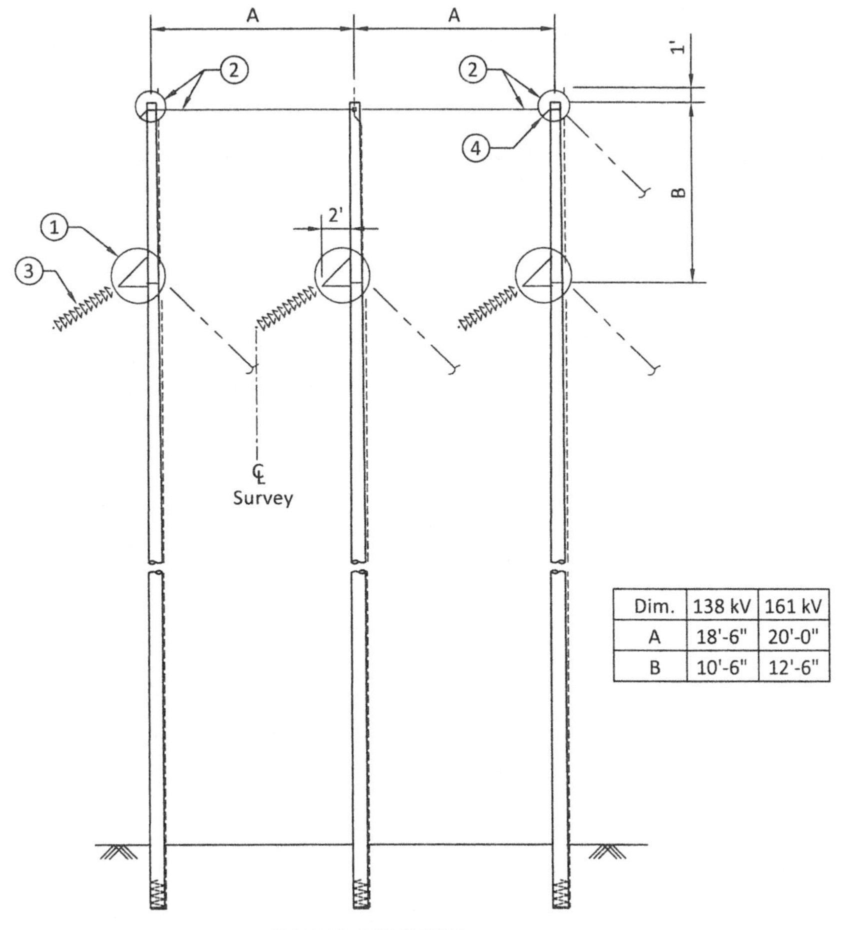

Dim.	138 kV	161 kV
A	18'-6"	20'-0"
B	10'-6"	12'-6"

[SOURCE : RUS / USDA]

Item	Description	Quantity
1	Bracket & Guy Attachment	3
2	Pole Tie, Large Angle	1
3	Insulator Assembly, Angle	3
4	OHGW Assembly, Angle	2

1 ft = 30 cm   1 in = 25.4 mm

Figure 3.32g 161 kV Medium Angle Structure.

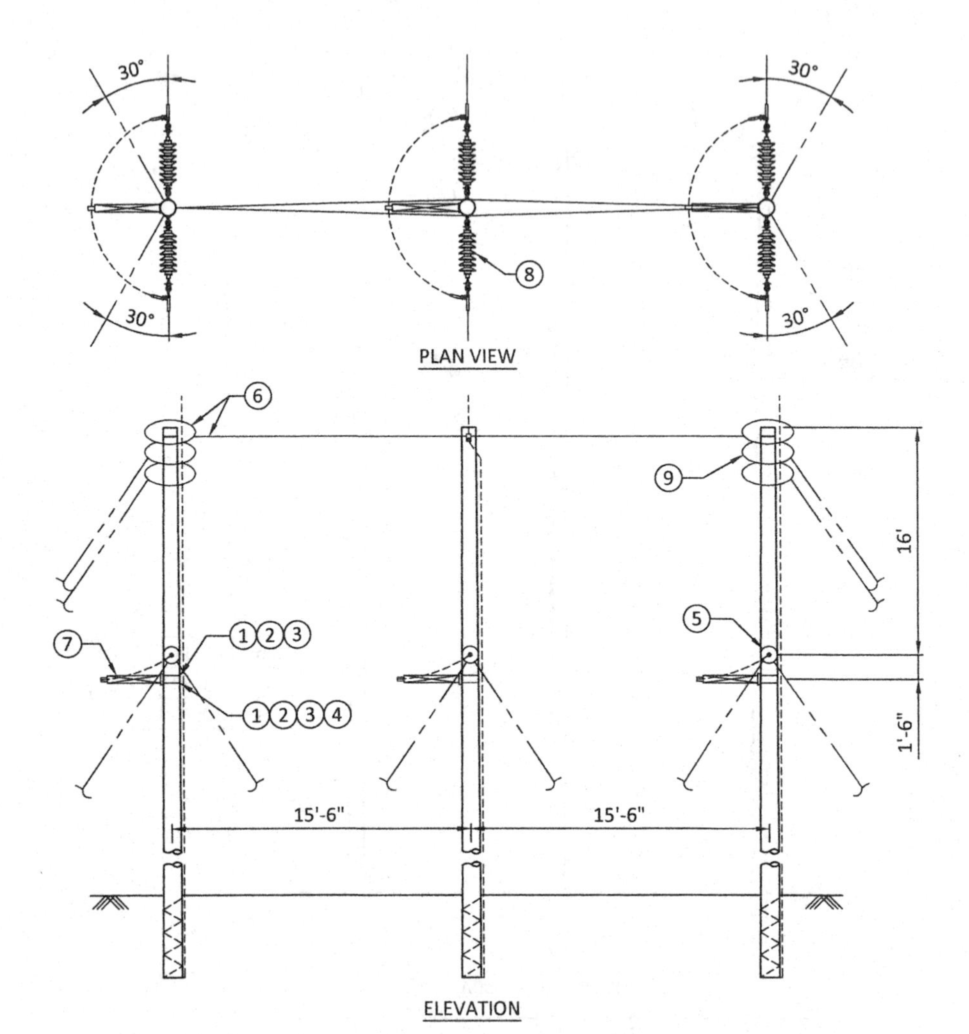

[SOURCE: RUS / USDA]

Item	Description	Quantity
1	3/4" Bolt	6
2	Curved Washer	6
3	3/4" Locknut	6
4	Spring Washer	3
5	Guy Attachment	3
6	Pole Tie, Guying	1
7	Insulator, Horizontal Post, w/Clamp	3
8	Insulator Assembly, Deadend	6
9	OHGW Assembly, Deadend	2

1 ft = 30 cm 1 in = 25.4 mm

Figure 3.32h 161 kV Tangent Deadend Structure.

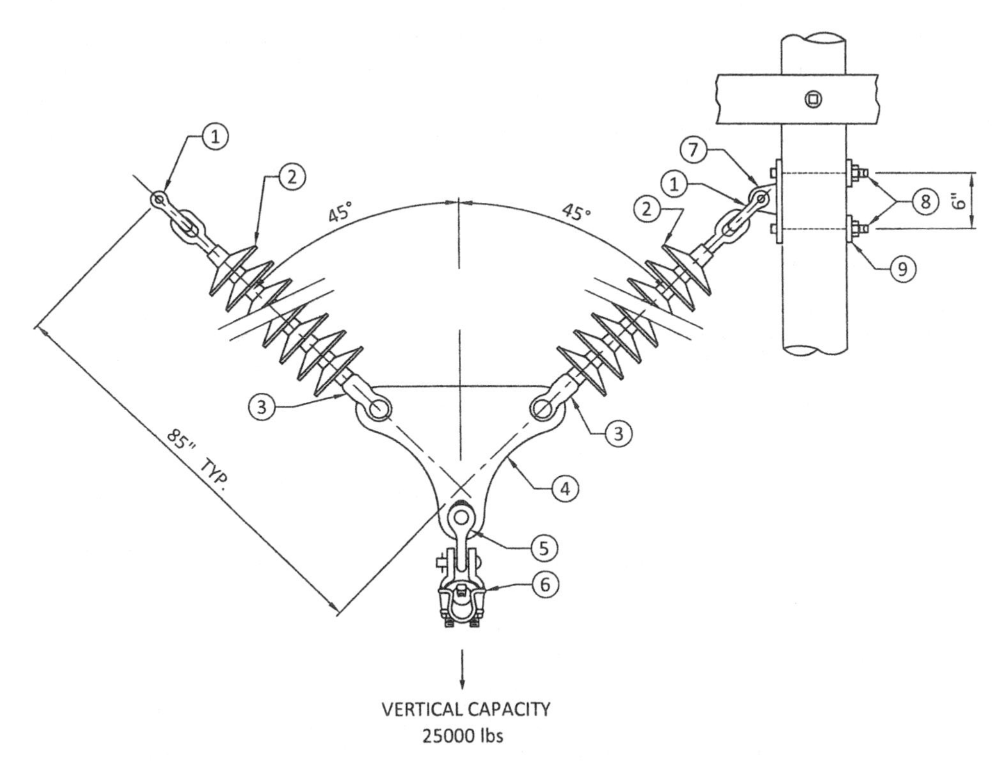

[Source : Allgeier Martin and Associates] [Modified]

Item	Description	Quantity
1	Anchor Shackle	2
2	Insulator, 161kV, Polymer, Suspension, Eye/ball	2
3	Socket Y Clevis	2
4	Yoke, 3/4" x 9" x 6"	1
5	Y-Clevis Eye	1
6	Suspension Clamp	1
7	Deadend Tee	2
8	Bolt, 7/8" W/Stud, Nuts, Spring Washer & Lock Nut	4
9	Washer, Flat, Round	4

1 ft = 30 cm 1 in = 25.4 mm 1 lb = 4.45 N

Figure 3.32i 161 kV Two-Part Conductor Support Assemblies.

The drawing in Figure 3.32h depicts what is known as a tangent-deadend with horizontal posts. Here the lines are strung with either same or different tensions on both sides, needing in-line guying to hold the tension differential. Note the large pole spacing which translates to larger ROW width. A special 2-part V-string insulator is shown on the last drawing, Figure 3.32i; this suspension system is often used on H-Frames or lattice towers (middle phase) where long spans and bundled conductors are supported.

3.5.3.3 345 kV structures

Figures 3.33a to 3.33g shows the assembly drawings for a full family of steel 345 kV structures with bundled conductors (two per phase). These drawings also illustrate the use of various special insulators handling multiple conductors. Note the use of three poles, large pole spacing, braced line posts (and yoke plates) for small angle locations and heavy-duty guying tees for medium angle structures. X-Braces for H-Frames are often custom-made since the large pole spacing requires longer brace lengths and therefore bulkier braces. Hughes Brothers (1953, 2000), for example, are one of the suppliers of specialty X-braces for large H-Frame systems; these heavy-duty braces are often made of steel tubular sections or laminated wood for higher buckling strength (see Figure 3.36c).

Figures 3.33d to 3.33g show several insulator assemblies that can be used for a typical bundled conductor application. The one visible difference between these assemblies and others is the presence of a corona ring which is generally used for all voltages above 115 kV. The last drawing (Figure 3.33g) also shows the use of a wire spacer yoke plate to maintain an 18 in. (45.7 cm) separation between wires for a bundled conductor application. This particular example shows a long (15 ft. or 4.57 m) strain string containing an extension link.

3.5.3.4 Distribution structures

Figures 3.34a to 3.34c show three distribution poles with post, pin and suspension strings with swing brackets. At lower voltages, say less than 14 kV, a neutral conductor is also employed with pin insulators. The spans of these lines are generally less than 300 ft. (91.4 m) with smaller conductors. Most framing drawings for distribution-level structures are available in RUS Bulletin 803 (1998) and Bulletin 804 (2005) supplemented Bulletins 150, 152 and 153 (all 2003).

3.5.3.5 Special structures

Figures 3.35 shows the framing details of a 3-way air break switch built with a laminated wood pole. Note the large phase spacing of 22 ft. (6.7 m).

3.5.3.6 Hardware

The eleven (11) drawings of Figures 3.36a to 3.36 k show various hardware items used in transmission line structures. These range from insulator strings, optical ground wire assemblies, X-braces for H-Frames, medium and heavy duty guying tees, swing brackets, davit arms, fiber optic deadend assemblies and grounding units. These drawings are only a representative sample of the type of hardware items used on transmission

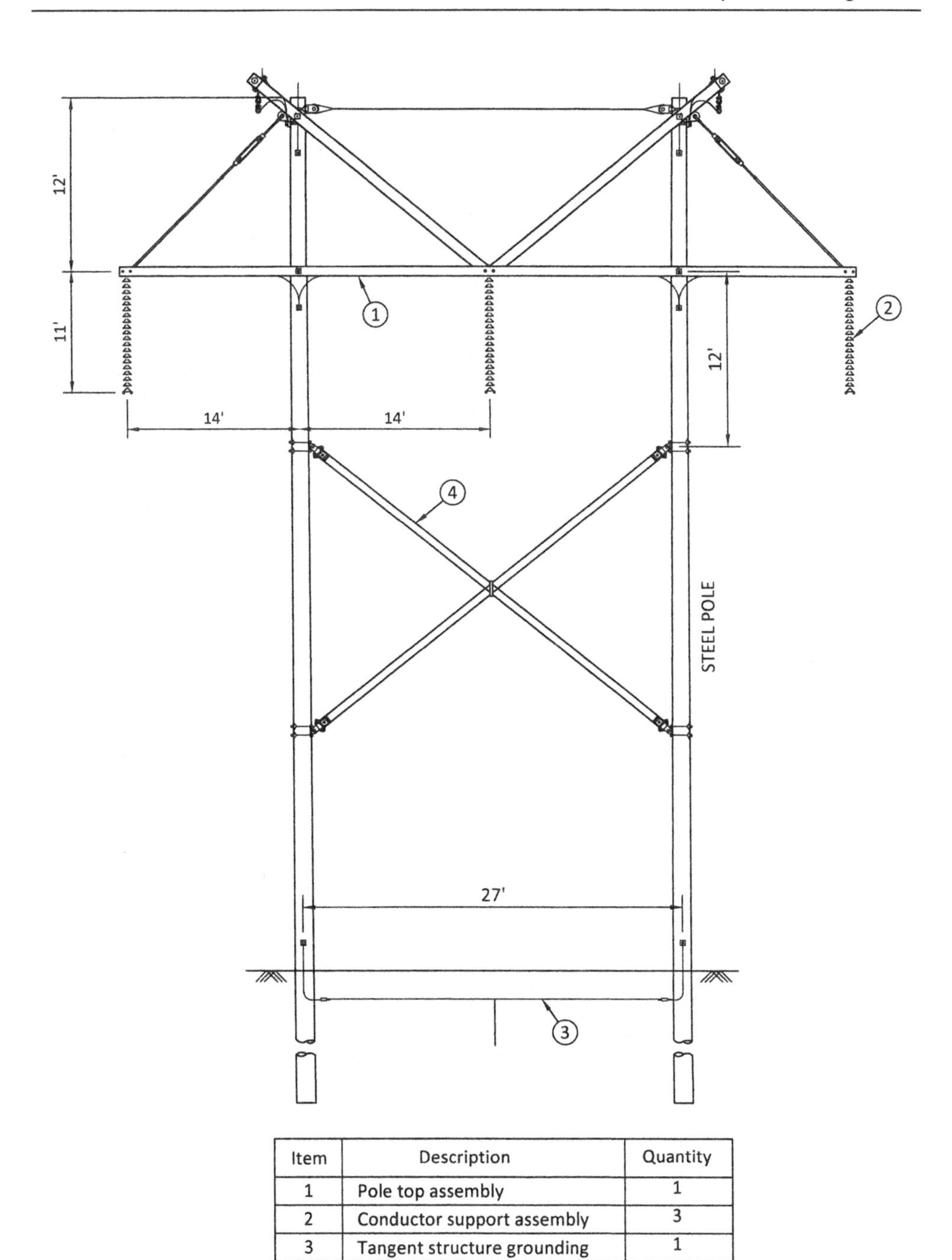

Item	Description	Quantity
1	Pole top assembly	1
2	Conductor support assembly	3
3	Tangent structure grounding	1
4	X-braces	1 pair

1 ft = 30 cm

Figure 3.33a Typical 345 kV Tangent Structure. [Source: Allgeier Martin and Associates] [Modified]

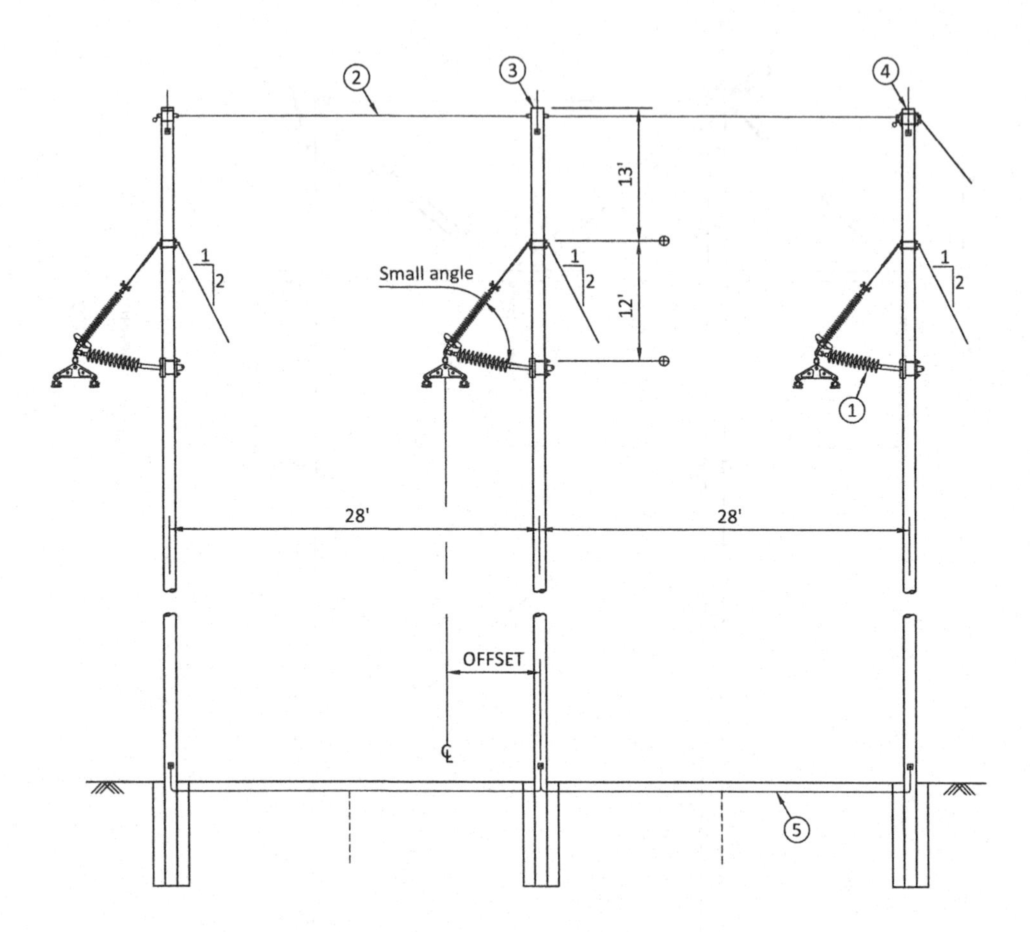

Item	Description	Quantity
1	Small angle conductor support	3
2	Pole Tie	2
3	Guy attachment	2
4	Guy attachment	1
5	Structure grounding	1

1 ft = 30 cm

Figure 3.33b Typical 345 kV Transmission Structure (Small Angle). [Source: Allgeier Martin and Associates] [Modified]

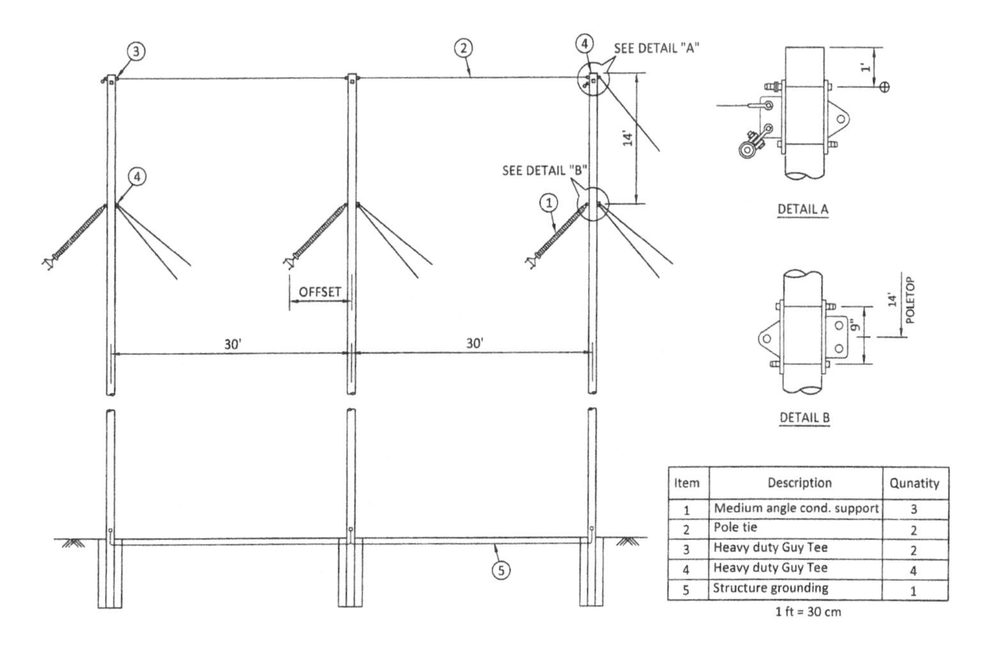

Item	Description	Qunatity
1	Medium angle cond. support	3
2	Pole tie	2
3	Heavy duty Guy Tee	2
4	Heavy duty Guy Tee	4
5	Structure grounding	1

Figure 3.33c Typical 345 kV Transmission Line Structure (Medium Angle). [Source: Allgeier Martin and Associates] [Modified]

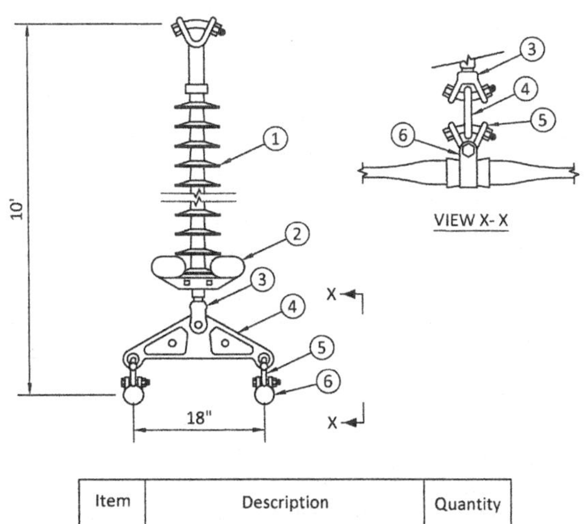

Item	Description	Quantity
1	Polymer Insulator, Y-Clevis Ball	1
2	Corona Ring	1
3	Socket Y-Clevis	1
4	Yoke Plate, 18" Spacing	1
5	Y-Clevis Eye 90	2
6	Clamp, Suspension, AGS for EHV	2

1 ft = 30 cm 1 in = 25.4 mm

Figure 3.33d Typical 345 kV Twin Conductor Support Assembly. [Source: Allgeier Martin and Associates] [Modified]

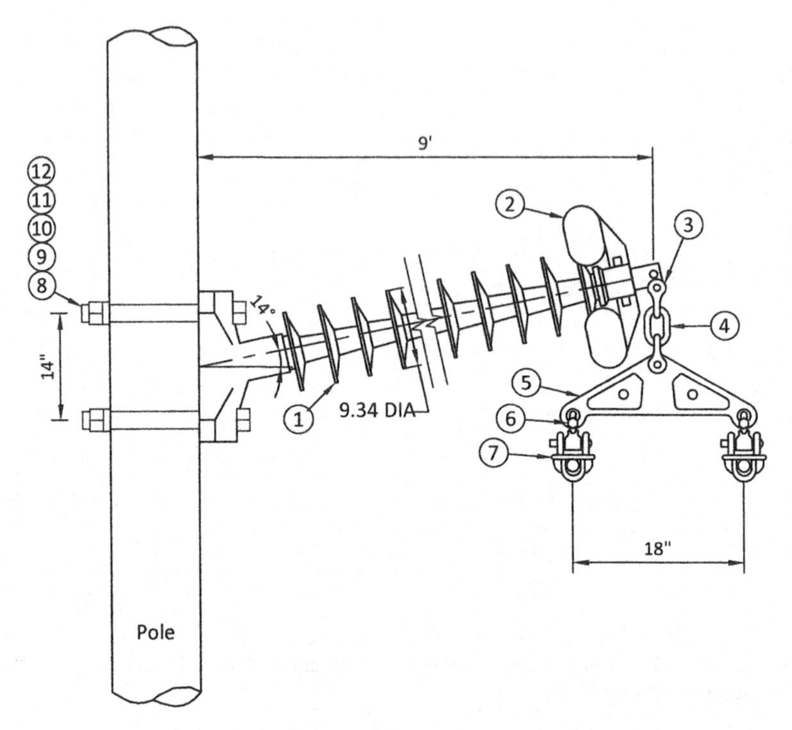

[Source : Allgeier Martin and Associates] [Modified]

Item	Description	Quantity
1	Polymer Line Post	1
2	Corona Ring, 15" for 3" Rod Insulator	1
3	Anchor Shackle	2
4	Chain Link	1
5	Yoke Plate, 18" Spacing	1
6	Y-Clevis Eye 90`	2
7	Clamp Suspension	2
8	Bolt, Machine	2
9	Washer, Square Flat	2
10	Locknut	2
11	Spring Washer for Bolt	As Reqd.
12	Clamp, Groundwire and Nut	1

1 ft = 30 cm 1 in = 25.4 mm

Figure 3.33e Typical 345 kV Twin Conductor Post Insulator Assembly.

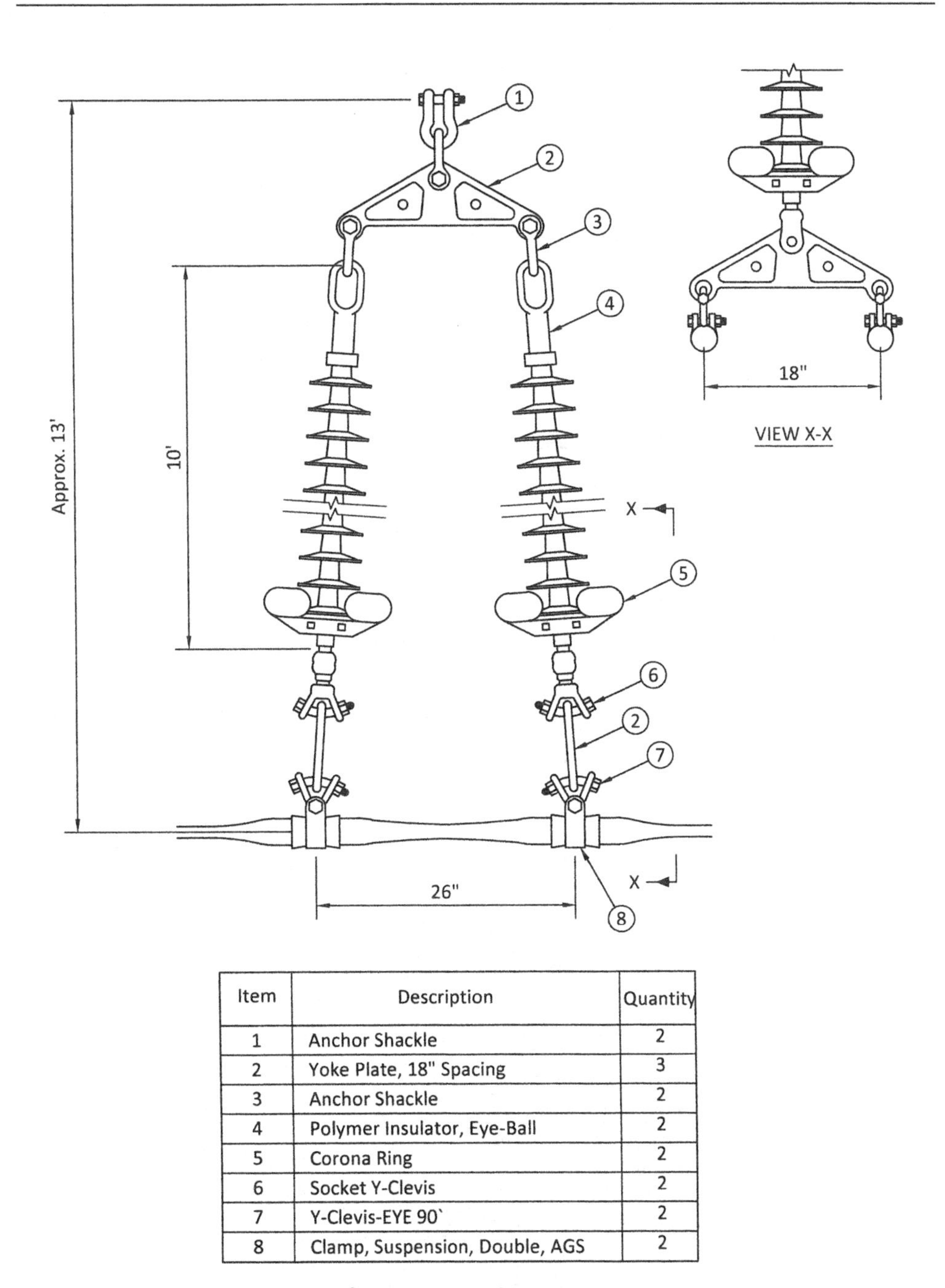

Item	Description	Quantity
1	Anchor Shackle	2
2	Yoke Plate, 18" Spacing	3
3	Anchor Shackle	2
4	Polymer Insulator, Eye-Ball	2
5	Corona Ring	2
6	Socket Y-Clevis	2
7	Y-Clevis-EYE 90`	2
8	Clamp, Suspension, Double, AGS	2

1 ft = 30 cm 1 in = 25.4 mm

Figure 3.33f Typical 345 kV Twin Conductor Support Assembly for Medium to Large Angles. [Source: Allgeier Martin and Associates] [Modified]

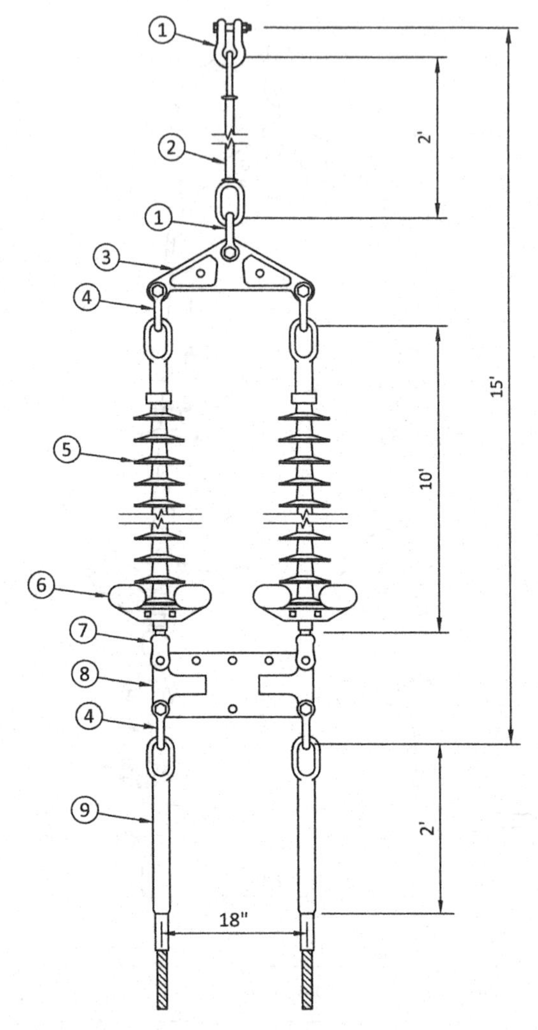

Item	Description	Quantity
1	Anchor Shackle	2
2	Extension Link, Eye-Eye, 90`, 21" Long	1
3	Yoke Plate, 18" Spacing	1
4	Anchor Shackle	4
5	Polymer Insulator, Eye-Ball	2
6	Grading Ring, 12" for Rod Insulator	2
7	Socket Y-Clevis	2
8	Rectangular Yoke Plate, 18" Spacing	1
9	Compression Deadend Single Tongue for EHV	2

1 ft = 30 cm 1 in = 25.4 mm

Figure 3.33g Typical 345 kV Twin Conductor Deadend Support Assembly. [Source: Allgeier Martin and Associates] [Modified]

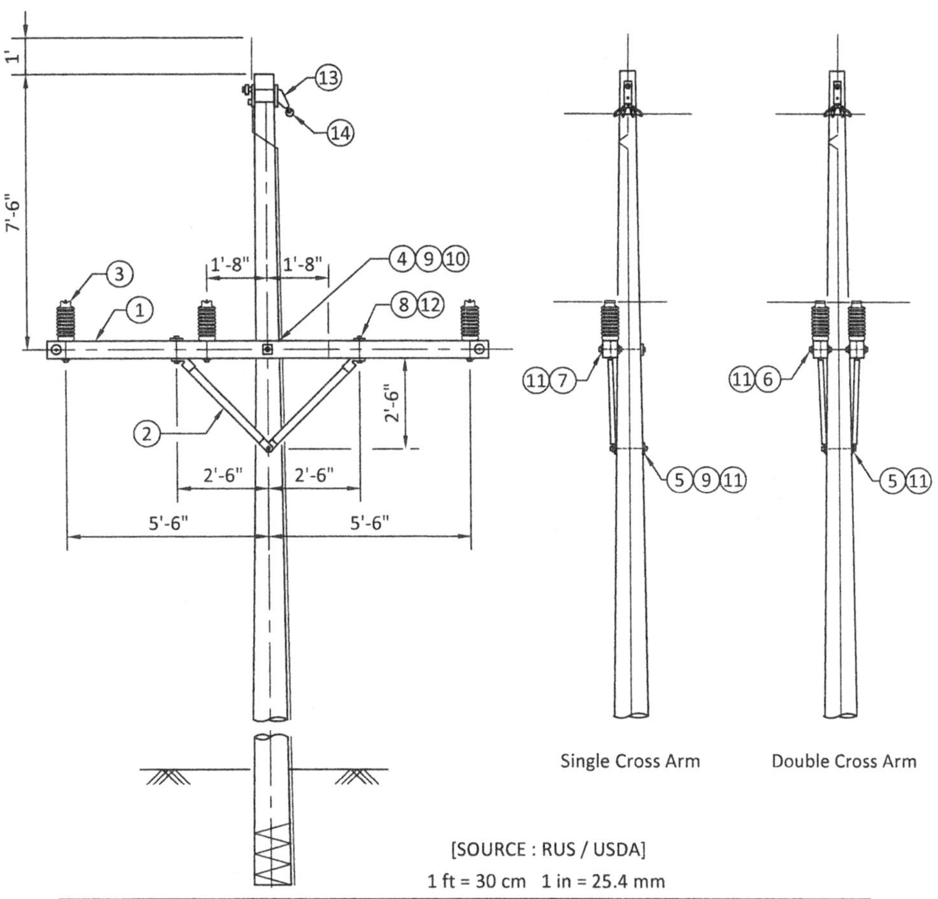

[SOURCE : RUS / USDA]

1 ft = 30 cm 1 in = 25.4 mm

Item	Description	Quantity	
		Single Cross Arm	Double Cross Arm
1	X-Arm, 4-5/8"x5-5/8"x12'-0", #28	1	2
2	Brace, X-Arm, 60"/30"	1pr	2pr
3	Insulator, Post Type	3	6
4	3/4" Bolt	1	1
5	5/8" Bolt	1	1
6	5/8" Bolt, Double Arming	-	2
7	5/8" Bolt, Washer Head	2	-
8	1/2" Bolt, Washer Head	2	4
9	Washer, Flat	3	2
10	3/4" Locknut	1	1
11	5/8" Locknut	3	9
12	1/2" Locknut	2	4
13	OHGW Support Assembly, Tangent	1	1
14	OHGW Assembly, Tangent	1	1

Figure 3.34a Medium Voltage Sub-Transmission Structure with Post Insulators.

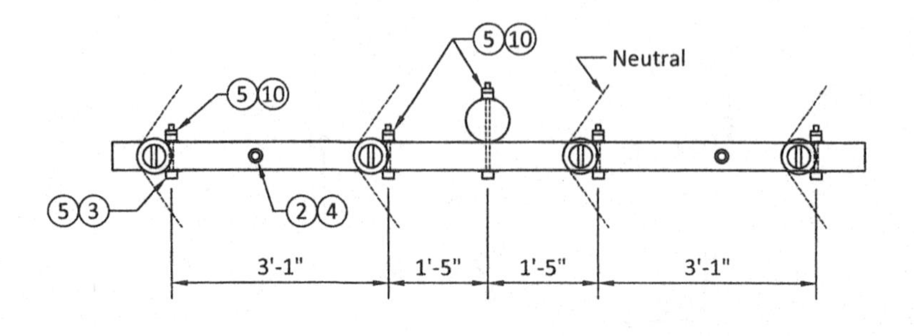

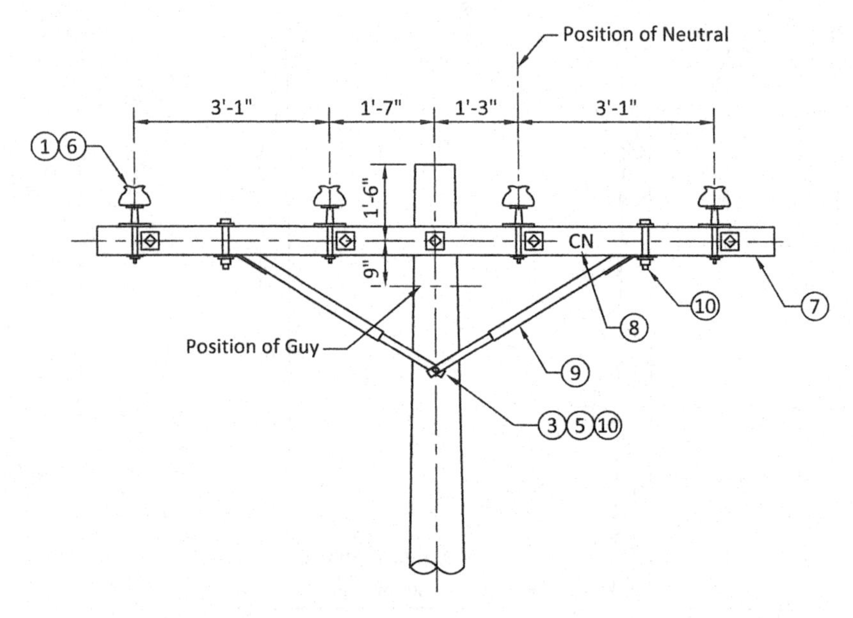

[SOURCE : RUS / USDA]

Item	Description	Quantity
1	15 kV Pin type White Insulator	4
2	1/2" Bolt	2
3	5/8" Bolt	6
4	Round Washer	2
5	Square Washer	11
6	Clamp type Steel Crossarm Pin	4
7	Crossarm, 3-5/8" x 4-5/8" x 10'	1
8	Letters	4
9	Brace, Wood, 60" span	1
10	Lock Nut	8

1 ft = 30 cm 1 in = 25.4 mm

Figure 3.34b Typical Distribution Structure with Pin Insulators (7.2 kV to 12.5 kV).

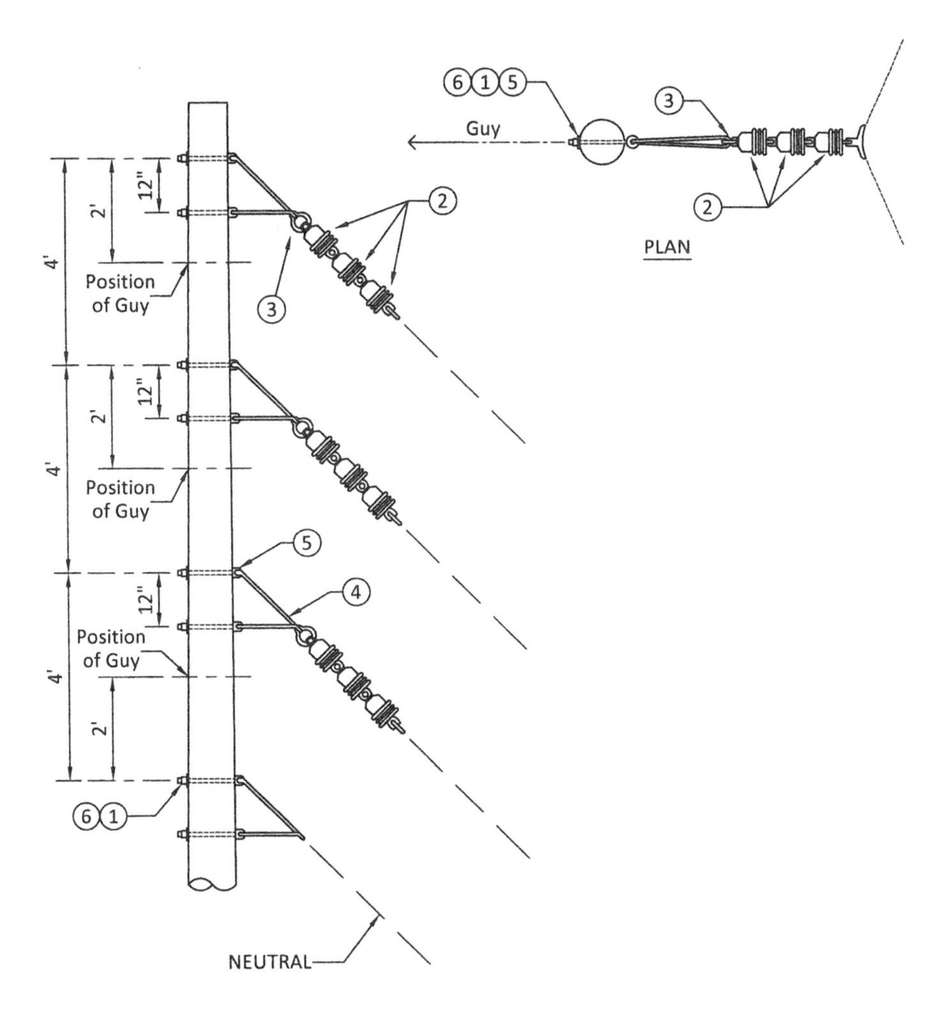

[SOURCE : RUS / USDA]

Design Parameters: Allowable Transverse load = 5000 lbs. / Conductor		
Item	**Description**	**Quantity**
1	Square Washer	8
2	Insulator, Suspension	9
3	Shackle, anchor	3
4	Bracket, angle	4
5	Bolt, clevis	8
6	Locknut	8

1 ft = 30 cm 1 in = 25.4 mm 1 lb = 4.45 N

Figure 3.34c Medium Running Angle Guyed Structure (7.2 kV to 12.5 kV).

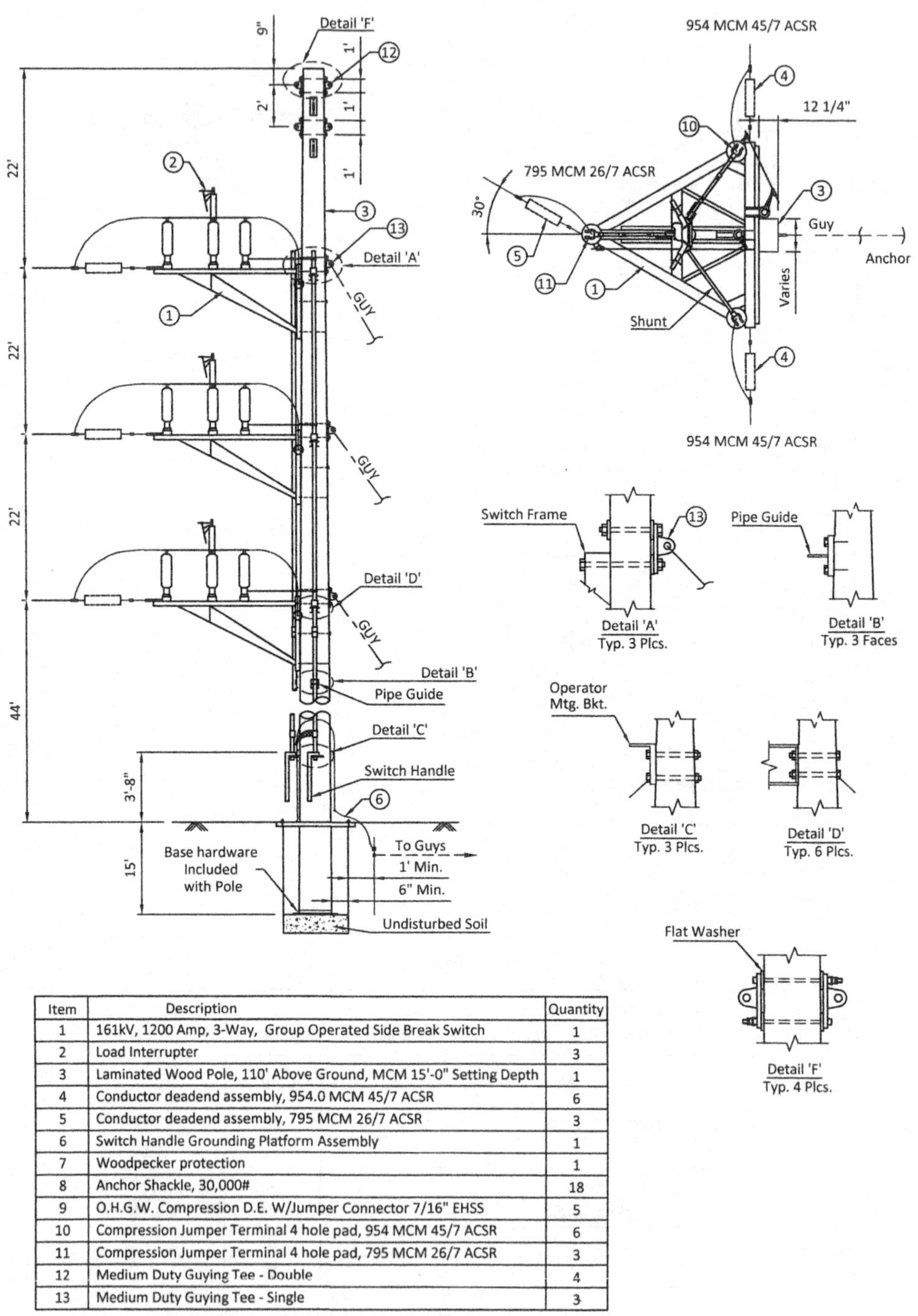

Item	Description	Quantity
1	161kV, 1200 Amp, 3-Way, Group Operated Side Break Switch	1
2	Load Interrupter	3
3	Laminated Wood Pole, 110' Above Ground, MCM 15'-0" Setting Depth	1
4	Conductor deadend assembly, 954.0 MCM 45/7 ACSR	6
5	Conductor deadend assembly, 795 MCM 26/7 ACSR	3
6	Switch Handle Grounding Platform Assembly	1
7	Woodpecker protection	1
8	Anchor Shackle, 30,000#	18
9	O.H.G.W. Compression D.E. W/Jumper Connector 7/16" EHSS	5
10	Compression Jumper Terminal 4 hole pad, 954 MCM 45/7 ACSR	6
11	Compression Jumper Terminal 4 hole pad, 795 MCM 26/7 ACSR	3
12	Medium Duty Guying Tee - Double	4
13	Medium Duty Guying Tee - Single	3

1 ft = 30 cm 1 in = 25.4 mm 1 lb = 4.45 N

Figure 3.35 Air-Break Switch 3-Way – 161 kV for 125 ft. Wood Pole. [Source: Allgeier Martin and Associates]

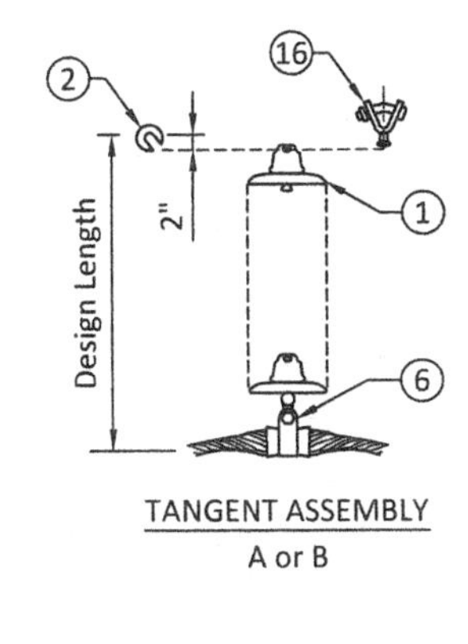

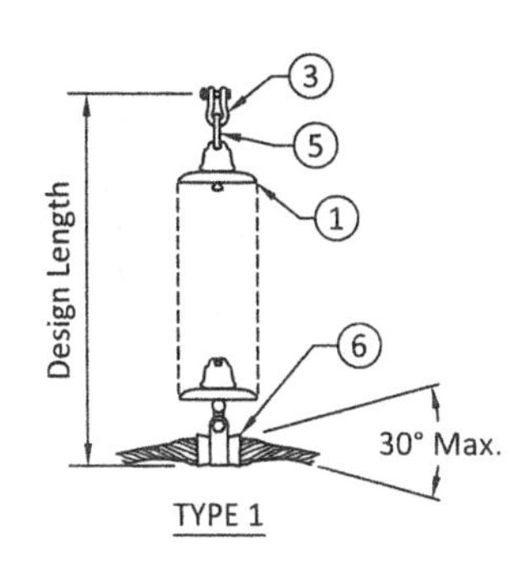

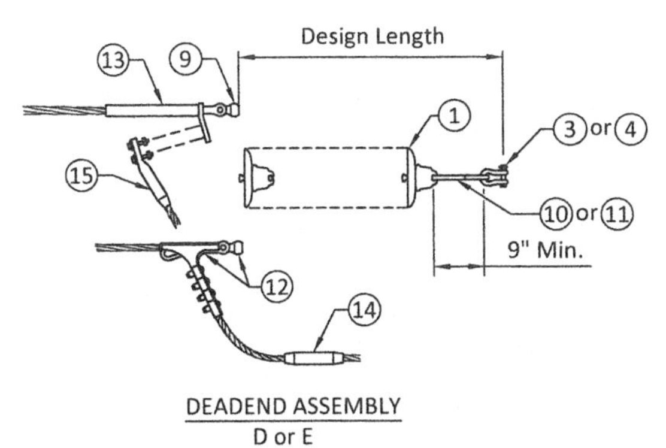

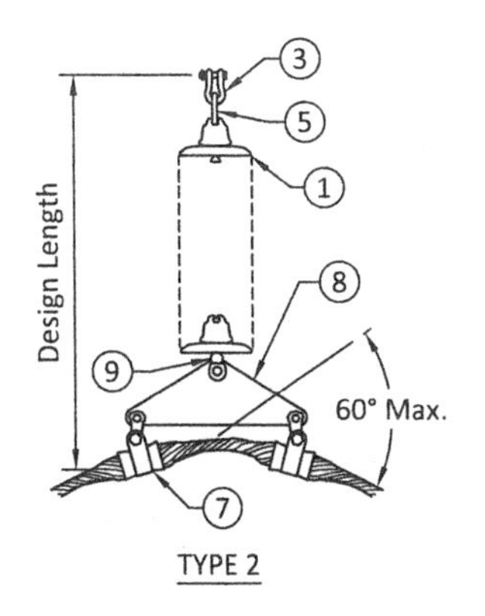

[SOURCE : RUS / USDA]

Item	Description
1	Insulator, Suspension
2	Hook, Ball, 30,000 lb.
3	Anchor Shackle, 30,000 lb.
4	Anchor Shackle, 50,000 lb.
5	Oval Eye Ball, 30,000 lb.
6	Clamp, Suspension & Socket Eye
7	Clamp, Suspension & Clevis Eye
8	Yoke Plate
9	Socket Adapter (Socket Eye or Clevis)
10	Link, Extension, Oval Eye Ball, 30,000 lb.
11	Link, Extension, Oval Eye Ball, 50,000 lb.
12	Clamp, Bolted D.E. & Socket Eye
13	Clamp, Compression D.E.
14	Jumper Connector, Compression
15	Jumper Terminal, Compression
16	30,000 lb. Ball Y-Clevis

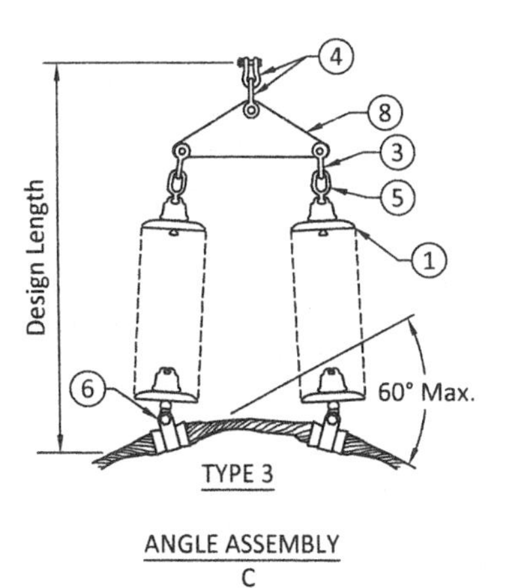

1 ft = 30 cm   1 lb = 4.45 N

Figure 3.36a Insulator Strings.

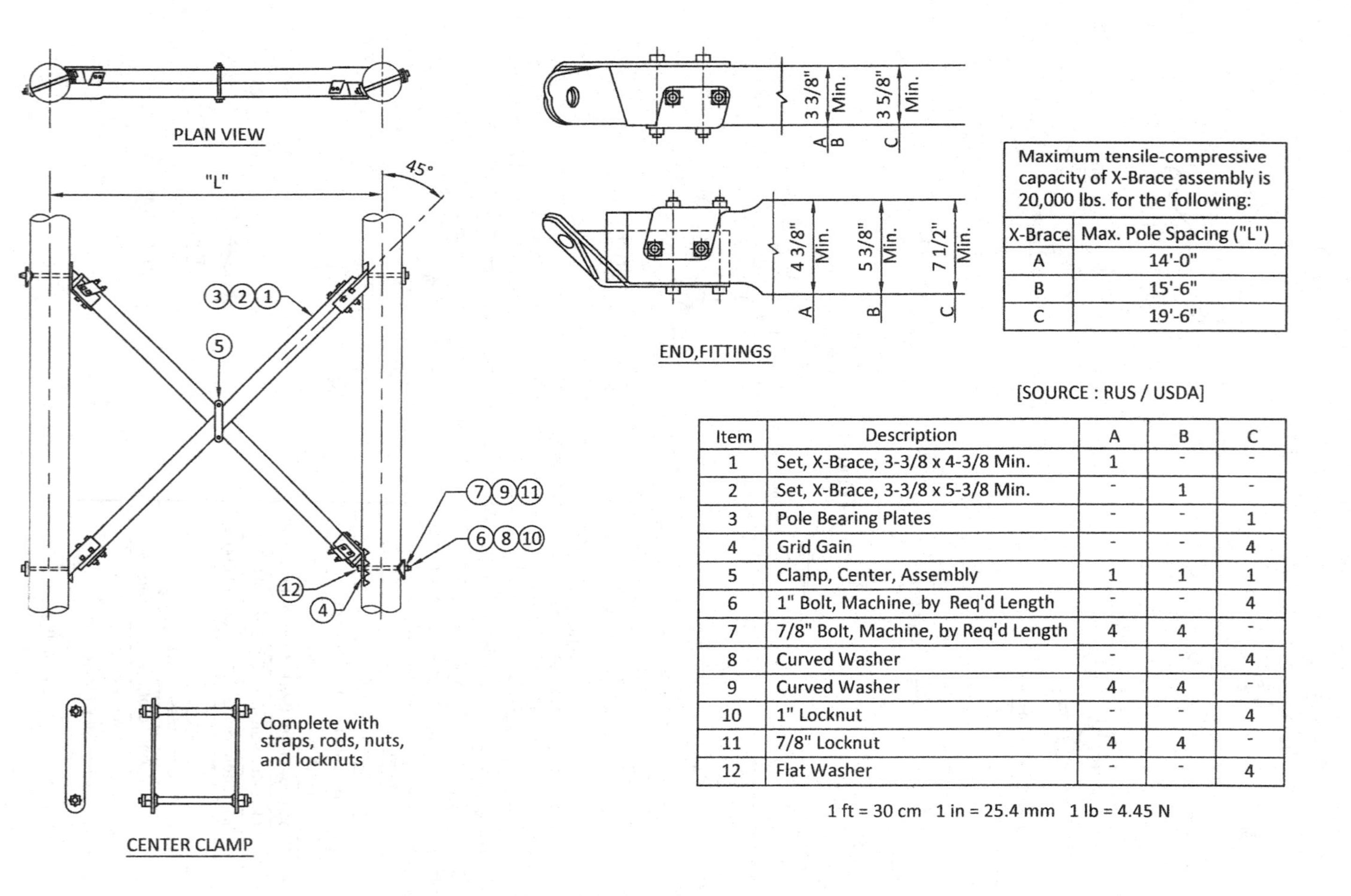

Maximum tensile-compressive capacity of X-Brace assembly is 20,000 lbs. for the following:	
X-Brace	Max. Pole Spacing ("L")
A	14'-0"
B	15'-6"
C	19'-6"

[SOURCE : RUS / USDA]

Item	Description	A	B	C
1	Set, X-Brace, 3-3/8 x 4-3/8 Min.	1	-	-
2	Set, X-Brace, 3-3/8 x 5-3/8 Min.	-	1	-
3	Pole Bearing Plates	-	-	1
4	Grid Gain	-	-	4
5	Clamp, Center, Assembly	1	1	1
6	1" Bolt, Machine, by Req'd Length	-	-	4
7	7/8" Bolt, Machine, by Req'd Length	4	4	-
8	Curved Washer	-	-	4
9	Curved Washer	4	4	-
10	1" Locknut	-	-	4
11	7/8" Locknut	4	4	-
12	Flat Washer	-	-	4

1 ft = 30 cm   1 in = 25.4 mm   1 lb = 4.45 N

Figure 3.36b Wood X-Braces.

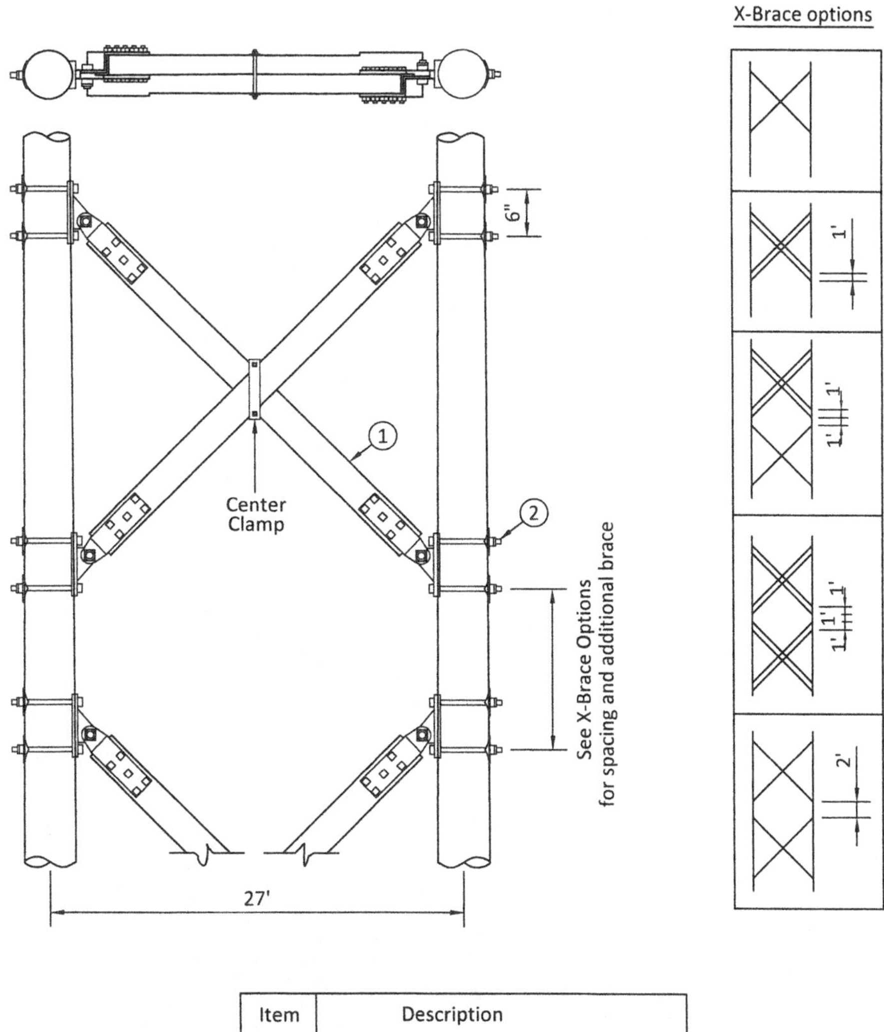

Item	Description
1	Heavy Duty X-brace with center clamp
2	End Fitting with Thru' Bolts

1 ft = 30 cm

Figure 3.36c Typical 345 kV X-Brace Assembly. [Source: Allgeier Martin and Associates] [Modified]

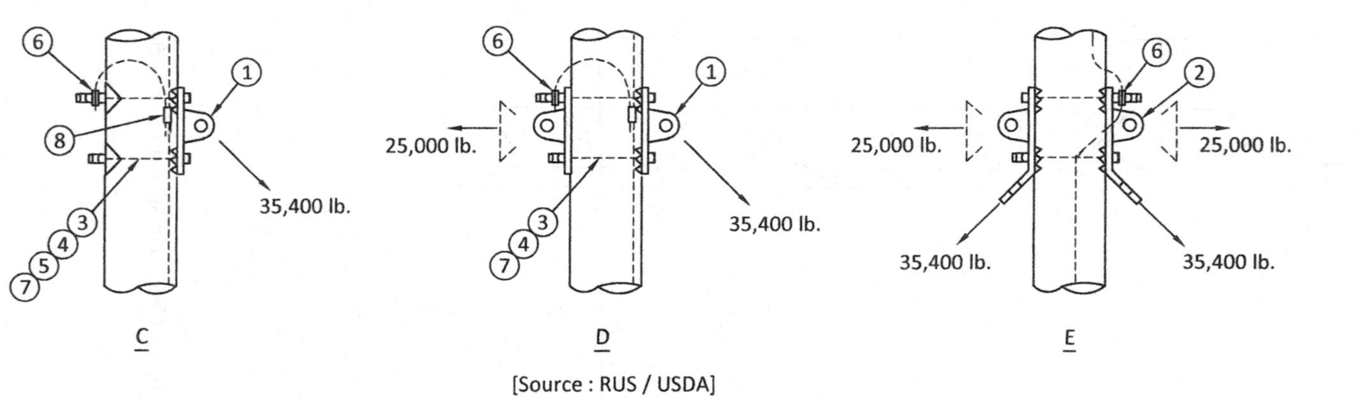

[Source : RUS / USDA]

Note: The indicated loads are design maximum

Item	Description	C	D	E
1	Tee, Deadend	1	2	-
2	Tee, Deadend & Guying	-	-	2
3	Grid Gain	2	2	4
4	7/8" Bolt	2	2	2
5	Washer, Curved	2	-	-
6	7/8" Clamp, Groundwire + 1 Nut	1	1	1
7	7/8" Locknut	2	2	2
8	Connector, Compression	1	1	-

1 lb = 4.45 N

Figure 3.36d Typical Heavy Duty Guying Tees.

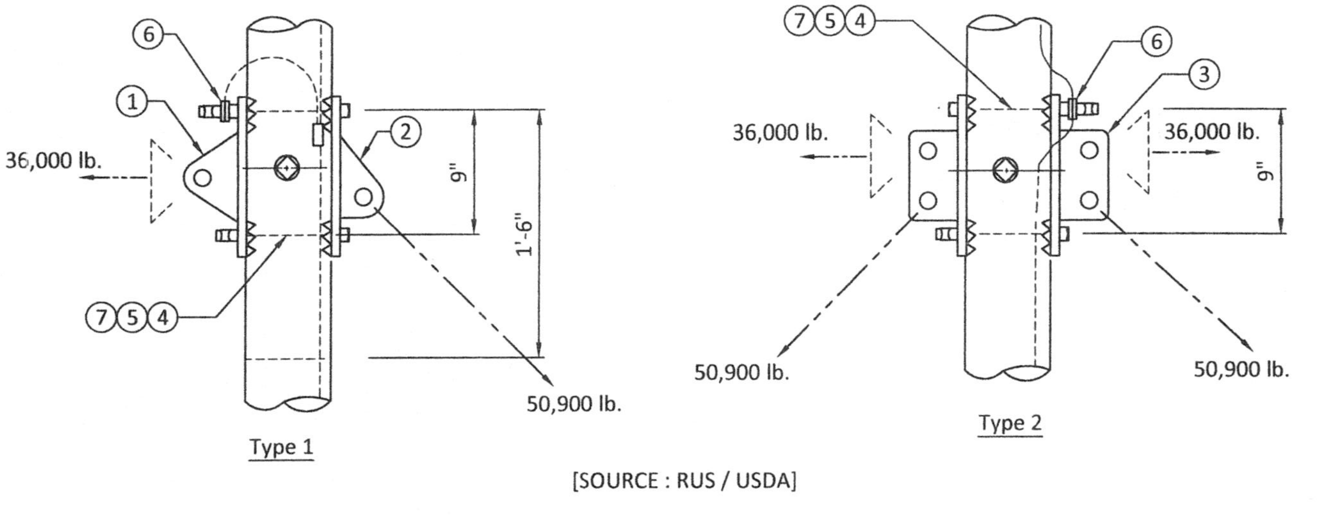

[SOURCE : RUS / USDA]

Item	Description	Quantity	
		Type 1	Type 2
1	Tee, Deadend	1	-
2	Tee, Droopy Deadend	1	-
3	Tee, Deadend & Guying	-	2
4	Grid Gain	4	4
5	1" Bolt	2	2
6	1" Clamp, Groundwire + 1 nut	1	1
7	1" Locknut	2	2

1 ft = 30 cm   1 in = 25.4 mm   1 lb = 4.45 N

Figure 3.36e Typical Medium Duty Guying Tees.

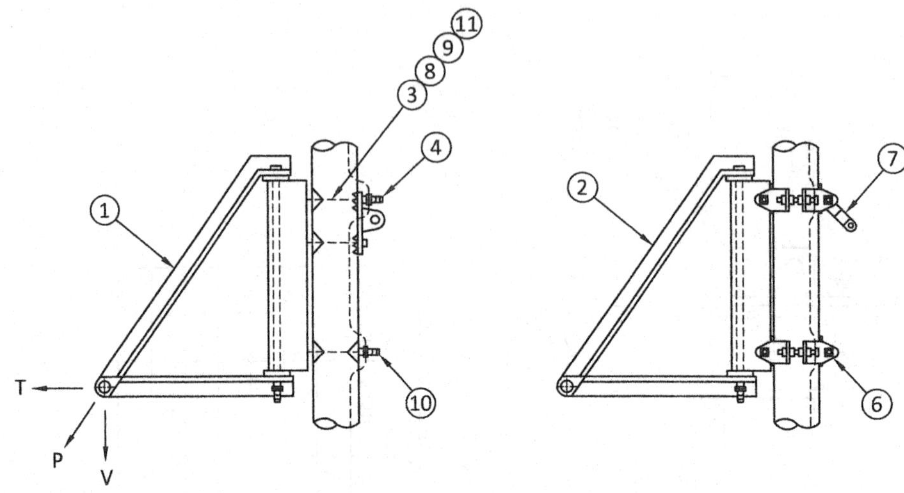

[Source : RUS / USDA]

Strength Limitations	
	Maximum Vertical Load, V = 5,000 lbs.
	Maximum Transverse Load, T = 10,000 lbs.
	Maximum Oblique Load, P = 7,000 lbs.

Item	Description
1	Bracket, Swinging Angle, Bolted Attach.
2	Bracket, Swinging Angle, Band Attach.
3	7/8" Mounting Hardware
4	Tee, Deadend
5	Plate, Pole Eye
6	Band, 4-Way Pole
7	Links, Connecting, 2: /8 x 2, w/ /16 Hole
8	Grid Grain
9	Washer
10	7/8" Clamp, Groundwire + 1 Nut
11	7/8" Locknut

1 ft = 30 cm   1 lb = 4.45 N

Figure 3.36f Bracket and Guy Attachment.

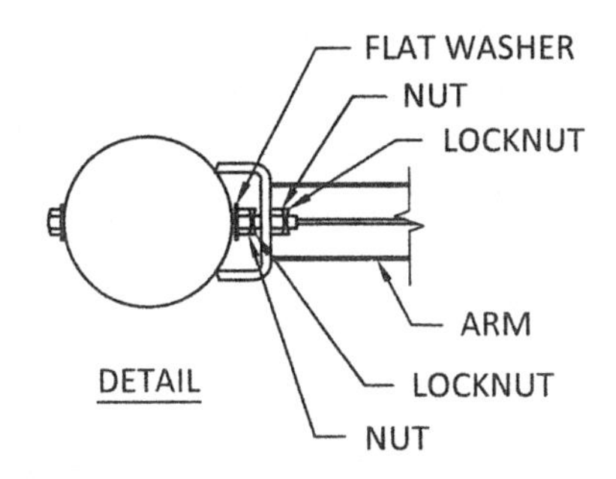

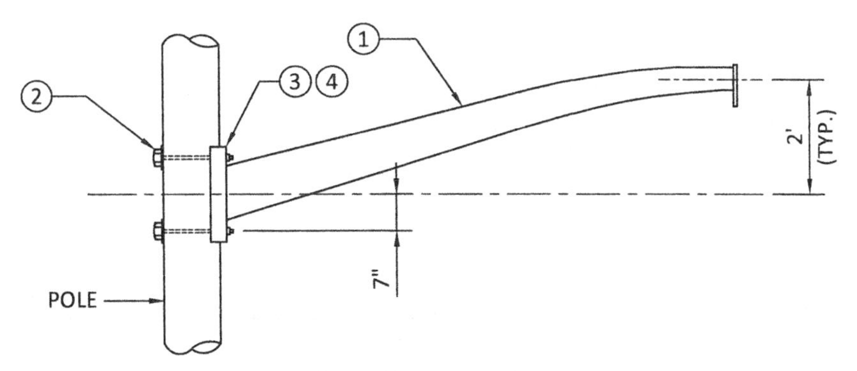

[Source : Allgeier Martin & Associates]

Item	Description	Quantity
1	8' Galv. Steel Suspension Arm	1
2	Machine Bolt	2
3	Square Washer	4
4	Locknut	4

1 ft = 30 cm 1 in = 25.4 mm

Figure 3.36g Steel Suspension Arm Assembly. [Source: Allgeier Martin and Associates]

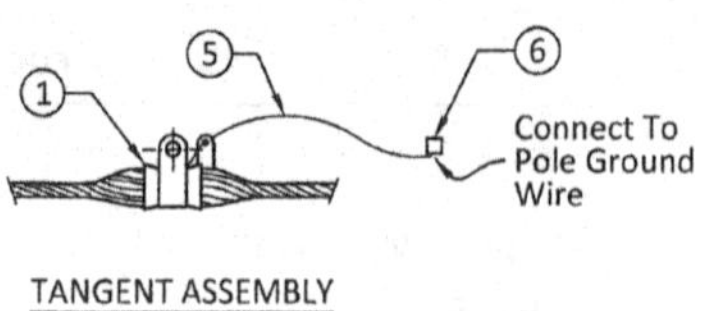

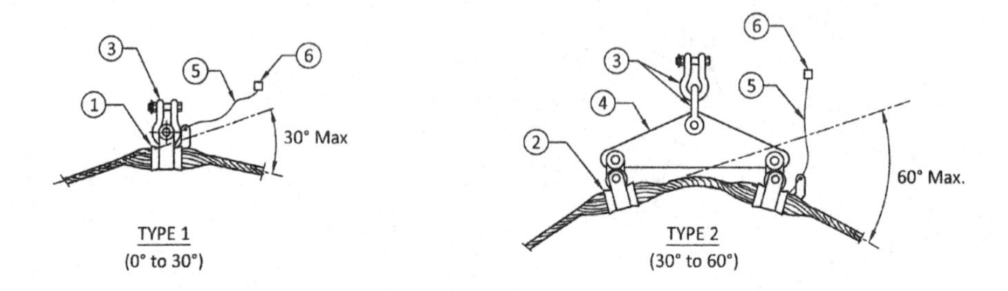

[SOURCE : RUS / USDA]

Item	Description
1	Clamp, Cushioned Suspension w/Ground Tab
2	Clamp, Double Cushioned Susp. & clevis Eye w/Ground Tab
3	Anchor Shackle, 30,000 lbs.
4	Yoke Plate
5	Ground Wire Assembly
6	Connector

1 lb = 4.45 N

Figure 3.36h Optical Ground Wire Assembly – Cushioned Suspension Clamp.

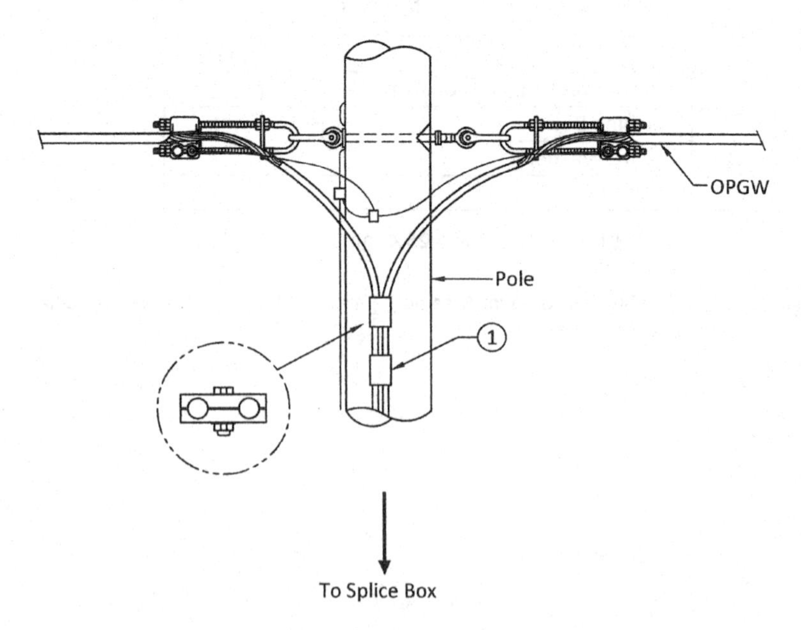

[Source : Allgeier Martin and Associates]

Item	Description	Quantity
1	Aluminum Fiber Optic Clamp	1

Figure 3.36i Fiber Optic Deadend Assembly – At Splice Box.

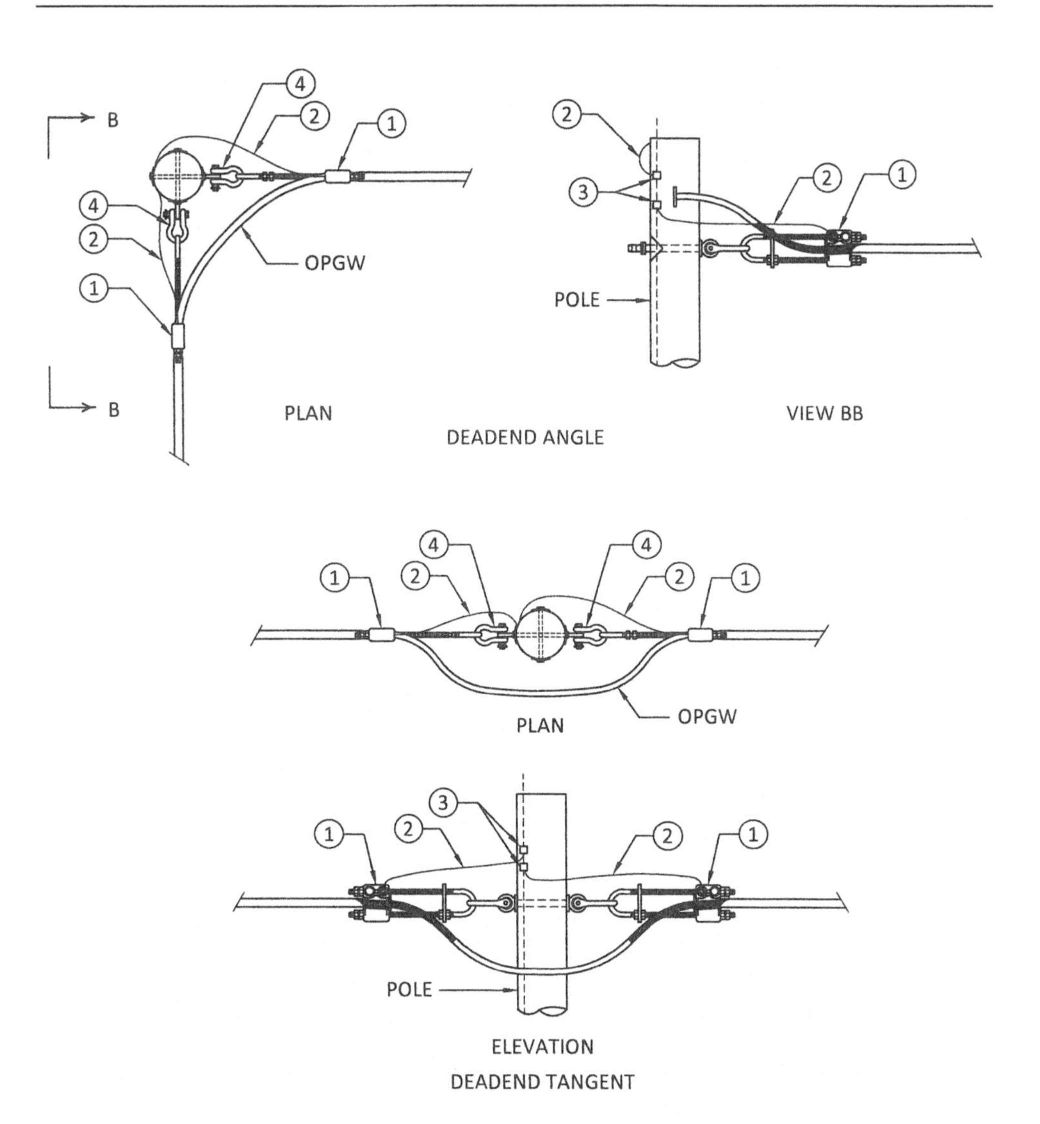

Item	Description	Quantity
1	Bolted Deadend for OPGW	2
2	Ground Wire Assembly	2
3	Connector	2
4	Anchor Shackle, 30,000# Capacity	2

1 ft = 30 cm 1 in = 25.4 mm 1 lb = 4.45 N

Figure 3.36j Fiber Optic Deadend Assembly. [Source: Allgeier Martin and Associates] [Modified]

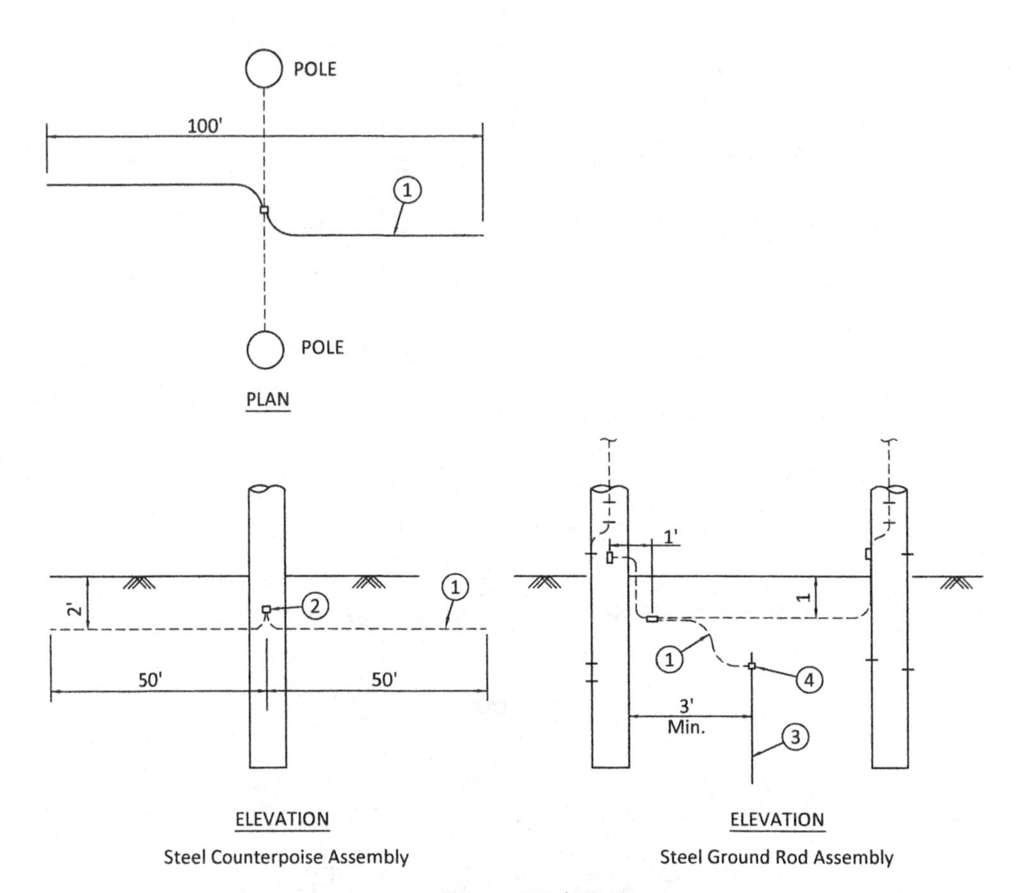

[Source : RUS / USDA]

Item	Description	Quantity	
		Steel Counterpoise Assembly	Steel Ground Rod Assembly
1	Steel, 3 Strand, 5/16" Dia., Soft Annealed wire	102'	5'
2	Connector, 3 Bolt Guy Clamp	1	
3	Ground Rod, Galv. Steel, 3/4" x 10'		1
4	Ground Rod Clamp for Steel Rod		1
5	Connector, 2 Bolt Guy Clamp		1

1 ft = 30 cm 1 in = 25.4 mm

Figure 3.36k H-Frame Grounding.

lines. While some details vary from manufacturer to manufacturer, the basic intent of all these items remains the same: to support the main components that comprise the structure carrying the conductors and ground wires.

Steel poles and lattice towers are often provided with climbing devices or ladders to facilitate worker movement during routine maintenance and repair operations. Steel poles usually contain step bolts while towers carry ladders. The responsibility for the fabrication and installation of these climbing provisions lies with the fabricator.

PROBLEMS

P3.1 Determine the RUS steel class designation for a Class H3 wood pole for (a) District Transverse Wind Load (b) Extreme Wind Load.

P3.2 Determine the maximum unbraced length of the crossing tower diagonals for the following data:
b = 7.5 ft. (2.29 m)
v = 12 ft. (3.66 m)
k = 2 ft. (0.61 m)

P3.3 Re-do Example 3.3 with each lateral wind load increased to 1000 lbs. (4.45 kN). All other data remains the same. RUS Standards apply.

P3.4 Re-do Example 3.9 for a 75 ft. (22.86 m) pole with a GL moment capacity of 184 kip-ft. (249.5 kN-m). Use pole top diameter of 8.7 in. (22.1 cm) and butt diameter of 16.9 in. (42.9 cm).

P3.5 For the pole shown in Example 3.4, use one-half the loads shown and determine the appropriate pole class. All other data remains the same.

P3.6 For the configuration in Example 3.2, determine the wire loads due to a line angle of 10° for the three situations shown.

P3.7 For the dual-use pole of Example 3.10 what will be the required pole height if the transmission circuit is for 230 kV instead of 161 kV? All other data remains the same.

P3.8 What is the design compressive force in the angle of Example 3.13 if it is 12 ft. (3.66 m) long?

P3.9 Re-do the problem of Example 3.14 assuming 5/8 in. (15.9 mm) diameter bolts.

P3.10 Re-do Example 3.15 assuming a 5/8 in. (15.9 mm) diameter bolts. All other data remains the same.

P3.11 Determine the design buckling capacity of the guyed wood pole of Example 3.5 if the phase spacing is 7 ft. (2.13 m) and if the lowest phase is 31 ft. (9.45 m) above the ground. All other data remains the same except the pole diameter at GL which is now 15 in. (38.1 cm).

P3.12 If the concrete strength in Example 3.16 is reduced to $f_c' = 8000$ psi (55.16 MPa), determine the cracking moment of the section.

P3.13 Determine the number of anchor bolts required in Example 3.11 if the pole diameter at ground line is 60 in. (152.4 cm). All other data remains the same.

P3.14 Re-do Example 3.8 if the pole is 55 ft. (16.76 m) of Douglas-Fir Class 1 with a GL moment capacity of 204.75 kip-ft. (277.6 kN-m). All other data remains the same. Compare results with that of Example 3.16. Assume RUS Standards apply.

Chapter 4

Foundation design

Prior to performing detailed foundation design, the engineer must finalize the alignment or route of the transmission line taking into consideration several constraining criteria, including the soil conditions at structure locations. The objective of detailed foundation design is to ensure that every structure is securely embedded or anchored into the ground, and the structure loads are effectively transferred to the ground strata below. To determine foundation requirements, the engineer must evaluate the nature and condition of the soil in the vicinity of the structure. The choice of foundation type also will depend on structure type, material, configuration, structure loads and the installation cost of the foundation.

This chapter discusses the design criteria for transmission structure foundations and presents numerical examples of detailed foundation designs.

4.1 GEOTECHNICAL DATA

From an engineering design perspective, soils are classified into two major groups: coarse-grained soils and fine-grained soils. Coarse-grained soils are also known as granular soils or cohesionless soils. Fine-grained soils are also known as cohesive soils.

While several geotechnical parameters influence the design of transmission structure foundations, the most important aspects are:

- Geotechnical properties of soil
- Soil classification
- Shear strength of the soils
- Geotechnical investigations

4.1.1 Geotechnical properties of soils

4.1.1.1 Index properties

The soil mass is a three phase system consisting of solid particles, water and air. The void space between the solid particles is occupied by water and air. When all the voids in a soil mass are filled with water, the soil is said to be saturated. When all the voids are filled only with air, the soil is called dry soil. Saturated soil and dry soil are examples of a two-phase system.

Geotechnical properties of soils are broadly categorized into two groups: Index Properties and Engineering Properties. The properties that are helpful in assessing the engineering behavior of soil and in determining its classification are termed index properties. Most index properties are determined by testing soil samples in the laboratory. Some index properties are listed below:

- Unit Weight
- Relative Density
- Consistency (Atterberg Limits)
- Soil Color
- Particle Size
- Void Ratio
- Specific Gravity
- Water Content
- Degree of Saturation
- Particle Size Distribution

Unit weight of soil

The ratio of total weight of soil to its total volume is termed bulk unit weight (γ). When the soil mass is completely saturated, then bulk unit weight becomes the saturated unit weight (γ_{sat}). The dry unit weight (γ_d) of soil mass is the ratio of the weight of soil solids to the total volume. Typical value of saturated unit weight of soil ranges between 100 to 130 lb/ft^3 (15.72 to 20.44 kN/m^3). For comparison, unit weight is water (γ_w) is 62.4 lb/ft^3 (9.81 kN/m^3) and unit weight of concrete is 150 lb/ft^3 (23.58 kN/m^3).

When the soil mass is submerged, its unit weight is termed submerged unit weight. The submerged unit weight (γ_{sub}) of a soil is the difference between the saturated unit weight and unit weight of water.

Relative density of cohesionless soils

The relative density is a very useful parameter for describing the strength and deformation behavior of cohesionless soils. A larger relative density indicates a higher strength and lower compressibility. The relative density (D_r) in % is given by the following the equation:

$$D_r = 100\left[\frac{\gamma_d - \gamma_{d\text{-min}}}{\gamma_{d\text{-max}} - \gamma_{d\text{-min}}}\right]\left[\frac{\gamma_{d\text{-max}}}{\gamma_d}\right] \quad (4.1)$$

where:
γ_d = in-situ dry unit weight
$\gamma_{d\text{-max}}$ = dry unit weight of soil in the densest state
$\gamma_{d\text{-min}}$ = dry unit weight of soil in the loosest state

The relative density ranges of cohesionless soils are summarized in Table 4.1.

Consistency

Consistency is a term used to indicate the degree of firmness of cohesive soils such as clays. The consistency of cohesive soils is expressed qualitatively by terms ranging from very soft state to hard state.

Table 4.1 Relative Density of Cohesionless Soils.

Description	*Relative Density (D_r) in %*
Very Loose	0 to 15
Loose	15 to 35
Medium	35 to 65
Dense	65 to 80
Very Dense	80 to 100

Table 4.2 Liquidity Index of Cohesive Soils.

Description	*Liquidity Index*
Liquid State	>1
Very Soft State	1
Very Stiff State	0
Semisolid State or Solid State	<0 (negative)

Moisture present in a cohesive soil impacts its strength. The moisture content or water content in a soil sample is given by the ratio of weight of water to the weight of solids. As the moisture content in dry cohesive soil is gradually increased the soil transitions successively from solid state to semi-solid, plastic and finally to liquid states. The moisture content at which soil changes from the liquid state to plastic state is called liquid limit (w_l) and the moisture content at which soil changes from the plastic state to semi-solid state is called plastic limit (w_p). When soil is close to its liquid state, it is very soft in consistency. In this state, it has low shear strength. However, when the soil is in its semisolid or solid state, it is very stiff and possesses high shear strength. One of the terms used to indicate consistency of soil is liquidity index (I_l) which is given by,

$$I_l = \frac{w_n - w_p}{w_l - w_p} \tag{4.2}$$

where:
w_n = natural moisture content of soil in undisturbed state
w_l = liquid limit
w_p = plastic limit

The qualitative relationship between consistency and liquidity index of cohesive soil is shown in Table 4.2. By determining the natural moisture content, liquid limit and plastic limit of a cohesive soil, one can indirectly assess its shear strength.

4.1.1.2 Engineering properties

The properties which directly determine the actual engineering behavior of soil are termed engineering properties. The most important engineering properties are:

- Shear strength

- Compressibility (popularly known as consolidation characteristics for cohesive soils)
- Permeability

Testing for engineering properties is generally time-consuming and expensive. For transmission line foundation design, the shear strength and compressibility properties are normally required.

4.1.2 Soil classification

Soil classification is the arrangement of soils into groups. Soils in a particular group exhibit similar properties and behavior. The two popular classification systems are:

- Unified Soil Classification System
- AASHTO System

The Unified Soil Classification System (Appendix 12) specified by ASTM D2487 (Appendix 13) is commonly used in transmission line structure foundation designs in the U.S. Highlights of this classification system are briefly described here. Gravels (G) and sands (S) fall into the category of coarse-grained soils. Inorganic silts (M), inorganic clays (C) and, organic silts and clays (O) fall into the category of fine-grained soils. In addition, peat, muck, and other highly organic soils (Pt) are considered in the Unified Soil classification.

Gravel size particles pass through a 3-in (75 mm) sieve and are retained on a No. 4 (4.75 mm) U.S. standard sieve. Sand size particles pass through a No. 4 (4.75 mm) U.S. standard sieve and are retained on a No. 200 (0.075 mm) U.S. standard sieve. Silts and clay size particles are called fines. Fines pass through No. 200 (0.075 mm) U.S. standard sieve. Soil particles finer than 2 microns (0.002 mm) are called clay size particles.

Coarse-grained soils are further classified into different groups. Group symbols are assigned based on gradation of particle size (W-well graded; P-poorly graded) and using plasticity chart. In case of fine-grained soils, the plasticity chart is used to classify the soils into different groups. The liquid limit of soil is represented on the X-axis and plasticity index (liquid limit-plastic limit) is represented on the Y-axis in the Plasticity chart. The A-line of the chart separates clays and silts. The liquid limit is used to indicate plasticity of fine-grained soils. Soils with liquid limit less than 50 are designed as low plastic (L) soils, whereas soils with liquid limit greater than 50 are designated as high plastic (H) soils.

As an example, well-graded sands and well-graded sands with gravel are designated by the group symbol SW. For the detailed list of group symbols, group names and complete classification procedure, the reader is referred to ASTM Standard D2487 or any Geotechnical Engineering text book.

4.1.3 Shear strength of soils

Foundations of transmission structures induce significant shear stresses in the ground. Failure occurs if the induced shear stresses exceed the shear strength of the soil. The Mohr-Coulomb failure criterion for shear strength of soils (τ) is expressed as,

$$\tau = c + \sigma \tan \phi \quad (4.3)$$

where:
τ = shear strength of soil
c = cohesion intercept
ϕ = angle of internal friction
σ = normal stress on the shear surface

Effective stress analyses

The portion of the total stress (σ) that is carried by the soil solids is the effective stress (σ'). In three phases of soil system, the shear strength in soil is primarily borne by the soil solids. The other two phases, water and air offer no shear resistance. Therefore, the shear strength is evaluated using effective stress σ'.

$$\tau = c' + \sigma' \tan \phi' \quad (4.4)$$

where:
τ = shear strength of soil
c' = effective cohesion
ϕ' = effective friction angle
σ' = effective normal stress on the shear surface

In both cohesionless and cohesive soils, typically, c' is zero. The non-zero c' occurs in heavily overconsolidated clays, cemented soils and partially saturated soils.

Total stress analyses

The total stress analysis is used when it is difficult to determine the effective stress directly.

$$\tau = c_T + \sigma \tan \phi_T \quad (4.5)$$

where:
τ = shear strength of soil
c_T = total cohesion
ϕ_T = total friction angle
σ = total normal stress on the shear surface

For saturated clayey soils under undrained conditions, $\phi_T = 0$. The shear strength of soil in this case is equal to c_T. It is also called undrained shear strength and is denoted by c_u or s_u. Here $\tau = c_T = c_u = s_u$. This analysis is widely used in saturated cohesive soils and it is known as "$\phi = 0$ analysis".

The shear strength parameters needed to design foundations depend on the rate and duration of loading and the rate of pore water pressure dissipation. When a saturated soil mass is subjected to additional stresses from foundation loads, excess pore water pressures build up and the pore pressures dissipate with time as the water drains out of voids. Drainage of water is a function of the permeability of soil. The coefficient of permeability of cohesionless soils such as gravels and sands is about a million times that of cohesive soils such as clays. If the rate of drainage is faster relative to rate of loading, then the pore water pressures dissipates in no time. Such conditions are known as drained conditions. Shear strength parameters should be determined simulating the drainage conditions occurring in the field under the applied load. Drained shear strength parameters are appropriate when the permeability of soil is relatively high or the rate of loading is relatively slow. Such an analysis is called effective stress analysis.

4.1.3.1 Sands and gravels

Effective stress analysis is common for coarse-grained/cohesionless soils such as sands and gravels. Both steady-state (long-term) and transient (short-duration, except earthquake) loads normally result in drained behavior. Since excess pore pressures dissipate rapidly in cohesionless soils, the pore water pressures in these soils equal hydrostatic pressures, and are easy to compute. Effective stresses, and therefore the shear strength can be accurately estimated in these soils. In this analysis, the shear strength parameters used are c' (effective cohesion) and ϕ' (effective angle of internal friction). The effective stress on the shear surface, used in the equation of shear strength, is σ' which is calculated with effective unit weight $\gamma_{sat} - \gamma_w$ in saturated soils, where γ_{sat} is saturated unit weight of soil and γ_w is unit weight of water.

In clean sands and gravels, it is a good practice to use $c' = 0$. Therefore, for pure sands and gravels, the shear strength is expressed in terms of angle of internal friction (ϕ'). In the rest of this book, the angle of internal friction is denoted by ϕ instead of ϕ'. However, it must be noted that it is an effective angle of internal friction determined under drained conditions. Laboratory tests such as triaxial tests (e.g.: consolidated, drained triaxial strength test) or direct shear tests can be used to determine the friction angle. This strength parameter can be also determined using correlations with N values determined from Standard Penetration Test (SPT).

4.1.3.2 Clays and silts

In case of saturated fine-grained soils such as clays and silts, transient or short-duration loads result in an undrained response. Due to low permeability, excess pore pressures develop during and immediately after transient loading. For effective stress analysis, determining the excess pore water pressures developed under undrained conditions is a difficult task. Therefore, for transient loads in cohesive soils, a total stress analysis is performed using undrained shear strength parameters. The shear strength parameters in this case are c_T (total cohesion) and ϕ_T (total angle of internal friction).

For saturated clays under undrained conditions, the shear strength is just equal to c_T and $\phi_T = 0$. In this case, the shear strength determined is called undrained shear strength. In this textbook, the undrained shear strength determined under total stress approach for saturated clays is denoted by s_u, with $\phi = 0$ condition. Some geotechnical engineering textbooks commonly use c_u in place of s_u.

Typically, undrained shear strength governs the design of foundations in cohesive soils with transient loading conditions. The extreme wind loads used in the design of transmission line structures in the US are short-term loads (3-second gust winds) and are thus considered transient loads. Even the extreme ice loading that typically lasts 3–5 days can be considered as a transient load for design purposes.

Table 4.3 Loading Conditions and Shear Strength Parameters.

Soil Type	*Drained/Undrained Behavior*	*Loading Condition*	*Shear Strength Parameter*
Coarse-grained Soils (Sands, Gravels)	Drained Behavior	Both Steady State and Transient Loads	Angle of Internal Friction (ϕ) for pure sands and gravels (Effective Friction Angle)
Fine-grained Soils (Clays, Silts)	Undrained Behavior	Transient Loads	Undrained Shear Strength s_u with $\phi = 0$ for pure clays
	Drained Behavior	Steady State Loads	Effective Shear Strength Parameters*

Steady state or long-term loads can control foundation design in some situations.

The steady-state loads result in drained behavior in cohesive soils and drained shear strength parameters have to be used in these situations. For example, the long-term (drained or effective strength) uplift capacity of a shallow footing is less than the short-term (undrained) uplift capacity and it can be as low as 50% of short-term capacity. At the same time, the ratio of wire tensions on angle/pure deadend transmission line structure resulting from every day loads to transient loads (e.g. extreme wind) is typically less than 50%. Therefore, the drained behavior may not govern foundation design in this situation.

However, for spread foundations such as grillages installed in fissured clays with relatively low embedment depths, drained conditions under long-term loadings can control foundation design. Similarly, in case of overconsolidated stiff clays, drained/effective shear strength parameters under sustained long-term loads can control size of drilled shafts used for self-supporting steel poles at angle and pure deadend locations. A summary of the controlling states for coarse and fine grained soils is presented in Table 4.3.

4.1.4 Geotechnical investigations

Subsurface exploration is the process of identifying soil strata and the characteristics of the strata underneath a proposed transmission structure. Geotechnical studies are routinely performed along the alignment of the transmission line to determine the types of soils and variation in the soil profile at a given location. Exploration of the line route includes several steps such as collection of preliminary information, field reconnaissance and detailed design investigation.

During the first step of gathering preliminary information, the data collected includes site geology and existing geotechnical information. Various sources such as topographic and geologic maps, aerial photos, soil reports, manuals published by Agricultural and Highway departments, previous construction and soil boring information along the line route provide useful preliminary information. During the field reconnaissance stage, visual inspection of line route is performed by the geotechnical engineer or the engineering geologist. The number of soil borings required along the line route can be minimized using the information gathered during the first two stages.

Table 4.4 Classification of Soils Based on Simple Field Tests – Cohesive Soils.

Term	*Field Test*
Very soft	Squeezes between fingers when fist is closed
Soft	Easily molded by fingers
Firm	Molded by strong pressure of fingers
Stiff	Dented by strong pressure of fingers
Very Stiff	Dented only slightly by finger pressure
Hard	Dented only slightly by pencil point

(Source: RUS/USDA.)

Table 4.5 Classification of Soils Based on Simple Field Tests – Cohesionless Soils.

Term	*Field Test*
Loose	Easily penetrated with a 1/2 in (13 mm) reinforcing rod pushed by hand
Firm	Easily penetrated with a 1/2 in (13 mm) reinforcing rod driven with a 5 lb (22 N) hammer
Dense	Penetrated 1 ft with a 1/2 in (13 mm) reinforcing rod with a 5 lb (22 N) Hammer
Very Dense	Penetrated only a few inches with a 1/2 in (13 mm) reinforcing rod driven with a 5 lb (22 N) hammer

(Source: RUS/USDA.)

A few simple field tests to classify the soils for wood transmission lines are given in Tables 4.4 and 4.5.

During the detailed design stage, site specific information needed for the design and construction of foundations for line structures are collected. Various tasks such as drilling soil borings, collecting soil samples, field testing (including geophysical surveys), and laboratory testing are performed. Soil borings are obtained by using different methods such as auger boring, wash boring, rotary drilling and percussion drilling. Both rotary and percussion drilling can be used in rocky strata.

The number of borings required for a transmission line depends on the voltage of the line, data collected during preliminary stages, variability of geotechnical data along line route, foundation and structure type. In general, higher voltage lines require relatively more borings. Drilled shaft foundations typically require more site specific borings than direct embedment foundations. More borings are required if the geotechnical data varies along a line.

It is also a common practice to perform soil borings at each angle and deadend structure locations especially in case of drilled shaft foundations. Some utilities perform one or two borings per mile for tangent structures; other utilities perform a soil boring at each structure location. Locations with poor soil conditions require soil borings due to the risk associated with foundation failure. Locations with variable rocky strata below ground level require soil borings so that an economical foundation type and foundation depth can be determined. For locations with high water table and unstable soil conditions, the information from soil borings can help the engineer select proper construction equipment. Ultimately, the number of borings performed for a given transmission line should be based on balancing cost and risk.

4.1.4.1 Field tests

During the detailed design stage, different types of field tests are conducted. The common in-situ tests include Standard Penetration Test (SPT), Cone Penetration Test (CPT), Vane Shear Test (VST), and Pressuremeter Test (PMT).

Standard Penetration Test (SPT) is very routinely performed to characterize the subsurface to aid in transmission line foundation design. This test is performed in a bore hole using a split spoon sampler. The sampler is penetrated into soil by a total depth of 18 inches (450 mm) using standard hammer weight and drop and the blow counts for each 6-inch (150 mm) increment are recorded. The number of blows required for the last 12 inches (300 mm) of penetration is recorded as the SPT resistance value, N. The units of N are number of blows per foot.

Per Peck et al. (1974), the correlation between SPT "N" value and the effective triaxial compression friction angle for cohesionless soils is shown in Table 4.6. An approximate correlation between SPT N value and undrained shear strength of cohesive soil per Terzaghi and Peck (1967) is presented in Table 4.7.

Table 4.6 Empirical Relationship Between SPT N Value and Friction Angle of Cohesionless Soils.

N value	*Relative Density*	*Approximate Friction Angle, ϕ (degrees)*
0 to 4	Very Loose	<28
4 to 10	Loose	28 to 30
10 to 30	Medium	30 to 36
30 to 50	Dense	36 to 41
>50	Very Dense	>41

(With permission from EPRI, Report EL-6800, 1990.)

Table 4.7 Empirical Relationship Between SPT N Value and Undrained Shear Strength of Cohesive Soils.

N value	*Consistency*	*Approximate Undrained Shear Strength, s_u in ksf (kPa)*
0 to 2	Very Soft	<0.25 (<12)
2 to 4	Soft	0.25 to 0.5 (12 to 25)
4 to 8	Medium	0.5 to 1.0 (25 to 50)
8 to 15	Stiff	1.0 to 2.0 (50 to 100)
15 to 30	Very Stiff	2.0 to 4.0 (100 to 200)
>30	Hard	>4.0 (>200)

(With permission from EPRI, Report EL-6800, 1990.)

The Cone Penetration Test (CPT) is a simple test that provides data for continuous soil strata. A hole is not bored in a CPT Test. In a cone test, the cone is penetrated in to the ground at a steady rate and the resistance to penetration is measured. Both the end resistance to cone and frictional resistance along the sleeve are measured. Soil properties obtain from established correlations. The Vane Shear Test (VST) is another test which is very useful for determining the in-situ undrained shear strength of cohesive soils.

The Pressuremeter Test is used for determining the pressuremeter modulus, an important parameter for calculating the deflection and rotation behavior of laterally-loaded drilled shaft foundations and direct embedment foundations under moment loads. Table 4.8 shows a brief summary of different methods used for determination of soil parameters for transmission foundation design.

4.1.4.2 *Strength and deformation parameters of rock*

Rock core samples are collected for rocky soils to establish the soundness of rock and to perform unconfined and high-pressure triaxial tests in the laboratory. If rock is encountered at shallow depths, it is imperative to determine if it is bed rock or a suspended boulder. Geological knowledge of rock formations is very helpful in determining the extent of rock. The Rock Quality Designation (RQD) is one of the metrics used to determine quality of a rock mass. The RQD is the ratio of the lengths of intact pieces of core greater than 4 inches (100 mm) over length of core advance. The higher the RQD value, the better is the quality of rock. If the RQD value is close to 100%, rock quality is considered excellent. If the RQD value is less than 25%, the rock is considered poor quality rock (Peck et al., 1974).

The most important rock parameters that quantify the strength of rock are the effective stress friction angle, the effective stress cohesion, and the modulus of deformation of the rock mass. The Rock Mass Rating (RMR_{76}) is one of the most popular classification systems for rocks and is an excellent metric that is used to quantify both

Table 4.8 Design Parameters and Methods.

	Method Used	
Design Parameter	*Direct Measurement*	*Correlation with Tests/Indexes*
Friction Angle for Cohesionless Soil	Triaxial Compression Test Direct Shear Test	Soil Relative Density Standard Penetration Test Cone Penetration Test
Undrained Shear Strength for Cohesive Soil	Unconsolidated Undrained Triaxial Compression Test Unconfined Compression Test	Soil Liquidity Index Standard Penetration Test
Unit Weight of Soil	Undisturbed Soil Sample Direct Measurement	Standard Penetration Test
Deformation Characteristics of Soil	Pressuremeter Test	Standard Penetration Test

strength and deformation parameters of rock. The RMR_{76} is one of the systems used by transmission line design engineers in the design of rock-socketed drilled shafts as well as in directly embedded pole foundations.

The RMR_{76} is derived from Geomechanics Classification System developed by Bieniawski (1973, 1976). The system considers five parameters in classifying a rock mass: Strength of the intact rock material, Rock Quality Designation (RQD), Spacing of joints, Condition of joints and Groundwater. A numerical value is assigned for each of the parameters and the values are then summed to result in the RMR_{76}. An adjustment for joint orientations is applied. The RMR_{76} of a rock layer can vary range from 0 to100. The higher the RMR_{76} value, the better is the rock in terms of strength and deformation behavior. Detailed discussions on correlation of strength and deformation parameters with RMR_{76}, and application to design rock-socketed transmission foundations under moment loads, are found in research performed by Rose et al. (2001).

4.1.4.3 Geotechnical report

A standard Geotechnical Report for transmission lines will feature:

- Project Description – brief overview of the type and nature of construction (wood, steel or concrete); magnitude of loads expected and equipment supported by line structures
- Site Description – brief overview of site conditions, topography, slopes, regional geology and presence of water bodies
- Field Exploration – details of sub-surface explorations, location of bore holes, depths of investigation, soil types encountered
- Laboratory Testing – index and engineering properties of soil samples
- Boring Logs – details of information gathered from each borehole in a graphical form which includes the soil types, index properties such as water content, liquid limit, plastic limit, SPT N values, and water table location.
- Geotechnical Recommendations – design data for shallow and deep foundations, site preparation, excavations, structural fill
- Earthquake and Seismic Design – USGS NEHERP ground motion values for the area

4.1.5 ASTM standards

ASTM Standards are commonly used as a Quality Control tool for soil investigations and laboratory tests performed for foundation design. Governing ASTM Standards for soil analysis and testing are furnished in Appendix 13.

4.2 DESIGN PHILOSOPHY

The optimization techniques for transmission line structure foundations differ significantly from foundations of other structures such as buildings and bridges. Transmission line structures and their foundations are constructed over a route spanning several miles. Subsurface soil conditions change from location to location along the route. It is impractical to investigate every structure location for geotechnical parameters. A few standard foundation types, especially for tangent structures, are designed for the worst soil conditions and these designs are used at other locations. Therefore, optimization of geotechnical investigations and foundation designs for families of structures will have a significant impact on cost.

Geotechnical investigations for transmission line structures tend to be less sophisticated compared to those of buildings or bridges. Human occupancy is not involved for transmission lines, and therefore, the serviceability requirements are less stringent. Allowable criteria for settlement calculations under compressive loads, upward foundation movement under uplift loads and deflections and rotations under moment loads vary widely among utilities.

4.2.1 Basic types of foundations

The types of foundations used for the transmission line structures are summarized in Table 4.9. The types of foundations include direct embedment, concrete drilled shafts, steel grillages, pile foundations, concrete spread footings and anchors. In poor soils, special foundations (i.e.) culverts, helical piles and vibratory caissons are used. Recently, the foundations such as micropiles are getting popular in challenging site conditions.

Direct embedment foundations are commonly used on wood pole tangent, guyed angle and deadend structures. Direct embedment of steel tangent poles is also common. If the soil conditions are poor and the design loads are significant, a concrete pier foundation may be used for single tangent steel poles. Concrete pier foundations are commonly required for self-supported angle and deadend single steel pole structures. Grillages, concrete drilled shafts and spread footings are commonly used on lattice towers.

Table 4.9 Types of Foundations for Transmission Line Structures.

Type of Foundation	*Application*
Concrete Drilled Shaft Foundation	Single poles (heavy angle/deadend structure) Tangents and Heavily-loaded guyed structures Lattice towers
Directly Embedded Pole Foundation	Single poles and H-frame structures (Tangent/Lightly-loaded structure) Guyed structures Culverts for single pole foundations in poor soils
Steel Grillage Foundation	Lattice towers
Anchor Foundation	Guyed structures
Pile Foundation	Single poles and H-frame structures in poor soil conditions Vibratory caissons in poor soils (driven piles) Helical piles for lattice towers and pole structures Micropiles in special situations for lattice towers and pole structures
Concrete Spread Footing	Lattice towers

4.2.2 Foundation design loads

In general, the two major classes of loadings encountered by transmission structures and foundations are steady state loads and transient loads. Loads that are sustained on transmission line structures for a longer duration are called steady state or long-term loads. Examples of steady-state loads include:

1. Vertical loads (dead weight of the structure, conductors/shield wire/OPGW, insulators and hardware)
2. Transverse loads (transverse pull due to line angle on angle structures)
3. Longitudinal loads (differential line tension in adjacent spans)

Terminal structures at substation ends (or deadends with slack tension spans) are subjected to sustained loading due to differential wire tensions. The foundations on these structures are subjected to steady state loads.

Loads that act on a structure for a short time period are known as transient loads. Examples of these loads include:

1. Vertical loads (ice on conductors, shield wires and structure, broken insulator string of a suspension structure)
2. Transverse loads (extreme wind loads on conductors, shield wires, and the structure with/without ice)
3. Longitudinal loads (broken wire loads and stringing loads caused by conductor jam in stringing blocs during conductor stringing)

Depending on the type of structure, different types of forces act on typical transmission line structures as shown in Figure 4.1. Controlling forces for different types of structure foundations are summarized in Table 4.10. These guidelines are general in nature and can vary depending on the type of structure configuration.

4.2.3 Structure and foundation reliability

The design of a transmission line is often broken up into the design of constituent elements such as structures, foundation, conductor, hardware, etc. The weakest element governs the failure of the line. The reliability desired for different components is carefully predetermined during design. It is generally preferable to design the foundation to be stronger (or in other words more reliable) than the structure. If a structure fails during an overload in an extreme event, it can be quickly replaced on the existing foundation (example: steel pole supported on a drilled shaft). However, if the foundation fails, it causes the structure to fail subsequently. The cost of replacing both the foundation and the structure is more compared to just replacing the structure.

An exception to this philosophy is made for directly buried pole foundations. For low voltage lines, some designers prefer the direct buried foundation to be weaker than the structure. The failure of the soil may cause a pole to lean over in an extreme event, and relieves load on the pole. After the extreme event, pole may be re-plumbed depending on the severity of deflection/rotations of the pole.

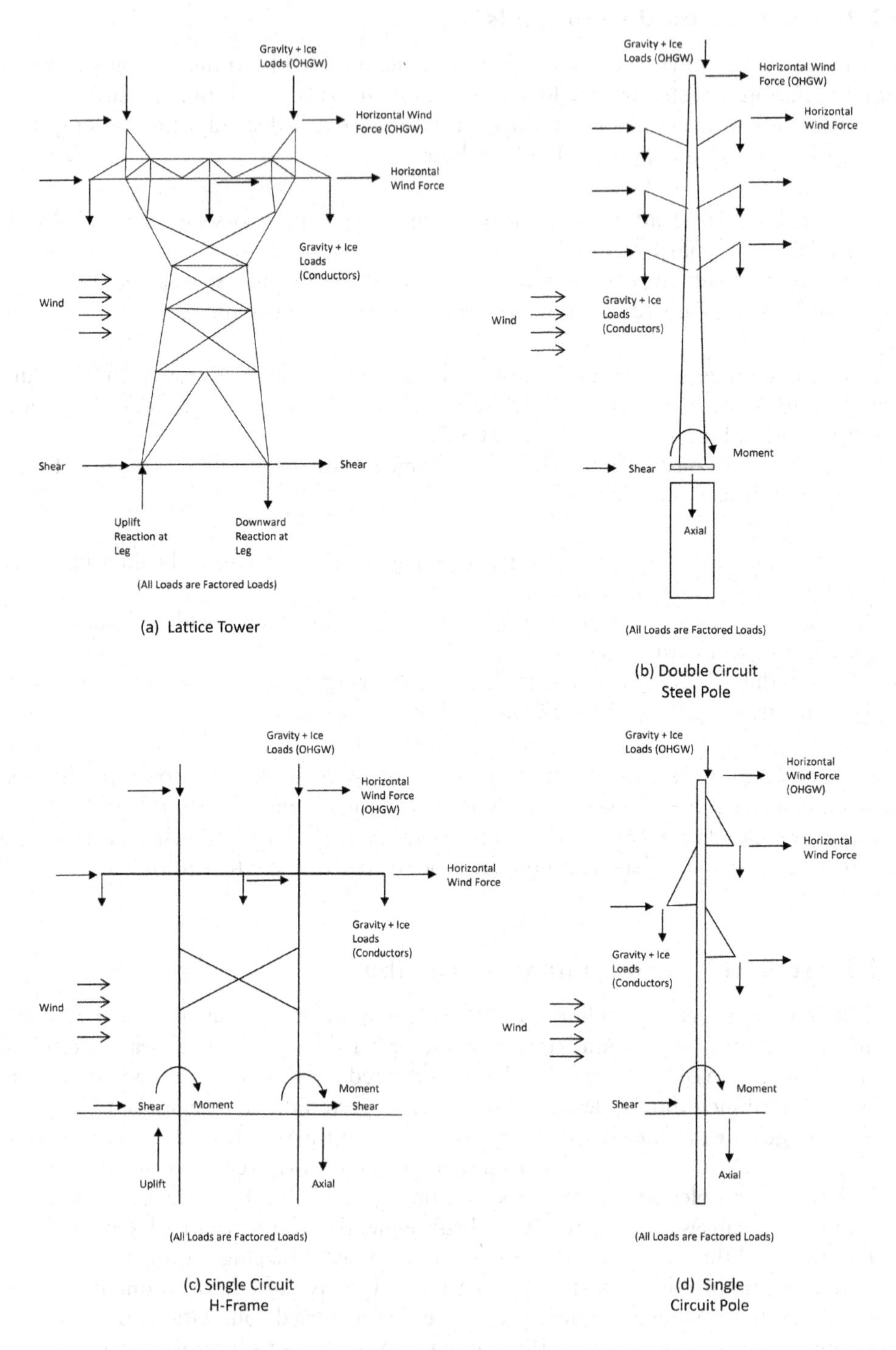

Figure 4.1 Design Loads and Reactions for Transmission Line Structures.

Table 4.10 Forces Acting on Various Types of Transmission Structure Foundations.

Type of Structure	*Foundation Reactions*	*Controlling Force (General)*
Single Pole Structure (Self-Supported/Unguyed)	Vertical (Compression) Horizontal (Shear) Torsional loads Overturning Moments	Overturning Moment
H-frame Structure	Vertical (Uplift and Compression) Horizontal (Shear) Overturning Moments	Vertical (Uplift/Compression) for H-frames with X-braces
Lattice Tower	Vertical (Uplift and Compression) Horizontal (Shear) Overturning Moments	Vertical (Uplift)
Guyed Structure	Vertical (Compression) Horizontal (Shear, at pole base) Pullout Force (Guys)	Vertical (Compression) at base Pullout Force (Guys)

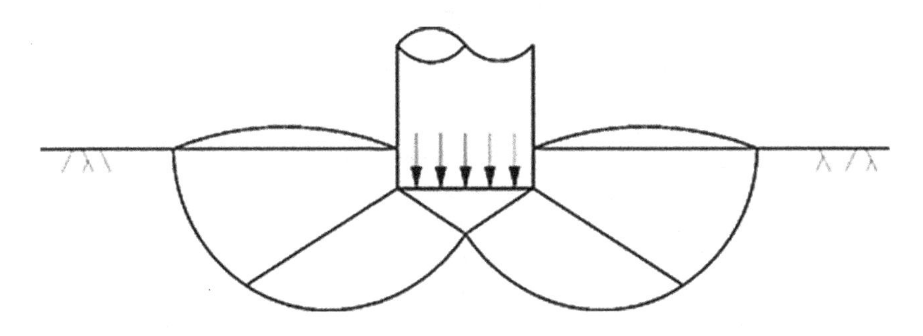

Figure 4.2 Bearing Capacity Failure in a Shallow Foundation.

4.2.4 Foundation design models

Foundations for transmission line structures are subjected to three predominant modes of loading as summarized in Table 4.10 and Figure 4.1.

- Compression
- Uplift
- Moment and/or Shear

The resistance offered by the soil under compression is known as the bearing capacity. Figure 4.2 illustrates a typical bearing failure of a shallow foundation. In case of a deep foundation, the compression load is resisted both by the skin friction and end bearing as illustrated in Figure 4.3.

The typical failure mechanism of shallow foundations under uplift loads is shown in Figure 4.4. The resistance offered by the soil is known as uplift capacity.

The uplift load is resisted by the skin friction and self-weight in deep foundations as shown in Figure 4.5.

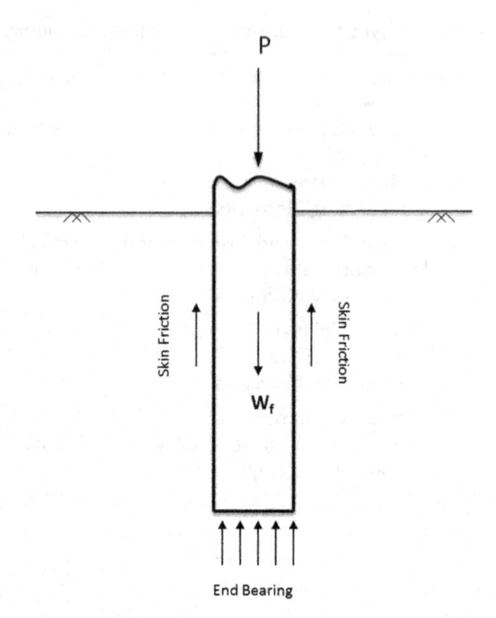

Figure 4.3 Deep Foundation Under Compression Load.

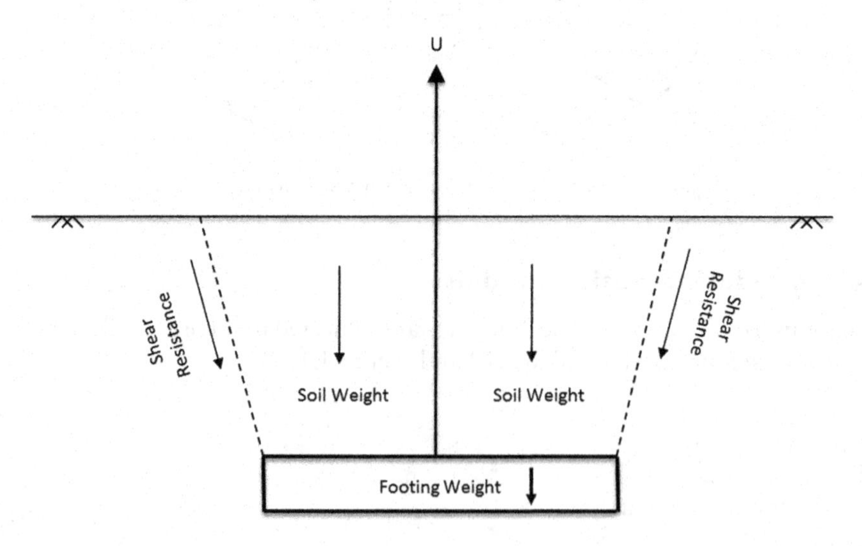

Figure 4.4 Uplift Failure in a Shallow Foundation.

Moment and shear loads are resisted by the lateral resistance of the soil in concrete drilled shafts and direct embedment pole foundations. Figure 4.6 shows a typical pattern of lateral soil resistance along a concrete drilled shaft.

Anchor foundations of guyed structures are subjected to pullout forces. The resistance to pullout is known as pullout capacity.

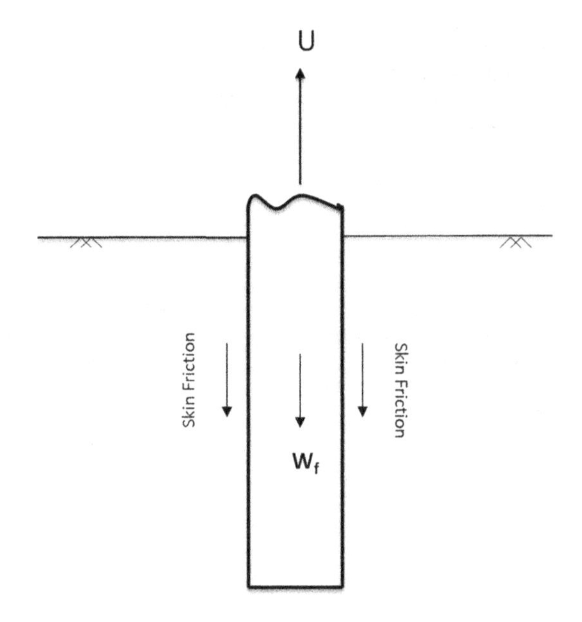

Figure 4.5 Deep Foundation Under Uplift Load.

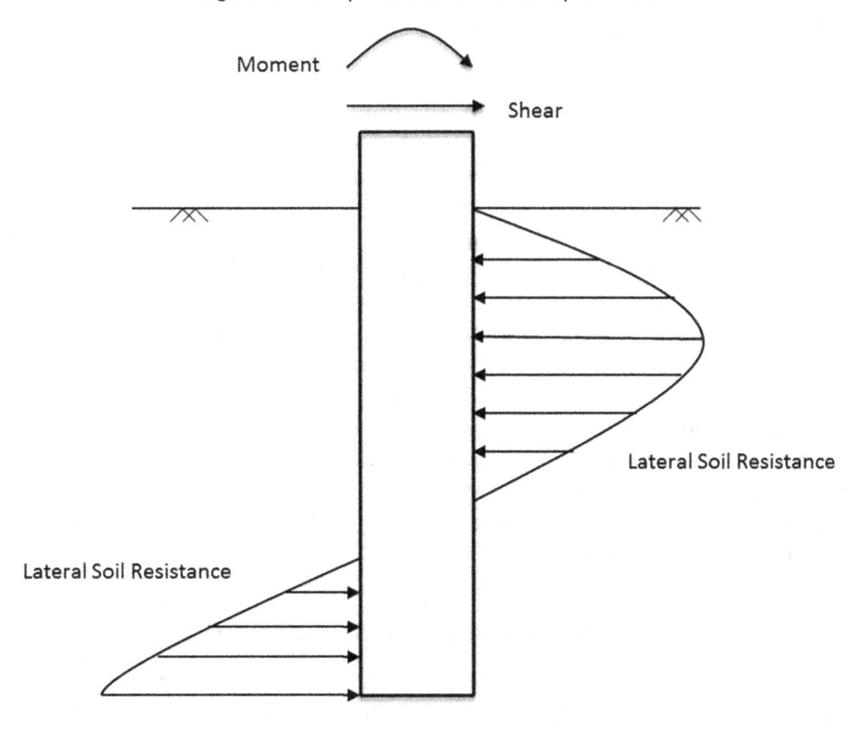

Figure 4.6 Concrete Drilled Shaft Foundation Under Moment and Shear Loads.

Table 4.11 Foundation Design Loads and Concepts.

Predominant Load	*Type of Foundation*	*Geotechnical Concept*	*Section*
Moment/Shear	Concrete Drilled Shaft	Ultimate Moment/Shear Capacities	4.4.1
Moment/Shear	Direct Embedment Pole Foundation	Ultimate Moment/Shear Capacities	4.4.2
Compression/ Moment	Spread Foundation	Ultimate Bearing Capacity (Shallow Foundations)	4.4.3
Uplift	Spread Foundation	Ultimate Uplift Capacity (Shallow Foundations)	4.4.4
Uplift	Concrete Drilled Shaft Foundations	Ultimate Uplift Capacity (Deep Foundations)	4.4.5
Uplift	Direct Embedment Pole Foundation	Ultimate Uplift Capacity (Deep Foundations)	4.4.5.1
Pullout Force	Anchor Foundation	Ultimate Pullout Capacity	4.4.6

Table 4.11 provides organization of different subsections of this chapter which provide basic design concepts and associated theories.

4.2.5 Structural and geotechnical designs

Foundation design typically involves two general stages: geotechnical design and structural design. Under geotechnical design, the foundation is designed to ensure safety against soil failure and deformations. The soil deformation leads to settlement or upward movement or deflection/rotation of foundation depending on the type of foundation and mode of loading. Based on soil data, the depth and plan dimensions of the foundation are calculated. In the structural design phase, which is commonly called material design phase, the foundation materials such as steel, concrete and wood are designed structurally to withstand the applied loads.

For example, in the case of design of a concrete spread footing for a lattice tower, the base dimensions and depth of the footing are determined in the geotechnical phase based on the bearing and uplift capacities of the soil. In the structural design phase, the shear and bending moments developed in the concrete pad are used to compute the thickness of the pad and the steel reinforcement required. Similarly, for a drilled shaft foundation of a steel pole, the depth and diameter of the pier are estimated under geotechnical design; the amount of steel, both longitudinal and transverse, is determined under structural design.

4.2.6 Deterministic and reliability-based designs

The allowable stress design is very commonly employed for designing transmission foundations. In this deterministic approach, foundation loads and capacities are assumed to be constant values; variability in loads and capacities is ignored. To account for the uncertainty in loads and capacities, a factor of safety is used. In the U.S., foundation design loads at the ground line are calculated from the factored forces acting on the structure and wires using the load cases from the National Electric Safety Code and internal utility standards. Foundation design capacities are calculated using theoretical geotechnical models. The factor of safety adopted by the US electric utilities depends on the experience and professional judgment of the engineer. In general, the factor of safety used for transmission line foundations is smaller than the factor of safety used in building and bridge foundations.

Although Reliability-Based Design (RBD) approaches are promising, they are not widely used in transmission structure foundation design. RBD design has recently been incorporated in FAD Tools (refer Section 4.5.2) using the Load and Resistance Factor Design (LRFD). In the RBD approach, both the load (Q) and resistance (R) are assumed to be random variables with their respective probability distributions. The strength factor times the foundation nominal resistance should be greater than or equal to the sum of the factored foundation design loads. The advantage of the RBD over the allowable stress is that the probability of failure and, hence, the reliability of foundation can be estimated. The strength factors used in the FAD tools for different foundation loading modes are calibrated using actual test data with 95% confidence (5% Lower Exclusion Limit). For example, the strength factor used on ultimate moment capacity of concrete drilled shafts is 0.63.

4.3 FOUNDATION TYPES

Foundations for transmission line structures can vary widely depending on the type of structural system, the soil profile and the cost of installation. In this section, the different types of foundations used for transmission line structures are discussed. These foundations include those supporting the main structure as well as anchors transmitting guy wire tensions to the ground.

4.3.1 Drilled shafts

Also known as pier foundations, concrete drilled shafts are ideal foundations for single pole structures. When the ground line moments and shears are small, direct embedment foundations are adequate. Poles subjected to large bending moments, shear and axial forces usually require a concrete drilled shaft. When a drilled shaft foundation is used, the steel pole is attached to the concrete drilled shaft by means of a base plate and anchor bolts as shown in Figure 4.7a, b and c.

Figure 4.7a and 4.7b show drilled shafts for round and square steel poles. Figure 4.7c shows a combination of a shaft and a mat footing, which is very common in situations where hard rock is encountered at shallow depths.

Drilled shaft foundations are also used for steel H-Frames and lattice towers. Transmission towers are normally anchored to the foundation by means of anchor bolt – base plate assemblies or by stub angles. In either case, the minimum diameter or dimension of the pier will be that required to accommodate the anchor bolt system or crimped stub angle. For steel pole structures, the diameter of the anchor bolt circle frequently controls the foundation shaft diameter. Accounting for the space required for vertical steel reinforcement bars, lateral ties and the concrete cover, a minimum required pier diameter will be established.

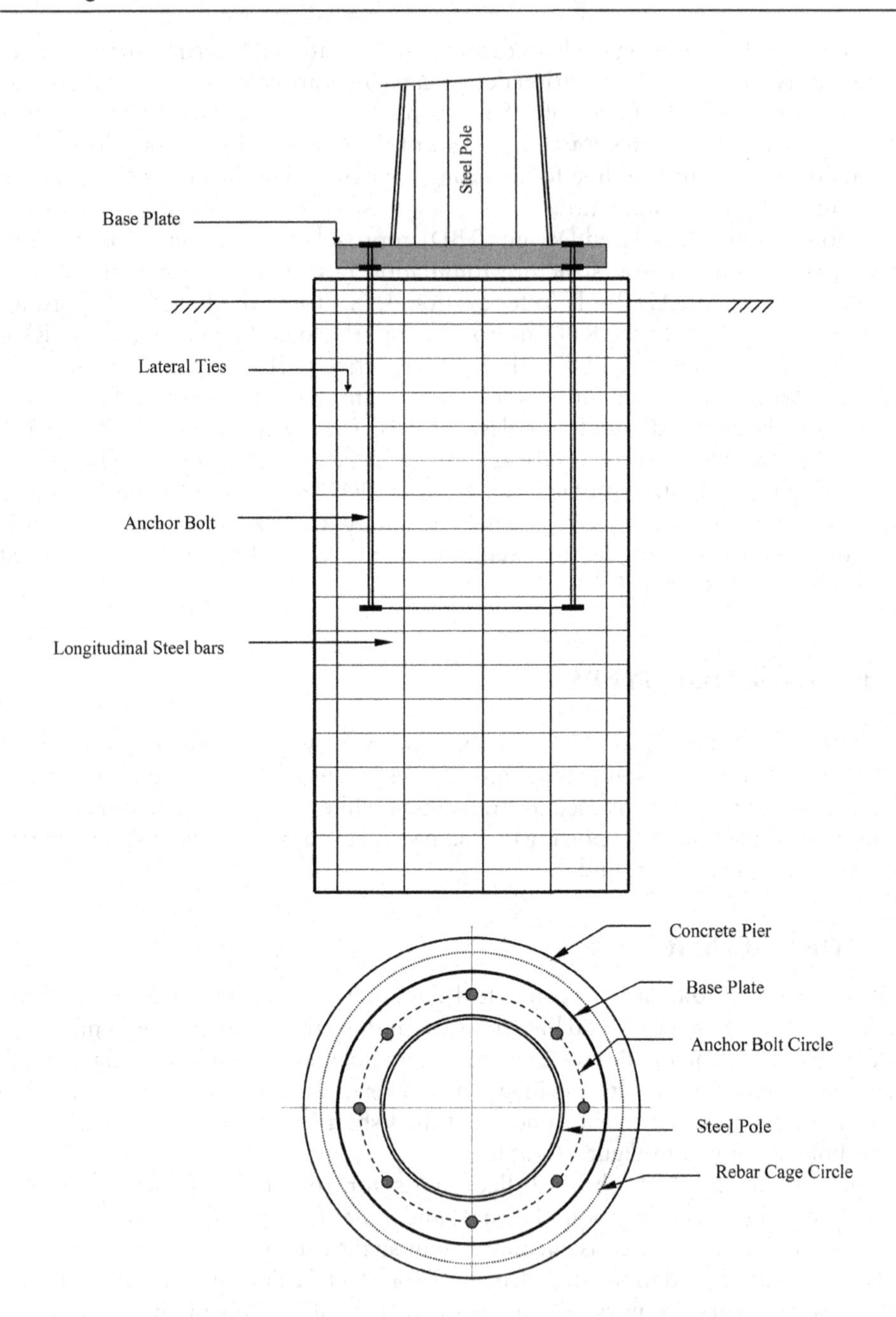

Figure 4.7a Round Steel Pole with a Drilled Shaft.

For construction of drilled shaft foundations, three different methods are used (1) The dry method (2) The casing method and (3) The wet method. In case of non-caving stable soils such as stiff clay, cemented granular soils and rock above the water table, the dry hole construction method is used. This is a quick and least expensive

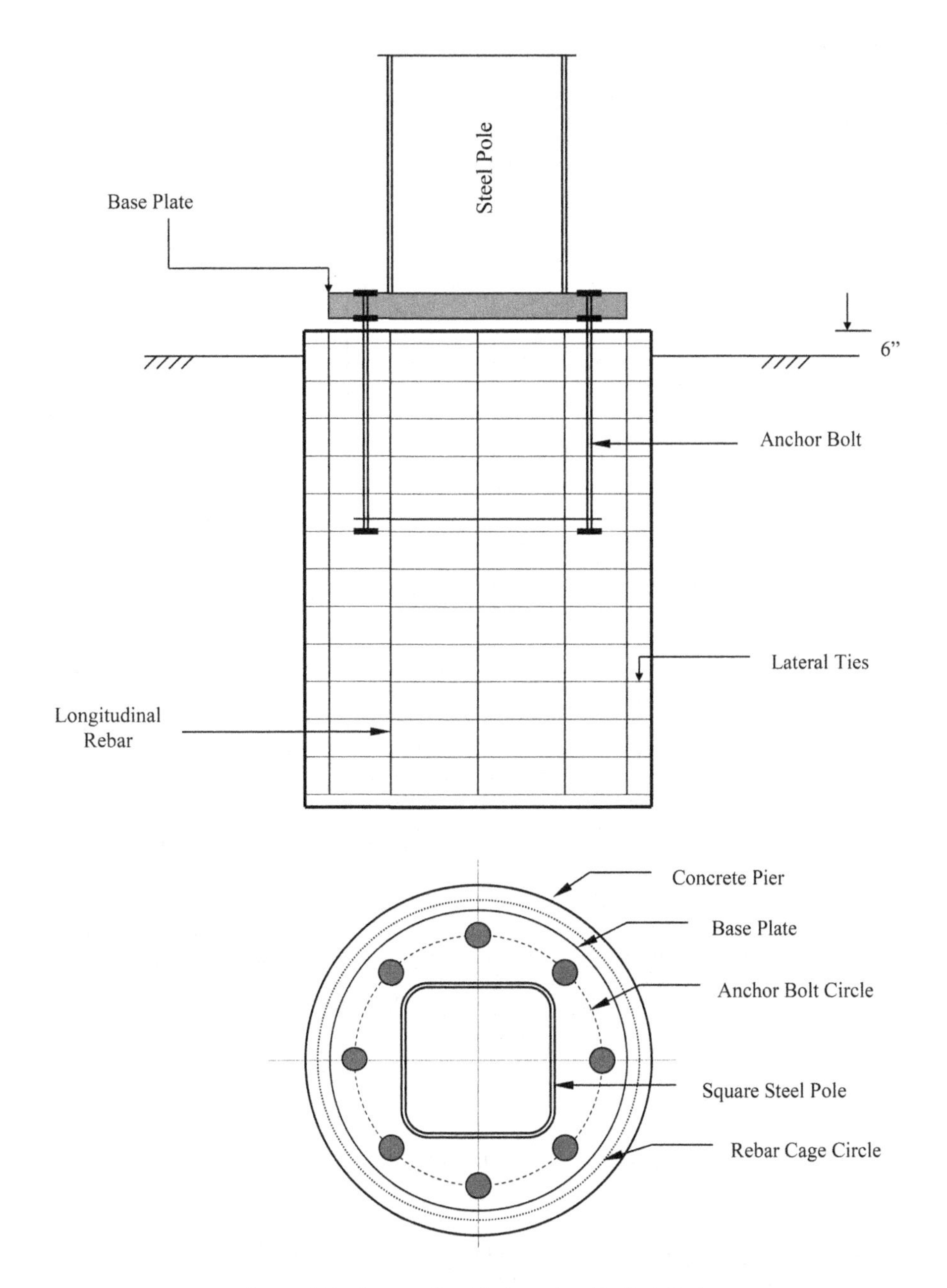

Figure 4.7b Square Steel Pole with a Drilled Shaft.

method. However, in case of caving soils, casing method is used to provide stability during the excavation of drilled hole. The wet method is normally used in granular soil areas and unstable cohesive soil formations with high ground water level. In this method, during entire operation of drilling the hole, placement of reinforcement and

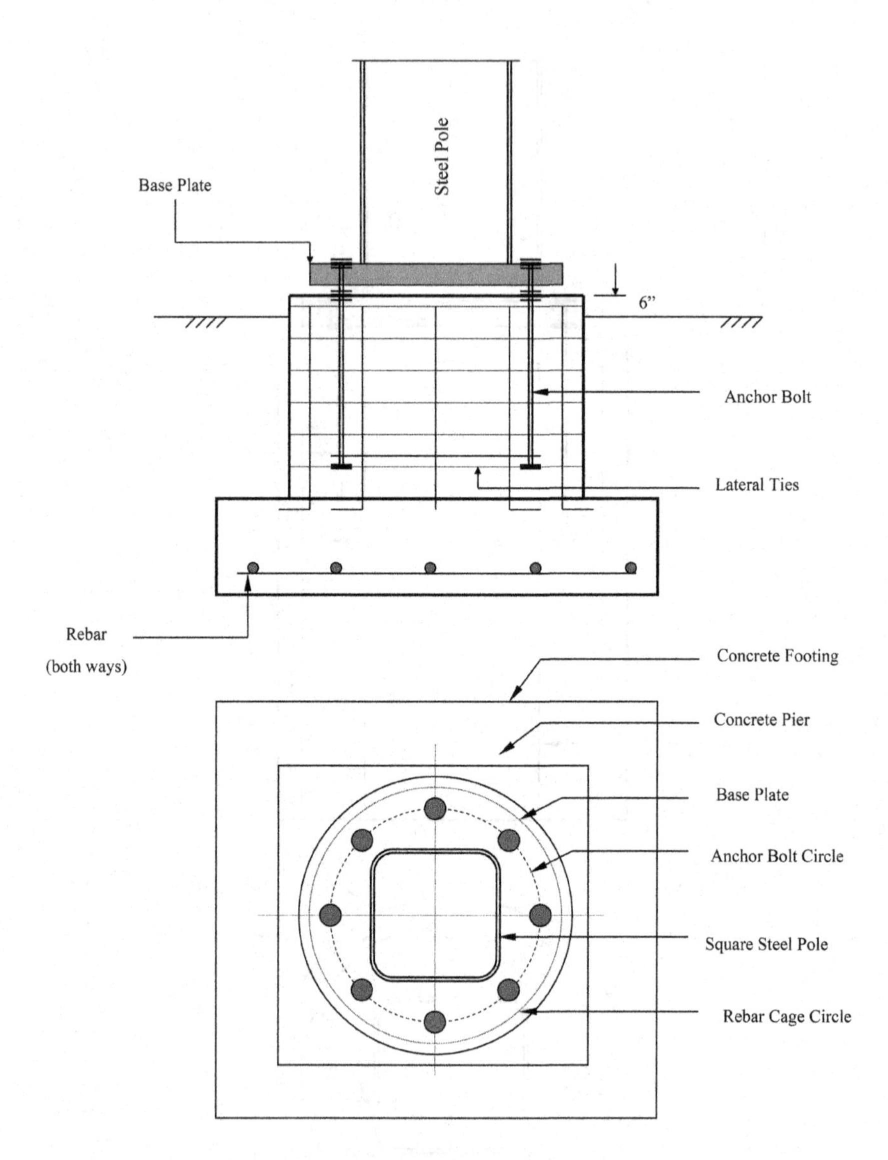

Figure 4.7c Short Shaft on a Mat Footing.

concreting, the excavated hole is kept filled with a drilling fluid with or without casing. The prepared slurry made with mineral bentonite or synthetic polymers are used as drilling fluids to maintain stability to the excavated hole. The seepage pressure exerted by the drilling fluid on the borehole wall provides stability to the excavated sidewall when the elevation of the slurry is above the elevation of the ground water level in the soil. In some stable excavations, water head alone can be used to prevent soil run-in at the bottom of the shaft. In many cases, depending on subsurface conditions, combinations of above three methods are employed.

Typical diameters of drilled shafts in transmission applications vary from 4 ft to 12 ft (1.2 m to 3.7 m). The depth of the piers vary anywhere from 15 ft to 70 ft (4.6 m to 21.3 m). The depth-to-diameter ratio ranges from 2 to 10 and typically is between 3 and 6.

The geotechnical design for a drilled shaft involves:

- Computing the required depth of the shaft for resisting the applied moment loads and verifying that the lateral deflection and rotation at the pier top are within acceptable limits for single pole structures
- Computing the required depth of the shaft for vertical/axial loads through bearing and/or skin friction for lattice towers and H-Frame structures.

The structural design includes determining the longitudinal and shear reinforcement; and the number, size and arrangement of anchor bolts. Most drilled shafts have 6 inches to 12 inches (15 cm to 30 cm) reveal or projection above ground. In high flood areas, a larger reveal is used.

Computer software programs such as MFAD™ (2014) require specifying what portion of the allowable deflection and rotation is permanent (inelastic) and recoverable (elastic). In cases where excessive lateral movement of the shaft is observed under large shears and moments, the shaft is often enclosed in a ½ inch to 1 inch (12.5 mm to 25 mm) thick steel casing to provide additional stiffness.

Base plate and anchor bolts

Structural members (pole or tower leg) supported on drilled shafts are often connected to the pier by means of a steel base plate and anchor bolts. Procedures for computing the thickness of the base plate are spelled out in ASCE 48-11 (2011), *Design of Steel Transmission Pole Structures*. Many steel manufacturers use internal design procedures to determine the thickness of the base plate. Standard anchor bolt material data is shown in Table A3.13 of Appendix 3. All anchor bolts are supplied with nuts and washers, and conform to ASTM Standards.

The length of embedment of anchor bolt is a function of several variables including the yield strength of the rod, the ultimate concrete strength, the diameter of the rod, and the number of threads per inch (for a threaded bolt). PLS-POLE™ (2012) is a commonly used computer program that is used for steel pole design. The program contains a provision for modeling base plates with specified anchor bolt patterns and arrangements; with this input, the program computes the minimum required thickness of the base plate.

4.3.2 Spread foundations

Spread foundations are used for lattice transmission towers. The common types of spread foundations are concrete foundations and steel grillages. Design of spread foundations is based on maximum vertical loads – both compression and uplift – and the associated shears. Uplift normally plays a critical role in determining the depth and the size of spread footings. Lattice towers can tolerate substantial foundation displacements under load; however, the possibility of excessive differential settlements of spread foundations in poor soils must be investigated by appropriate means.

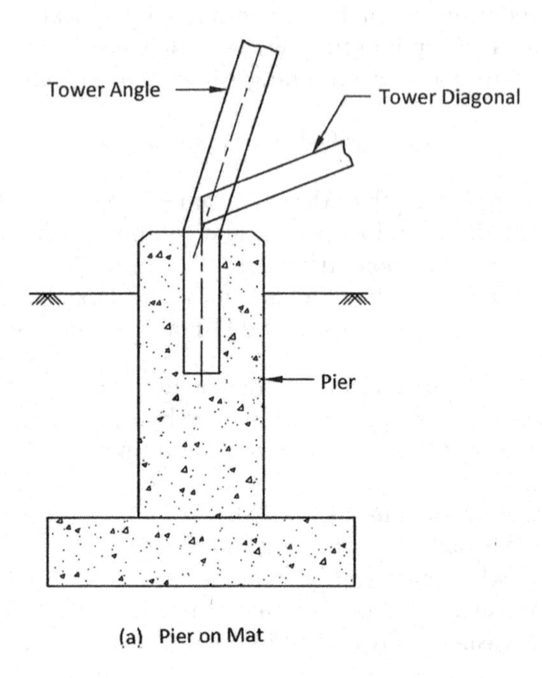

Figure 4.8a Concrete Foundations.

4.3.2.1 Concrete foundations

Concrete foundations are either cast-in-place or precast. The foundation consists of a base mat or pad and a square/cylindrical/stepped pier. The base pad is generally square or rectangular in plan. The base slab thickness for standard lattice towers ranges from 1.5 ft to 3 ft (45 cm to 90 cm). The maximum depth of these foundations is approximately 15 ft (4.57 m) and the ratio of depth to base pad width varies from 1 to 2.5. The slab must be of sufficient thickness to be able to accommodate the reinforcing bars and must provide for proper cover to the reinforcement. As shown in Figure 4.8a, the tower stub angle is bent and the directly embedded in the pier. The pier is centered on the base pad. A configuration that uses anchor bolts is shown in Figure 4.8b. A grillage encased in a concrete box is shown in Figure 4.8c; this configuration is commonly used for angle and deadend towers that sustain large uplift and compressive forces.

In situations where surface rock is present, the tower stub angle is directly embedded in the concrete as shown in the Figure 4.9. The base pad is not required in this situation.

4.3.2.2 Steel grillages

Grillage foundations consist of a tower stub leg that is connected to a horizontal grillage. The horizontal grillage is made up of standard structural steel sections below the ground line. The stub leg continues at the same slope as the lattice tower leg. The

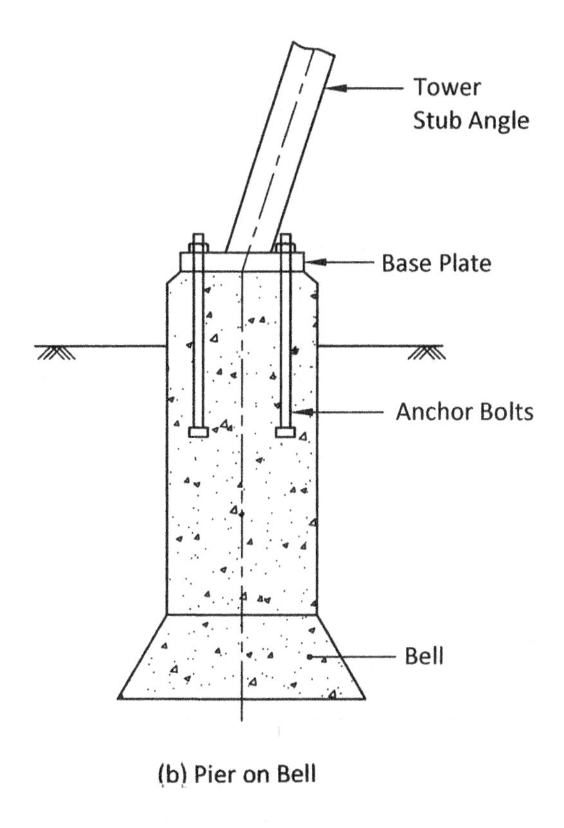

Figure 4.8b Concrete Foundations. (cont'd)

horizontal grillage is usually made of rolled steel sections such as I beams, channels and angles. The structural sections aid in transferring the compression forces to the soil below and uplift forces to soil above the base. In a pyramidal arrangement, the main leg stub is connected to the four smaller stub members, which in turn are connected to the horizontal grillage base. All the connections in the grillage foundations are bolted.

Some of the advantages of grillage foundations are:

- Foundations are made steel angle members which eliminates the need for heavy equipment such as concrete trucks which is a great advantage in remote and highly inaccessible areas
- The steel members can be carried in pieces to the remote areas and assemble at site
- Unless the excavated soil is poor such as peat or muck, the excavated material can be used for backfilling and thus eliminates the need for barrow backfill
- In case of single stub arrangement, it is easy to adjust the assembly which is a great advantage in a rolling terrain. However, in case of multi-legged arrangement, it is difficult.

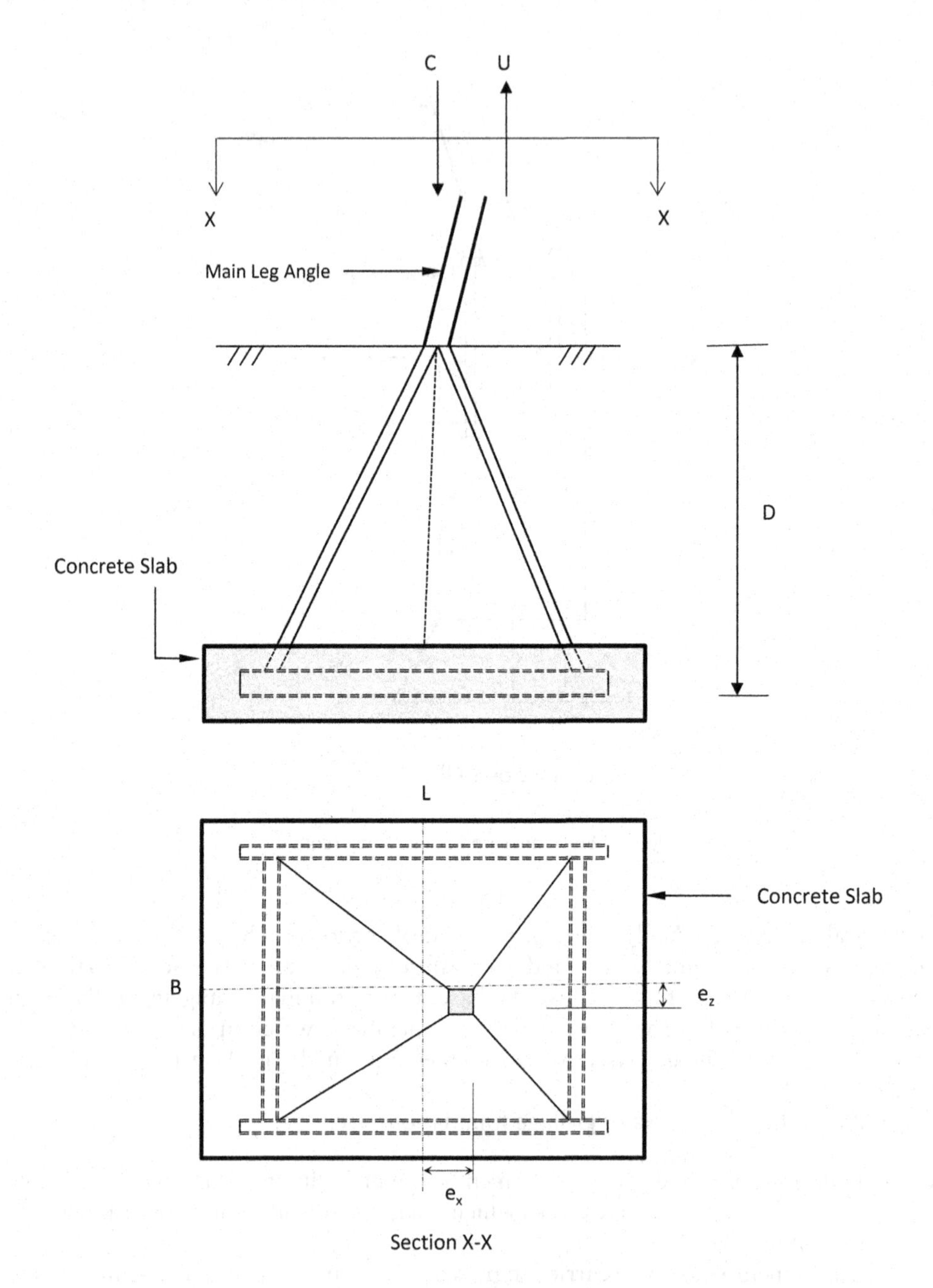

Figure 4.8c Grillage Foundation.

A grillage foundation is difficult to install and experienced crews are required for setting and leveling. Moreover, corrosion of steel members is a concern in aggressive corrosive environments. The following measures are adopted to prevent corrosion:

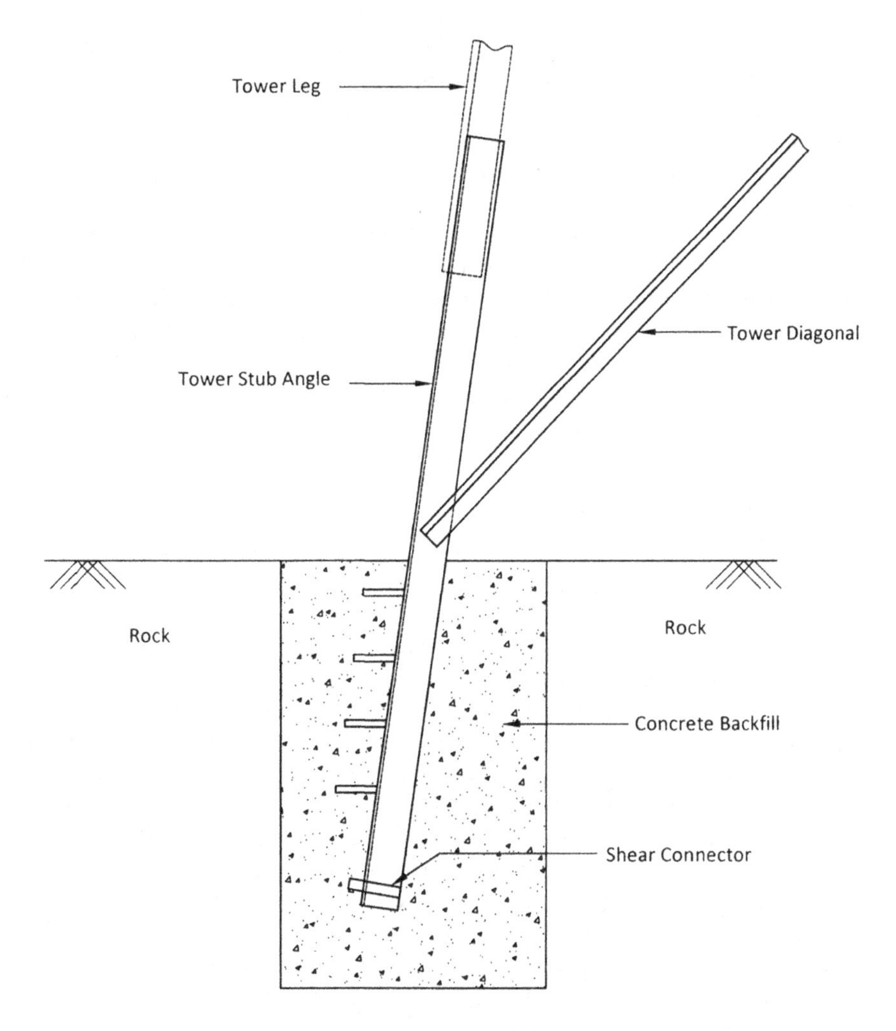

Figure 4.9 Foundation in Rock.

- Steel members are galvanized and further coated with coal-tar
- Grillage is encased in concrete
- Cathodic protections

4.3.3 Direct embedment

Direct embedment foundations are common for single poles as well as H-Frame structures. The pole is placed in a borehole with a diameter larger than the pole and the annular space is filled using either the excavated soil or select backfill material. The quality of the backfill, method of placement and degree of compaction greatly affect the strength and performance of the embedded pole foundation.

Wood, steel, concrete or composite poles in tangent structures are generally embedded directly into the ground. In these poles, the foundation (i.e.) the embedded portion of the pole, resists lateral forces consisting of ground line shears and overturning bending moments, in addition to the axial loads. The required embedment depth is determined from soil lateral resistance for single pole structures subjected to moment/shear forces. Vertical compressive loads are resisted by end bearing and skin friction. Vertical uplift loads are resisted by skin friction and self-weight of the foundation. Axial loads can control foundation depth for braced H-Frame structures.

Figure 4.10 show various configurations of directly embedded poles in wood. The concept is similar for concrete poles. In situations where uplift forces are encountered, wood poles are fitted with base shoes or bearing plates as shown in the figure. A concrete pad (precast or cast-in-situ) is often used at the bottom of poles to distribute large axial loads. The butt plate welded to the bottom of the steel pole helps distribute the axial load over a larger area. It also helps in generating uplift resistance by developing a frictional resistance over a larger surface area.

4.3.4 Pile foundations

Pile foundations are used in situations where the foundation must resist large loads in poor soils. Piles are long, slender members made of steel, concrete or wood. When poor soils are present in the top layers of the soil strata, these foundations are used to transmit large loads to underlying stronger soils or to bedrock. In case of expansive soils, pile foundations are extended beyond the active zone of swelling/ shrinkage to stable substrata. The disadvantage of pile foundations is that they are generally costly due to large equipment mobilizations costs. The engineer must explore the possibility of other alternatives before specifying a pile foundation for a project.

Pile foundations are very effective in resisting compression, uplift and lateral forces. They may be used in single or in group configurations with vertical and/or batter piles. When piles are installed at an angle to the vertical, they are called batter piles. Batter piles are very effective in resisting lateral loads. When grouped together, they can resist vertical loads as well.

Depending on the mode of installation, pile foundations are broadly classified into driven piles and cast-in-situ piles. Driven piles are made of steel, concrete or timber whereas the cast-in-situ piles are made of concrete. Cast-in-situ foundations eliminate vibration issues associated with the driven piles and are installed in a manner similar to drilled shaft foundations. Common types of driven piles are open ended steel and H- and I-shaped steel sections.

In poor soils such as wetlands and organic soil deposits, thin-walled, open-ended steel casings are installed using vibratory hammers. These foundations are also known as vibratory caissons. They are very effective and economical in poor soils. Typical diameters of these foundations vary from 3 ft to 5 ft (90 cm to 150 cm). Vibratory caissons are connected to the steel pole structures using flange, slip and socket connection.

In case of hard strata, the installation of these foundations is difficult and these foundations can be damaged during the installation. Therefore, the information of soil strata is very essential. Corrosion resisting measures such as coatings and cathodic protections are required for these foundations.

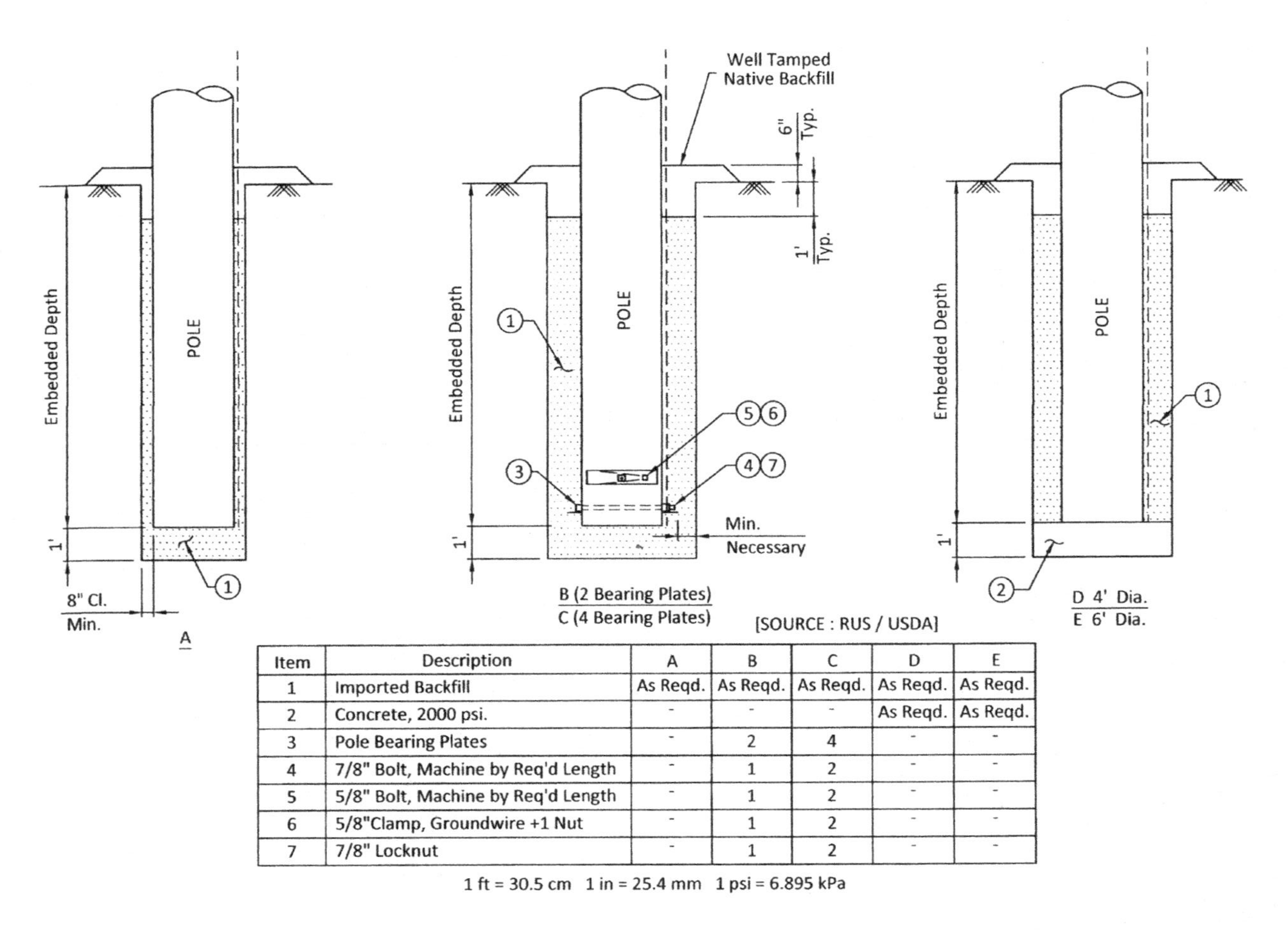

Item	Description	A	B	C	D	E
1	Imported Backfill	As Reqd.	As Reqd.	As Reqd.	As Reqd.	As Reqd.
2	Concrete, 2000 psi.	-	-	-	As Reqd.	As Reqd.
3	Pole Bearing Plates	-	2	4	-	-
4	7/8" Bolt, Machine by Req'd Length	-	1	2	-	-
5	5/8" Bolt, Machine by Req'd Length	-	1	2	-	-
6	5/8"Clamp, Groundwire +1 Nut	-	1	2	-	-
7	7/8" Locknut	-	1	2	-	-

1 ft = 30.5 cm 1 in = 25.4 mm 1 psi = 6.895 kPa

Figure 4.10 Wood Pole Embedment Configurations.

4.3.5 Micropiles

Micropiles are small-diameter, drilled and cement-grouted piles, usually less than 12 inches (30 cm) in diameter. Micropiles utilize steel reinforcing bar located at the center to transfer structural loads to a suitable soil or rock stratum. A schematic of a micropile is shown in Figure 4.11. As is shown in the figure, a permanent steel casing is typically used in the top portion of the micropile to enhance stiffness against lateral loading. AASHTO (2010) classifies micropiles into five different types A, B, C, D, E based on the grouting sequence.

Micropiles develop their axial capacity primarily through the bond between the grout body and soil or rock in the pile-bonded zone. This allows them to provide both tension and compression resistance. In case of larger loads, these piles are used in groups and tied together with a cap made of steel or concrete.

Micropiles can be installed using much of the same drilling and grouting equipment that is used for grouted rock anchors. They offer an ideal solution for transmission line foundation support in difficult subsurface conditions and at locations where access is restricted. They are also used in, in environmentally-sensitive areas and places where elevated groundwater is present. Due to their small size and type of materials used, micropiles do not require heavy construction equipment and can be constructed even in hard stratum. They can be used to provide additional foundation support for existing structures at times of retrofit.

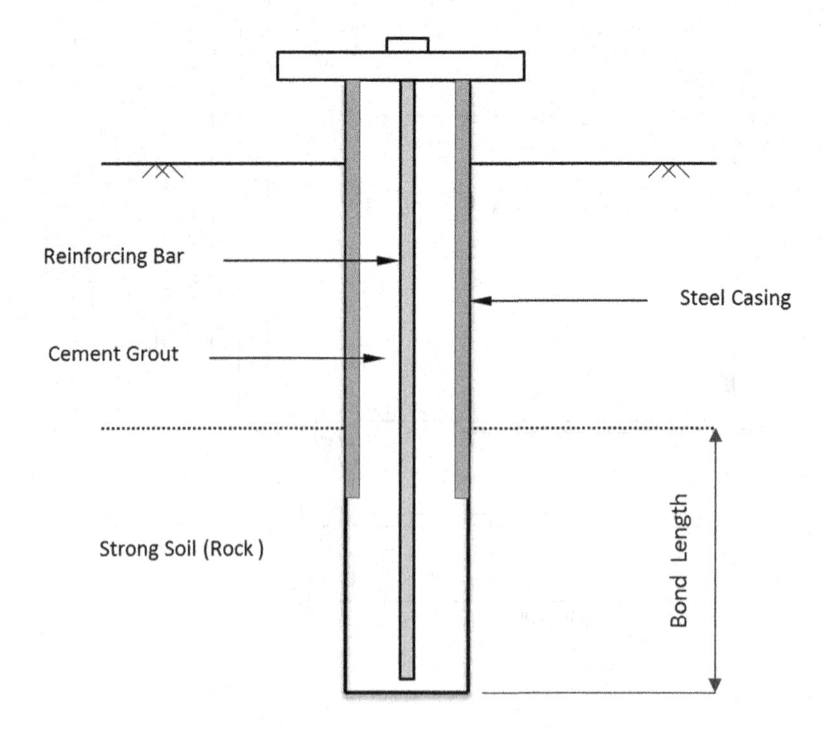

Figure 4.11 Micropile Foundation.

4.3.6 Anchors

Anchor foundations are mainly used for resisting the pullout force imposed by the guywire loads. Angle and deadend transmission structures are often guyed and their stability depends heavily on the proper design and construction of anchor foundations. The pullout of an anchor or a significantly displaced anchor under extreme events can lead to the failure of the structure.

Some anchors routinely used in guyed transmission line structures are listed below:

- Wooden log anchor
- Plate anchor
- Grouted soil anchor
- Grouted rock anchor
- Helical anchor

RUS log anchors are popular in the Midwestern US. Single rod and double rod modified log anchors are shown in Figures 4.12 and 4.13 respectively. As shown in these figures, a vertical trench is excavated and the log anchor is inserted and back-filled. Typical log anchors are 5 ft or 8 ft (1.5 m and 2.4 m) long and are 12 inches (30 cm) in diameter. Ultimate capacities vary from 16,000 to 50,000 lbs (71 to 222 kN), depending on the size of the rod and class of soil (see Appendix 12). Anchor rods are usually galvanized and vary from 5/8 inch to 1 inch (15.9 mm to 25 mm) in diameter, with rod strengths ranging from 16,000 lbs (5/8 inch rod) to 50,000 lbs (1 inch high strength rod).

A typical RUS steel cross plate anchor and relevant design data is shown in Figure 4.14.

In grouted soil anchors, a steel bar or steel cable is placed into a predrilled hole and the hole is filled with cement grout under pressure. Depending on the diameter of the bar, grouted soil anchors are classified into two major groups: large diameter and small diameter grouted anchors. Typical large diameter grouted soil anchors are shown in Figure 4.15.

Grouted soil anchors resist pullout forces by frictional resistance at the soil-grout interface. When a bell is used, it offers end bearing resistance. The diameter of the bell is typically larger than that of the initial drilled shaft. Different types of grouted soil anchors and soils in which they are typically used are shown in Tables 4.12 and 4.13.

Grouted rock anchors are employed where the guying wires must be anchored in rock. High strength grout is used to fill the hole. The most common grout used is pure cement-water mix in the ratio of 2.5:1 by weight, with an approved expansion agent added per the manufacturer's instructions.

While installing grouted rock anchors, the required diameters and lengths are determined by the strength of the rod, shear bond between the rod and grout and shear bond between grout and rock. Normally, anchor rods are high-strength reinforcing steel bars or threaded bars. The hole size for rock anchors is usually 1.5 to 3 times the diameter of the rod. Core drilling should be performed to explore rock quality and core testing must be performed to determine rock strength. A pull-out test may also be performed to confirm anchor capacity.

Helical anchors consist of a square or round steel shaft fitted with one or more helically deformed plates. The number of plates typically varies from one to four.

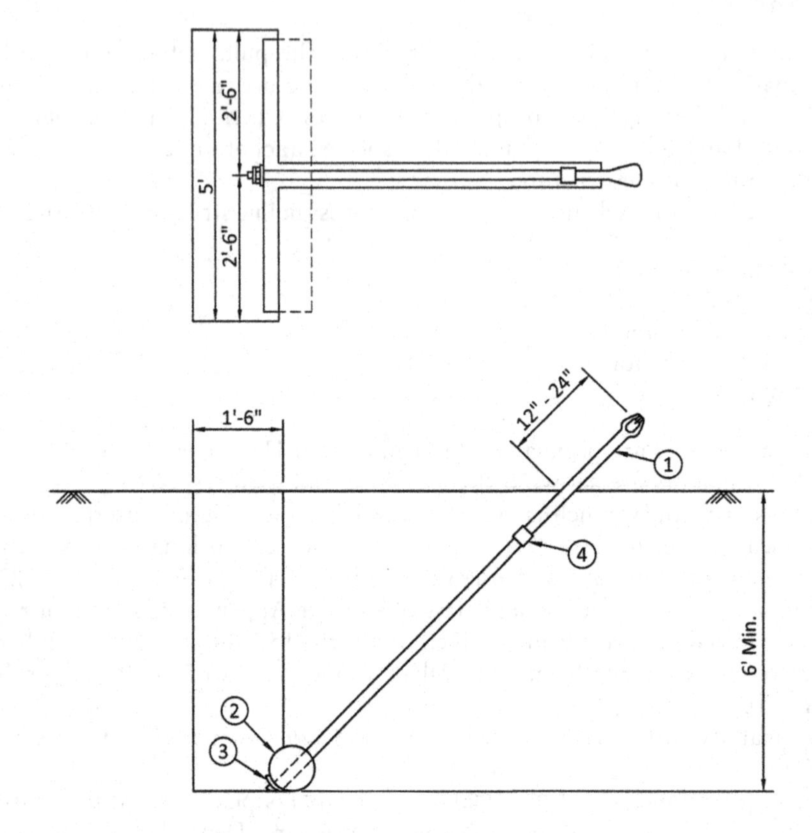

Capacity : 20,000 lbs. (ultimate) in Class 6 soil

Item	Description	Quantity
1	Anchor Rod, Twineye, Galvanized, 1" x 10'	1
2	Anchor, Log, 12" x 5'	1
3	Galvanized Curved Washer	2
4	Ground Clamp for Steel Rod	1

1 ft = 30.5 cm   1 lb = 4.45 N

Figure 4.12 Single Rod Log Anchor (Source: RUS/USDA).

Helical anchors are installed using a truck-mounted power auger by the simultaneous application of both torque and axial load. They offer pullout resistance through the bearing of each helical plate on soil. If the plates are closely spaced, a cylindrical failure surface will form along the periphery of top and bottom plates. The frictional resistance along this failure surface contributes to pullout resistance. End bearing resistance is obtained at the top plate. In the US, helical anchors are very common types of anchors for transmission applications. Besides guy applications, helical foundations are also used for pole and tower foundations. These are normally called helical piles. Depending

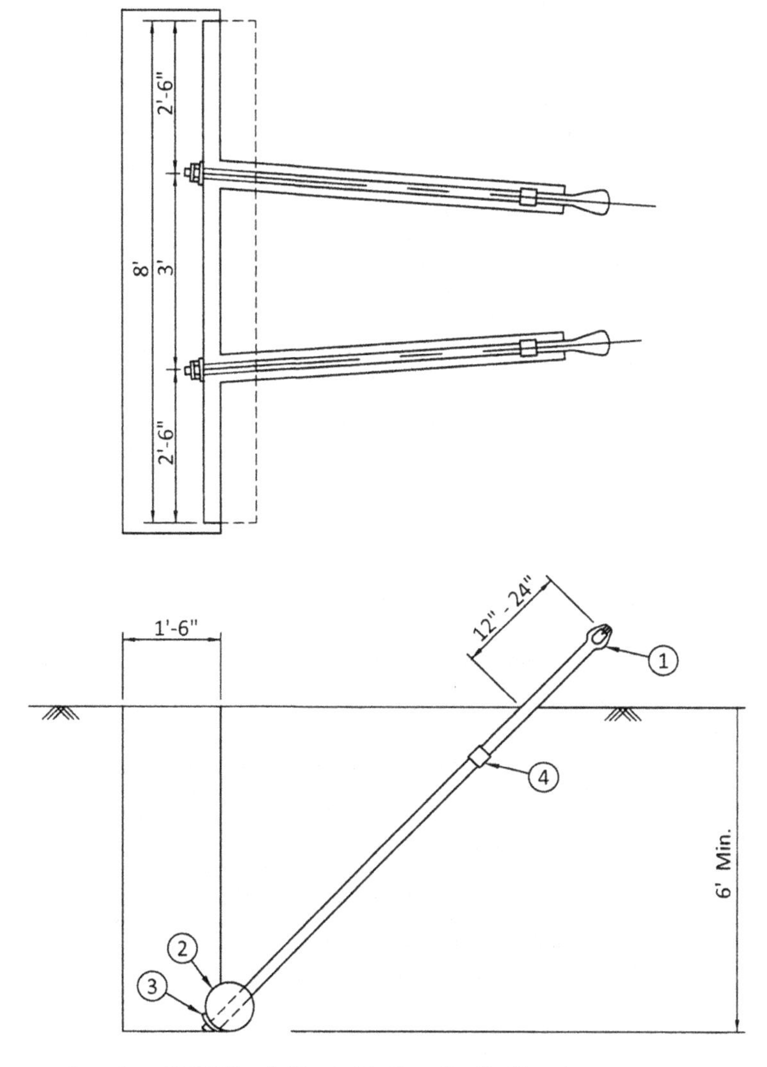

Capacity : 40,000 lbs. (ultimate) in Class 6 soil - 1" rod
36,000 lbs. (ultimate) in Class 6 soil - ¾" rod

Item	Description	Quantity
1	Anchor Rod, Twineye, Galvanized, 1" x 10'	2
2	Anchor, Log, 12" x 8'	1
3	Galvanized Curved Washer	4
4	Ground Clamp for Steel Rod	2

1 ft = 30.5 cm 1 in = 25.4 mm 1 lb = 4.45 N

Figure 4.13 Double Rod Log Anchor (Source: RUS/USDA).

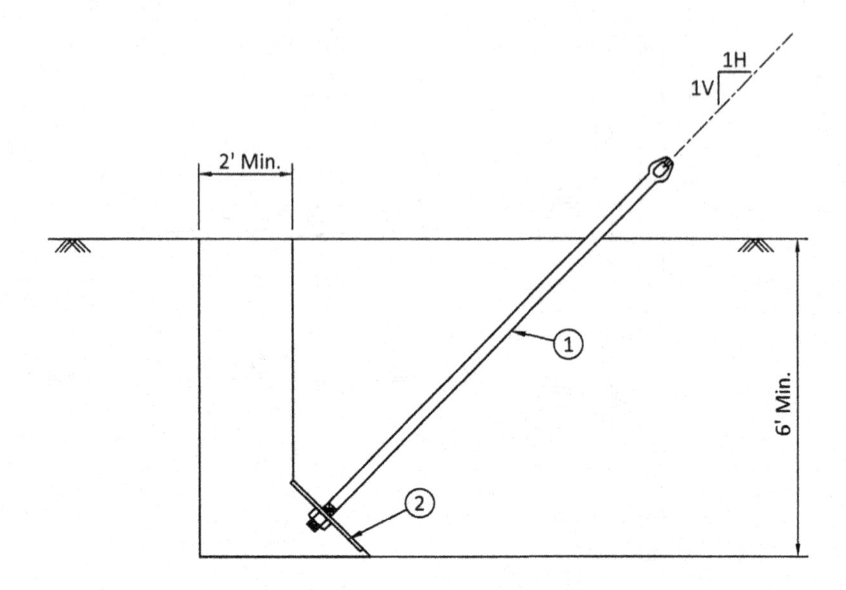

For guy slope 2H : 3V, minimum depth shall be 7'

Item	Description	Quantity
1	Anchor Rod, Twin Eye, Nut	1
2	Anchor Plate	1

Anchor Units and Capacities			
Unit No.	Soil Class	Ultimate Capacity (lbs)	Anchor Rod Size
1	5	16,000	3/4" dia x 8'
2	4	24,000	1" dia x 10'

1 ft = 30.5 cm 1 in = 25.4 mm 1 lb = 4.45 N

Figure 4.14 Steel Cross Plate Anchor (Source: RUS/USDA).

on structure type, these foundations can be single or grouped. These foundations work well in wetlands where disturbance needs to be minimized.

4.4 DESIGN MODELS

The prerequisite for efficient foundation design for transmission structures is the understanding of how line and structure loadings, along with soil data, influence foundation performance and reliability. Interpretation of soil data is critical as soil properties

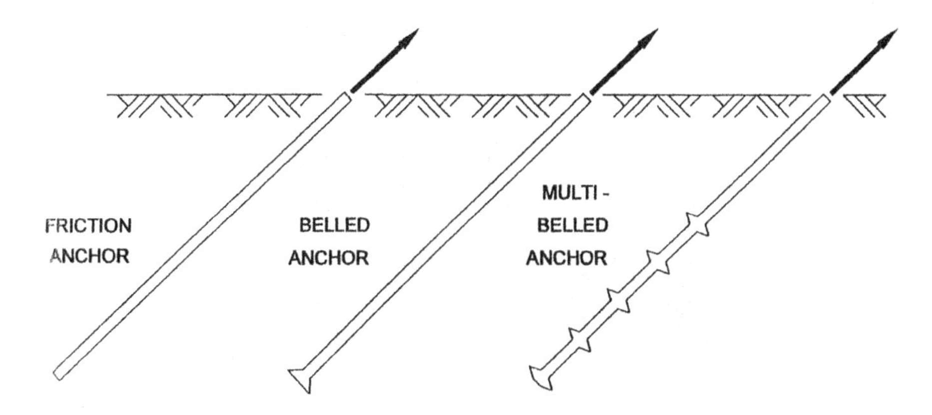

Figure 4.15 Typical Large Diameter Grouted Soil Anchors (with permission from ASCE).

vary significantly along a transmission line. Construction variables such as installation techniques and backfill compaction also influence overall foundation performance. Several other factors that influence the selection of a foundation type are rotation and deflection limitations, equipment availability, site accessibility and environmental concerns.

Analysis and design models of transmission line foundations are discussed in various books, reports and manuals. Conventional design models used by utilities are discussed in detail in several subsections.

4.4.1 Drilled shafts under moment and shear

A concrete drilled shaft/pier foundation of a self-supported steel pole structure is subjected to significant moment loads in addition to the horizontal (shear) and vertical (axial) loads. Therefore, utility engineers need to estimate the ultimate moment carrying capacity as well as the ground line deflection and rotation of the shaft. The effect of applied loads on the foundation shafts must be less than the ultimate capacity of the soil. Deflections and rotations of the foundation must be within the permissible limits.

Computation of the ultimate lateral resistance of a shaft and deflection of a shaft as the load reaches its ultimate value is very complex and involves a study of the interaction between a semi-rigid structural element and the soil. Theories originally developed for pile foundations are extended to the drilled shaft foundations subjected to moment loads. Some of the original concepts developed for pile foundations under lateral loads are briefly discussed below.

4.4.1.1 Classification of laterally loaded piles

A pile can be classified as rigid (short) pile or flexible (long) pile based on its embedment depth. If the ratio of the embedment depth to the diameter of the pile is less than 10 to 12, the pile can be classified as rigid (Broms, 1964). The stiffness factor (Broms, 1964) and relative stiffness factor (Poulos and Davis, 1980) are more scientific ways of classifying short and long piles.

Table 4.12 Low Pressure Grouted Anchor Types.

Type of Anchor	*Diameter of Shaft in inches (cm)*	*Size of Bell, in inches (cm)*	*Gravity Concrete*	*Grout Pressure in psi (kPa)*	*Suitable Soils for Anchorage*	*Load Transfer Mechanism*
Straight Shaft Friction (Solid Stem Auger)	12 to 24 (30 to 60)	Not Applicable	Applicable	Not Applicable	Very Stiff To Hard Clays, Dense Sands	Friction
Straight Shaft Friction (Hollow-Stem Auger)	6 to 18 (15 to 45)	Not Applicable	Not Applicable	30 to 150 (200 to 1035)	Very Stiff To Hard Clays, Dense Sands, Loose To Dense Sands	Friction
Underreamed Single Bell At Bottom	12 to 18 (30 to 45)	30 to 42 (75 To 105)	Applicable	Not Applicable	Very Stiff To Hard Clays, Dense Sands, Soft Rock	Friction and Bearing
Underreamed Multi-Bell	4 to 8 (10 to 20)	8 to 24 (20 to 60)	Applicable	Not Applicable	Very Stiff To Hard Clays, Dense Sands, Soft Rock	Friction and Bearing

Source: IEEE 691-2001, IEEE Guide for Transmission Structure Foundation Design and Testing.
Reprinted with permission from IEEE.

Table 4.13 High Pressure – Small Diameter Grouted Soil Anchors.

Type of Anchor	*Diameter of Shaft in inches (cm)*	*Size Of Bell in inches (cm)*	*Gravity Concrete*	*Grout Pressure in psi (kPa)*	*Suitable Soils For Anchorage*	*Load Transfer Mechanism*
Not Regroutable[1]	3 to 8 (7.5 to 20)	Not Applicable	Not Applicable	150 (1035)	Hard Clays, Sands, Sand-Gravel Formation, Glacial Till or Hard Pan	Friction or Friction and Bearing in Permeable Soils
Regroutable[2]	3 to 8 (7.5 to 20)	Applicable	Not Applicable	200 to 500 (1380 to 3450)	Same Soils as Above and Stiff To Very Stiff. Clay Varied and Difficult Soils	Friction and Bearing

[1]Friction from compacted zone having locked in stress. Mass penetration of grout in highly pervious sand/gravel forms bulb anchor.
[2]Local penetration of grout form bulbs which act in bearing or increase effective diameter.
Source: IEEE Std 691-2001, IEEE Guide for Transmission Structure Foundation Design and Testing.

The failure mode of a rigid pile, with free head conditions, under a lateral load, applied at an eccentricity above the ground level, is shown in Figure 4.16.

A short rigid pile unrestrained at the top tends to rotate and causes passive soil resistance to develop in the soil. Eventually the rigid pile will fail by rotation when the passive resistance of the soil at the head and toe are exceeded. The failure is accompanied by a lot of movement to enable the passive resistance to develop fully. In many transmission applications, drilled shaft foundations are idealized to behave like a rigid pile.

4.4.1.2 *Ultimate capacity models*

Several theories are proposed for estimating lateral and moment capacities of piles and drilled shafts.

a) Broms' Method
b) Hansen's Method
c) MFAD™ Model
d) Prasad & Chari Method

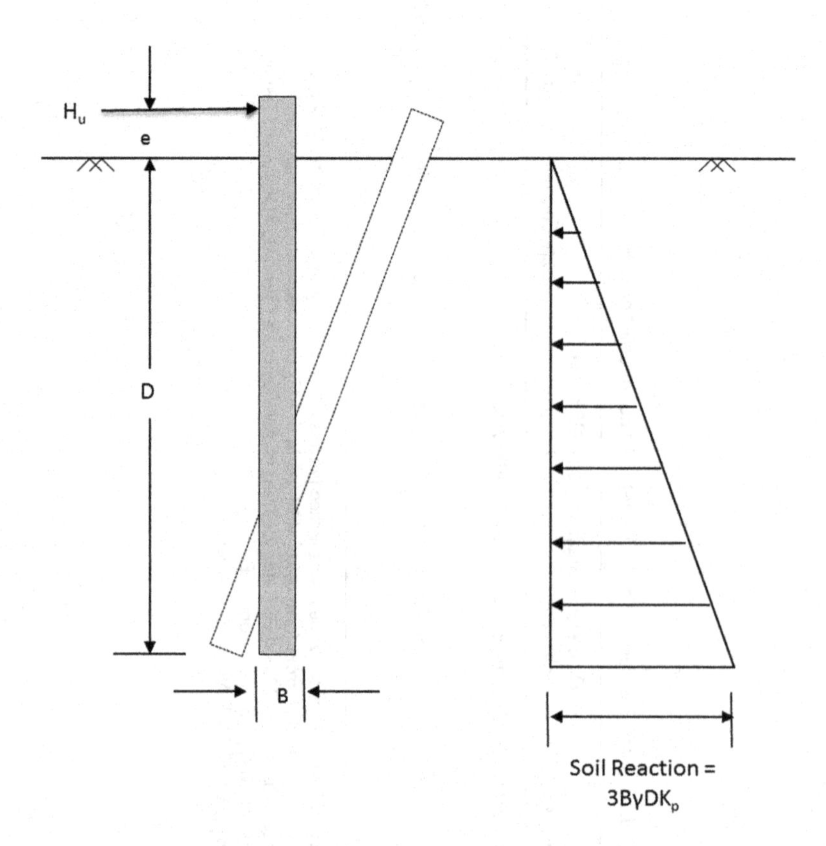

Figure 4.16 Rigid Pile Under Lateral Load – Cohesionless Soil (with permission from ASCE).

These methods are based on the theory of earth pressure on a rigid retaining wall. The theories are extended to rigid pile foundations with appropriate modifications.

Basics of Lateral Capacity of Rigid Pile Foundation
The passive pressure on the rigid wall at any depth, z in granular soils is given by:

$$p_{z\,Wall} = k_p \gamma Z \tag{4.6}$$

where:
k_p = passive earth pressure coefficient = $\tan^2(45 + \frac{\phi}{2})$
γ = unit weight of soil and
ϕ = angle of internal friction

The ultimate lateral resistance per unit width of a rigid pile is greater than that of a corresponding wall, because of the shearing resistance on the vertical sides of the failure wedges in the soil. Using the same distribution of earth pressure at failure of the pile as was assumed for the wall, the three-dimensional effect for a pile can be, approximately, determined by multiplying the net earth pressure on the wall by a shape factor.

Therefore, the passive earth pressure at any depth, z, on a rigid pile in granular soils is given by,

$$p_{z\,Pile} = S_F k_p \gamma Z \tag{4.7}$$

where:
S_F = shape factor to account for the three dimensional effect of soil resistance.

Different investigators assumed different values of the shape factors in their rigid pile theories. Broms (1964) assumed a constant shape factor of 3.0 which is independent of the embedment ratio and angle of internal friction of the soil. Petrasovits and Award (1972) assumed a value of 3.7. Meyerhof et al. (1981) suggested different values of the shape factors ranging from 1 to 10, depending on both the angle of internal friction of the soil and the embedment ratio of the pile.

Broms' method

Broms (1965) proposed a theory for determining the embedment depth by assuming a soil pressure distribution as shown in Figure 4.16 for cohesionless soils under drained conditions. For such soils, the maximum lateral earth pressure at the base of the shaft is three times the Rankine's ultimate passive pressure. The high magnitude of passive pressure developed behind the tip is approximated by concentrated force acting at the tip of the pile. According to this method, the ultimate lateral capacity, H_u is given by:

$$H_u = \frac{0.5 \gamma D^3 K_p B}{(e + D)} \tag{4.8}$$

where:
γ = unit weight of the soil
D = depth of drilled shaft
K_p = Rankine's earth pressure coefficient = $\tan^2(45+\frac{\phi}{2})$
ϕ = angle of internal friction of the soil

e = eccentricity of horizontal load
B = diameter of the drilled shaft

Broms' theory for cohesive soils is presented in the Figure 4.17. As shown in the figure, the theory neglects the soil resistance in the top region up to a depth of 1.5 times the diameter of pile. The soil resistance is given by $9s_uB$ where s_u is the undrained shear strength of cohesive soil.

The total shaft depth using Broms' theory is given by the following equation.

$$D = 1.5B + k + g \tag{4.9}$$

The k and g are as shown in the figure and they are determined using following two equations.

$$k = \frac{H_u}{9s_uB} \tag{4.10}$$

$$H_u(e + 1.5B + 0.5k) = 2.25s_uBg^2 \tag{4.11}$$

Example 4.1 A self-supported angle steel pole is installed on a concrete drilled shaft foundation. Determine the ultimate lateral and moment capacity of foundation using Broms' theory and compare them to loads acting on the foundation using following data:

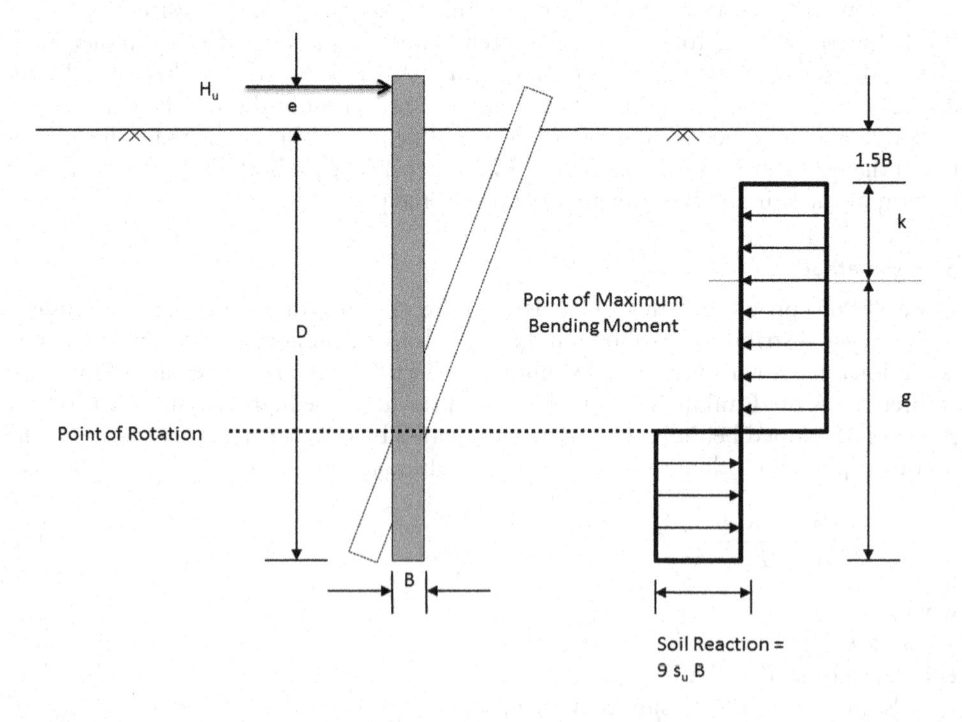

Figure 4.17 Rigid Pile Under Lateral Load – Cohesive Soil (With permission from ASCE).

Drilled shaft dimensions:
Diameter = B = 7 ft (2.13 m)
Depth = D = 20 ft (6.1 m)

Ground line loads:
Moment = M = 5000 kip-ft (6779.66 kN-m)
Horizontal load = H = 60 kips (266.91 kN)
Vertical load = V = 8 kips (35.59 kN)

Soil Properties:
γ = unit weight of in-situ soil = 110 pcf (17.293 kN/m^3)
ϕ = Angle of internal friction = 35°

Solution:

According to Broms' theory in cohesionless soil, the ultimate lateral capacity, H_u is given by:

$$H_u = \frac{0.5\gamma D^3 K_p B}{(e + D)}$$

$\gamma = 110$ pcf
$B = 7$ ft
$D = 20$ ft
$K_p = \tan^2(45 + \phi/2) = \tan^2(45 + 35/2) = 3.69$
$e = \frac{M}{H} = \frac{5000}{60} = 83.3$ ft (25.39 m)
$H_u = \frac{0.5(110)(20^3)(3.69)(7)}{(83.3 + 20)} = 110022$ lbs $= 110$ kips (489.32 kN)
$M_u = H_u e = 110\,(83.3) = 9163$ kip-ft (12424.41 kN-m)
Ultimate lateral capacity = 110 kips (489.32 kN)
Lateral load on foundation = 60 kips (266.91 kN) OK
Ultimate moment capacity = 9163 kip-ft (12424.41 kN-m)
Moment on foundation = 5000 kip-ft (6779.66 kN-m) OK

Example 4.2 A drilled shaft foundation is used for a self-supported single steel pole deadend in uniform clayey soil. The shaft diameter is 7.5 ft (2.286 m). The unit weight of in-situ soil is 100 pcf (15.721 kN/m^3) and the undrained shear strength = 1 ksf (47.88 kN/m^2). Determine depth of shaft required for a lateral resistance of 90 kips (400.36 kN) and moment resistance of 7500 kip-ft (10169.49 kN-m).

Solution:

$$k = \frac{H_u}{9 s_u B}$$

$H_u = 90$ kips
$s_u = 1$ ksf
$B = 7.5$ ft

$$k = \frac{90}{9(1)(7.5)} = 1.33\,\text{ft}\ (0.4054\,\text{m})$$

$$e = 7500/90 = 83.3\,\text{ft}\ (25.39\,\text{m})$$

Solving for g using the following equation yields, $g = 22.54$ ft (6.87 m)

$$H_u(e + 1.5B + 0.5k) = 2.25s_uBg^2$$

$$90(83.3 + 1.5(7.5) + 0.5(1.33)) = 2.25(1)(7.5)g^2$$

Therefore, $g = 22.54$ ft

Total foundation depth $= D$ is given by,
$D = 1.5B + k + g = 1.5(7.5) + 1.33 + 22.54 = 35.12$ ft (10.705 m)

Use 36 ft (11 m)

Hansen's method

Hansen's (1961) method can be used to calculate the ultimate lateral and moment capacity of rigid shafts. The advantage of this method is that it can be applied to layered soils in addition to uniform soils. In this method, the ultimate unit lateral resistance of an element (p_z) of at a depth Z below the ground surface is given by

$$p_z = qK_q^z + cK_c^z \tag{4.12}$$

where:
q = effective overburden pressure at depth Z
K_q^z = earth pressure coefficient for overburden pressure at depth Z
K_c^z = earth pressure coefficient for cohesion at depth Z
c = cohesion at depth Z

The equations for calculating K_q and K_c are provided in the RUS Bulletin 205 (1995). For cohesive soils under undrained conditions ($\phi = 0$), $K_q = 0$. The equation then simplifies to:

$$p_z = s_uK_c^z \tag{4.13}$$

where s_u is undrained shear strength of soil.

Using the unit pressure, the total lateral resistance at a given depth can be calculated and the lateral soil resistance distribution can be established over the entire length. Using the lateral soil resistance distribution, the ultimate lateral capacity can be determined using the horizontal force and moment equilibrium criteria for a rigid body. In this analysis, the point of rotation below the ground is established by the process of trial and error.

MFAD™ model

MFAD™ is an acronym for Moment Foundation Analysis and Design software developed by the Electric Power Research Institute (EPRI). This software uses the semi-empirical theoretical model for ultimate capacity and the non-linear load deflection response of drilled shaft foundations originally developed by Davidson (1981).

For pile foundations subject to lateral load and moment, the lateral resistance offered by the soil is the main component considered in the theoretical models. In the case of a drilled shaft whose diameter is much larger compared to a pile foundation, the contributions of various other forces such as vertical side resistance (friction), shear force acting on the base of the shaft, and the base normal force are significant enough to be included in the moment capacity calculation. MFAD™ software uses a "four-spring" nonlinear subgrade modulus model, where the continuous nature of the soil medium is idealized by four different types of non-linear springs. The four springs represent lateral resistance, vertical side shear, base shear and base normal force/moment. Hansen's theory discussed in the previous section is used in the calculation of ultimate lateral resistance in this model. The work done by Ivey (1968) is also incorporated in the four-spring ultimate capacity model.

DiGioia (1985) provides a summary of the four-spring theoretical model used for drilled shafts. DiGioia et al. (1989) also performed statistical analysis by comparing several theoretical model predictions with full scale load test data. MFAD™ predictions compared very well with test data of several drilled shafts.

Prasad and Chari method

Most of the ultimate capacity models are based on the assumption that the interaction between the pile and the soil can be characterized by net lateral soil pressures acting on the pile. The variation among these available theoretical methods is due to the assumptions made in the pressure distribution pattern on the pile at its ultimate load. An extensive review of assumed pressure distributions in cohesionless soils by different investigators is presented by Prasad and Chari (1999). Laboratory tests were conducted on fully instrumented scaled rigid pile model 4 ft (1.22 m) long and 4 inches diameter (10.2 cm). The theoretical model developed for cohesionless soils and the shape of the pressure distribution assumed is shown in the Figure 4.18.

The value of $p_{0.6x}$ is given by the following equation:

$$p_{0.6x} = (10^{(1.3\tan\phi+0.3)}\gamma(0.6x)) \tag{4.14}$$

where:
$p_{0.6x}$ = lateral resistance of soil at a depth of $0.6x$
x = depth of point of rotation (point where the rigid pier rotates below the ground).

A factor of 0.8 is applied on the peak pressure to account for the non-uniform pressure distribution across the pile section. From force and moment equilibrium, the following equations were developed.

$$H_u = \int_0^D 0.80 p_z B dz \tag{4.15}$$

$$H_u e = \int_0^D 0.80 p_z BZ dz \tag{4.16}$$

where p_z = lateral resistance of soil at a depth of Z.

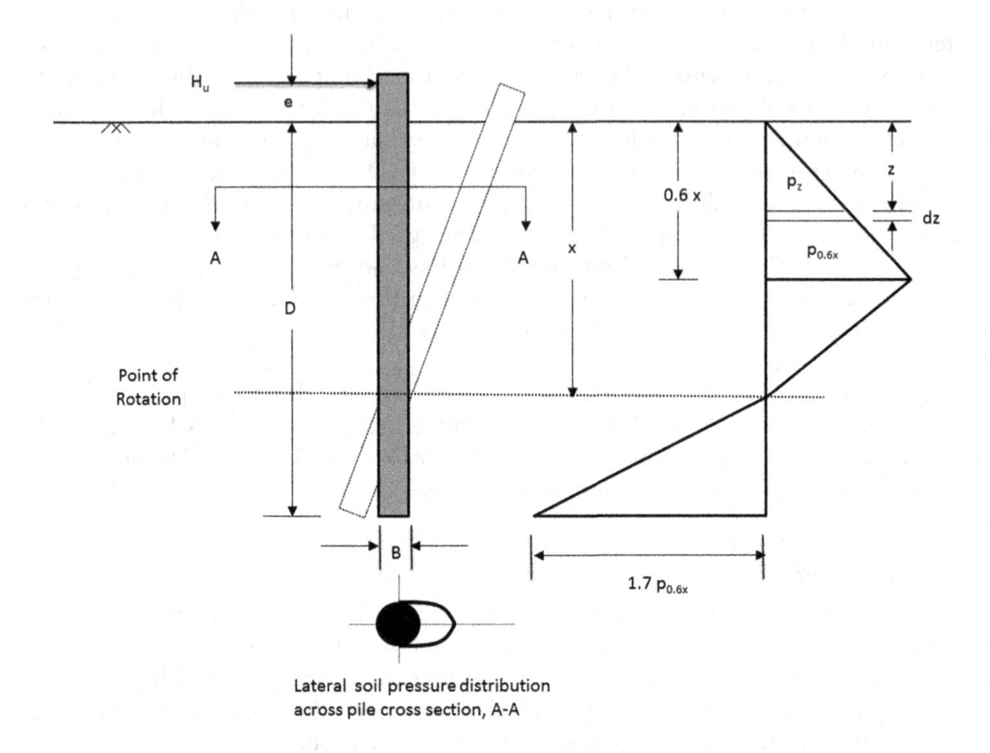

Figure 4.18 Pressure Distribution Under Lateral Load.

Knowing the drilled shaft diameter and soil properties, the point of rotation x and the ultimate load H_u can be determined using the following equations.

$$x = \left[\frac{-(0.567D + 2.7e) + (5.307D^2 + 7.29e^2 + 10.541eL)^{0.5}}{2.1996} \right] \tag{4.17}$$

$$H_u = 0.24(10^{(1.3\tan\phi + 0.3)})\gamma x B(2.7x - 1.7D) \tag{4.18}$$

The measured pressure distributions and model and field test capacities are compared with several existing theories. The proposed method predicts pressure distribution and ultimate lateral capacities better compared to other theories.

Example 4.3 A self-supported angle steel pole is installed on a concrete drilled shaft foundation. Determine the ultimate lateral and moment capacity of foundation using Prasad and Chari Method and compare them to loads acting on the foundation using following data:

Drilled shaft dimensions:
Diameter = B = 7 ft (2.134 m)
Depth = D = 20 ft (6.096 m)

Ground line loads:
Moment = M = 4000 kip-ft (5423.73 kN-m)
Horizontal load = H = 48 kips (213.52 kN)
Vertical load = V = 6 kips (26.69 kN)

Soil Properties:
γ = unit weight of in-situ soil = 125 pcf (19.651 kN/m^3)
ϕ = Angle of internal friction = 40°

Solution:

The depth of point of rotation is calculated using the following equation.

$$x = \left[\frac{-(0.567D + 2.7e) + (5.307D^2 + 7.29e^2 + 10.541eD)^{0.5}}{2.1996}\right]$$

$D = 20$ ft

$$e = \frac{M}{H} = \frac{4000}{48} = 83.3 \text{ ft } (25.39 \text{ m})$$

$$x = \left[\frac{-(0.567D + 2.7e) + (5.307D^2 + 7.29e^2 + 10.541eD)^{0.5}}{2.1996}\right]$$

$$x = \left[\frac{-(0.567(20) + 2.7(83.3)) + (5.307(20)^2 + 7.29(83.3)^2 + 10.541(83.3)(20))^{0.5}}{2.1996}\right]$$

$$= 13.1 \text{ ft } (3.993 \text{ m})$$

$$H_u = 0.24(10^{(1.3\tan\phi + 0.3)})\gamma x B(2.7x - 1.7D)$$

$\gamma = 125$ pcf

$B = 7$ ft

$\phi = 40°$

$x = 13.1$ ft

$$H_u = 0.24(10^{(1.3\tan 40 + 0.3)})(125)(13.1)(7)(2.7(13.1) - 1.7(20))$$

$$= 92691.664 \text{ lbs} = 92.692 \text{ kips } (412.33 \text{ kN})$$

$$M_u = H_u e = 92.692(83.3) = 7721.2 \text{ kip-ft } (10469.424 \text{ kN-m})$$

Ultimate lateral capacity = 92.69 kips (412.33 kN)
Lateral load on foundation = 48 kips (213.52 kN) OK
Ultimate moment capacity = 7721.2 kip-ft (10469.42 kN-m)
Moment on foundation = 4000 kip-ft (5423.73 kN-m) OK

The above methods show the basis behind various theoretical approaches for determining the ultimate lateral/moment capacities. Uniform soil conditions are assumed in the design problems to explain the concepts. Calculations using uniform soil conditions are not useful for stratified soils. In reality, stratified soils are a very common occurrence and, computer programs must be used for the analysis and design of drilled shaft foundations. Popular software programs used in US utility industry for design of drilled shafts include MFAD™, CAISSON™ and LPILE™.

4.4.1.3 *Design aspects of drilled shafts*

Drilled Shaft Diameter
The minimum diameter of the drilled shaft depends on the steel pole base diameter. Minimum shaft diameter,

$$B = ABC_o + 2s_c + 2d_b + 2d_{tb} + 2c_c \tag{4.19}$$

where:
ABC_o = Anchor bolt circle outside diameter
s_c = clear spacing between the anchor bolt and vertical or longitudinal reinforcement cage
d_b = vertical or longitudinal reinforcement bar diameter
d_{tb} = tie bar diameter
c_c = concrete clear cover

Anchor bolt circle diameter is given by the steel pole supplier and is determined by the steel pole base diameter. # 18 bar (2.25 in or 57.15 mm diameter) is commonly used for anchor bolts in the U.S. utility industry. The clear spacing between the anchor bolt cage and vertical or longitudinal reinforcement cage shall be more than aggregate size; a minimum spacing of 2 in to 3 in (50 mm to 75 mm) clear spacing is recommended. The typical concrete clear cover is 3 in to 4 in (75 mm to 100 mm).

The number of vertical steel bars depends on diameter of bar and loading on pier. Typical bar sizes range from # 8 (1 in or 25.4 mm) to # 11 (1.41 in or 35.81 mm), although larger bars are occasionally used. Tie bar size depends on the diameter of the longitudinal bar. Typically, # 3 (0.375 in or 9.52 mm) to # 5 (0.625 in or 15.87 mm) are used.

Factors impacting drilled shaft capacity

Several variables impact the lateral/moment capacity of drilled shaft foundations in granular soils. These include:

- diameter
- depth
- unit weight of soil
- friction angle of soil

The parameters that influence the capacity of the drilled shaft when Broms' theory is used in cohesionless soils are shown in Table 4.14.

Effect of diameter of shaft

Keeping all other parameters in the Table 4.14 constant, the diameter of shaft is increased from 5 ft (1.52 m) to 7.5 ft (2.29 m) and then to 10 ft (3.05 m). The capacity doubled when the diameter is doubled as shown in Table 4.15. The capacity linearly increased with the diameter as expected from Equation 4.8.

Table 4.14 Variables in Parameter Study.

Variable	*Unit*	*Value*
D	feet (meters)	25 (7.62)
ϕ	degrees	30
k_p	Non-dimensional	3
e	feet (meters)	60 (18.29)
ϕ	lbs/ft^3 (kN/m^3)	120 (18.85)
B	feet (meters)	5 (1.52)
H_u	kips (kN)	165.4 (735.70)
M_u	kip-ft (kN-m)	9926.5 (13458.53)

Table 4.15 Effect of Pier Diameter.

B in feet (meters)	*M_u in kip-ft (kN-m)*
5.0 (1.52)	9926.5 (13458.53)
7.5 (2.29)	14889.7 (20187.72)
10.0 (3.05)	19852.9 (26916.92)

Table 4.16 Effect of Pier Length.

D in feet (meters)	*M_u in kip-ft (kN-m)*
25 (7.62)	9926.5 (13458.53)
30 (9.14)	16200.0 (21964.25)
40 (12.19)	34560.0 (46857.07)
50 (15.24)	61363.6 (83197.87)

Effect of depth of shaft

The capacity increased almost six times when the depth is doubled (see Table 4.16) from 25 ft to 50 ft (7.62 m to 15.24 m). The depth of the drilled shaft has a larger impact on the moment capacity than the diameter of shaft. This aspect has practical implications in steel pole – drilled shaft optimization.

Effect of unit weight of soil

Keeping all other parameters in the Table 4.14 constant, the unit weight of soil increased from 120 to 140 lbs/ft^3 (18.87 to 22.01 kN/m^3). The capacity linearly increased in proportion to unit weight of soil as expected from the Equation 4.8. The effect of water table also has a significant impact on capacity of drilled shafts in cohesionless soils; it can reduce the capacity by almost 50% (see Table 4.17).

Table 4.17 Effect of Unit Weight of Soil.

γ in lbs/ft^3 (kN/m^3)	M_u in kip-ft (kN-m)
120 (18.85)	9926.5 (13458.53)
130 (20.42)	10753.7 (14580.06)
140 (21.99	11580.9 (15701.59)
57.6 (9.05)	4764.7 (6460.07)

Table 4.18 Effect of Friction Angle of Soil.

ϕ in degrees	M_u in kip-ft (kN-m)
30	9926.5 (13458.53)
35	12209.6 (16553.99)
40	15220.6 (20636.36)
45	19290.4 (26154.27)

Effect of friction angle of soil

The capacity linearly increased in proportion to Rankine's earth pressure coefficient, k_p. The capacity almost doubled when the soil changed from loose sand ($\phi = 30$ deg) to very dense sand ($\phi = 45$ deg) as shown in Table 4.18.

Depth versus diameter of drilled shaft

During design stage, optimization of shaft size is an important consideration and can significantly impact the project cost. The engineer has a choice of adjusting either the diameter, B or the depth, D of the shaft. From the above parametric study in cohesionless soils, it can be seen that the depth of pier will have a larger impact on the moment capacity compared to the diameter. Another consideration to be taken into account is the volume of concrete in a drilled shaft.

$$V = \frac{\pi}{4} B^2 D \tag{4.20}$$

The volume of concrete increases as a function of the square of diameter compared to a linear increase with depth. We can thus conclude that compared with the diameter, the pier depth increases capacity significantly while minimizing the concrete volume. Assuming no adverse soil conditions and no extra costs for going deeper, increasing the depth of the shaft provides a more economical solution relative to increasing the diameter.

The shaft diameter depends on steel pole design as discussed previously. Larger diameter steel poles require larger diameter drilled shafts. For a given set of loading conditions, steel pole manufacturers balance the diameter of pole and the thickness of steel. In general larger diameter steel pole gives a lighter and hence an economical steel pole. Therefore, from steel pole manufacturer perspective, larger diameter pole

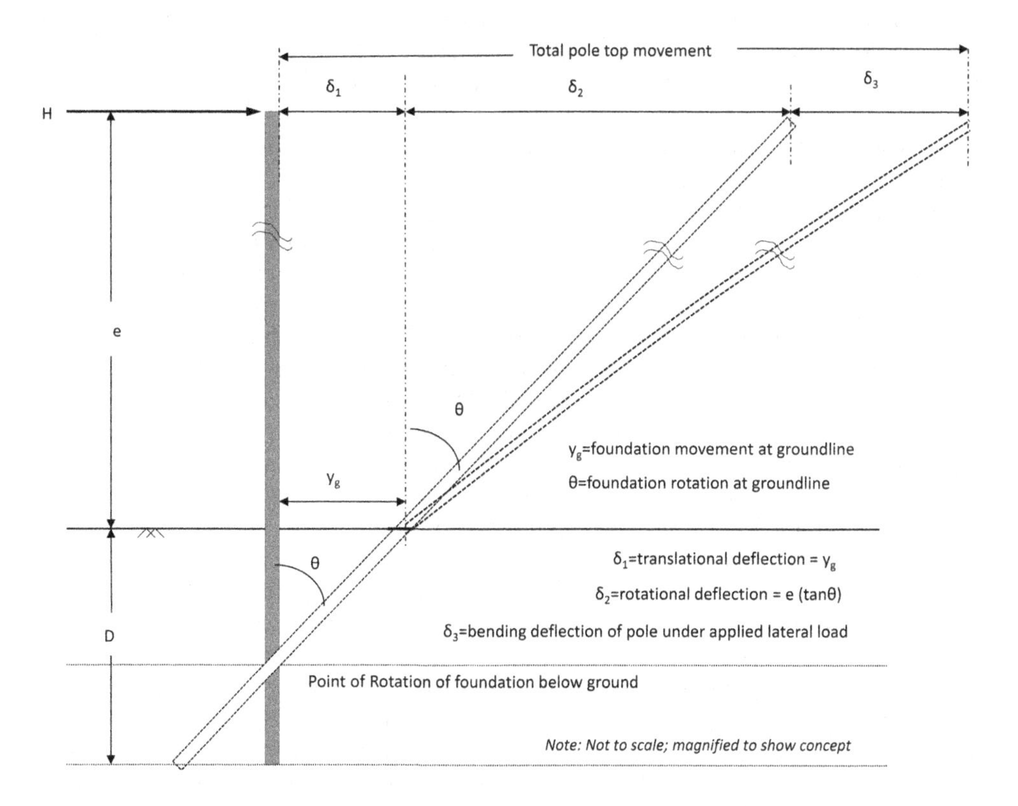

Figure 4.19 Deflection and Rotation of a Pole Foundation.

is preferable. However, this leads to a larger diameter shaft which is not economical from foundation perspective. The cost of steel pole and drilled shaft foundation work in opposite direction. The total cost (steel pole and drilled shaft) should be reviewed together to decide the cheapest pole and shaft options.

4.4.1.4 Deflection and rotations

For pole structures, even a small angular rotation at ground level can induce large displacements at the top of the pole (see Figure 4.19). The increased displacement in turn leads to reduced horizontal and vertical clearances. Further, the deflections will induce additional pole stresses due to what is known as the P-Delta (P-Δ) effect. Additionally, deflections may affect aesthetics of the structure. Despite this significance, few utilities ignore deflection and rotation limits in the design of drilled shaft foundations.

There are no universal standards for deflection and rotation limits of pole foundations. The total rotation and deflection at ground line include both recoverable (elastic) and non-recoverable (inelastic) components. The total rotation varies from 1 to 3 degrees and a 2 degree design requirement is commonly used. It is must be noted that the original EPRI research observed 2 degrees of rotation at the top of shaft to correspond approximately with the ultimate moment capacity (Davidson, 1981). The non-recoverable rotations vary from 0.5 degrees to 1.5 degrees. The total ground line deflection varies from 2 inches to 6 inches (50.8 mm to 152.4 mm). The non-recoverable portion varies from 1 inch to 3 inches (25.4 mm to 76.2 mm). These deflection and rotation limits are important design considerations for angle and dead-end transmission structures for which drilled shaft foundations are commonly used. Since these structures are more critical than tangent structures, their foundations must be designed for higher reliability levels.

In the recent study on performance criteria for short drilled shaft foundations, Kandaris et al. (2012) recommended more stringent rotation/deflection limits. During the initial iteration of design phase, the total rotation recommended by this study is 1 degree and the corresponding total deflection is 4% of shaft diameter. The one degree total rotation will generally keep rigid shaft movement within the elastic deformation range. Drilled shafts typically move towards more plastic behavior of soil above this level. Due to smaller values of limits, typically, these performance criteria may control the drilled shaft foundation design over the ultimate moment capacity.

Several methods are available to calculate the deflections and rotations of drilled shaft foundations. Some of these methods include Broms (1964), Poulos and Davis (1980), p-y curves (Reese and Matlock, 1956), Bhushan et al. (1981), MFAD™ and Prasad (1997). These methods use either the subgrade reaction approach or the continuum approach. In the subgrade reaction approach, the continuous nature of the soil medium is ignored and the shaft reaction at any point is considered to be directly proportional to deflection at that point. To include nonlinearity in the subgrade reaction approach, an iterative procedure is normally used. The iterative procedure incorporates nonlinear soil p-y curves for various depths in the computational approach. In the continuum approach, soil medium is treated as a continuous elastic medium. In the US, MFAD™ and LPILE™ software programs incorporate p-y curve concepts to calculate deflections and rotations of concrete shafts subjected to lateral loads and moments.

4.4.2 Direct embedment foundations

A single pole structure foundation is usually subjected to large moment loads along with relatively small vertical and shear loads. To resist vertical loads, the base of a direct buried steel pole is welded to a wider plate that increases the bearing area. In case of wood poles, bearing shoes are affixed near the bottom of the pole to transfer vertical loads safely to the soil while reducing bearing pressure. General principles of bearing capacity can be applied to these foundations. It must be noted that for unguyed single poles, moment loading usually controls the foundation design.

4.4.2.1 Failure modes

In general, three modes of failure may occur in directly embedded pole foundations.

(i) Flexural failure of pole – The pole may fail due to bending above or below the ground line. Failure due to insufficient flexural strength is common in wood poles.

(ii) Functional failure of transmission pole – The pole may fail due to excessive rotation and deflection at the ground level. This type of failure depends to a great extent on the deformation characteristics of the soil. If the pole deflects significantly, it will reduce electrical clearances as well as adversely impact aesthetics of the structure.

(iii) Collapse of pole due plastic failure of soil – Complete collapse of the pole can occur when the soil becomes sufficiently plastic. The capacity of the pole at this condition is known as ultimate moment capacity. In that situation, the rear end of the pole loses contact with the soil and a big gap forms behind the pole.

4.4.2.2 Calculation methods for embedment

Some of the methods and theories currently used to determine pole embedment are:

- Rule of Thumb Method
- RUS Method
- Broms' Method
- Hansen's Method
- CEA Method
- MFAD Method

Rule of thumb method

This widely used empirical approach is popularly known as the $10\% + 2'$ rule. If the length of the pole is L (feet), the required depth of embedment (D_e) in feet is given as:

$$D_e = 0.10 * L + 2 \quad (4.21)$$

This equation implies that the depth of embedment of the pole is solely a function of length of the pole. The obvious drawback of this method is that it yields the same depth of embedment for both poor and good soils. This method was originally suggested for good soil conditions; but is being used routinely on all types of soils.

Limitations

Keshavarzian (2002) analyzed various classes of wood poles in different soil conditions and presented an excellent discussion on limitations of $10\% + 2'$ rule. Additional embedment depths over $10\% + 2'$ were recommended by the RUS Bulletin 200 for the following conditions:

- Low areas near streams, rivers, or other bodies of water where a high water table or a fluctuating water table is likely. Poles in a sandy soil with a high water table may "kick" out.
- Areas where the soil consists of soft clay, poorly compacted sand, pliable soil, or soil that is highly organic in nature.
- Locations where higher safety is desired. Typical locations include unguyed small angle structures where a portion of the load is relatively permanent in nature, and structures at river, line, or road crossings.
- Locations where poles are set adjacent to or on steep grades.
- Locations where more heavily loaded poles are used.
- Locations where underground utilities such as water or sewer will be located next to the pole.

Internal standards of some utilities recommend the above rule of thumb, with an adder different than 2′ (0.61 m). Depending on the soil conditions, voltage level, type of structure (single pole or H-Frame), the adder could vary from 1.5 ft (0.46 m) to 10 ft (3.05 m).

Example 4.4 A transmission pole 80 feet (24.4 m) long is to be installed in homogeneous soil deposits of (a) loose sand ($\phi = 30°$) and (b) dense sand ($\phi = 40°$) using direct embedment foundation. Determine the depth of embedment using the 'Rule of Thumb' method.

Solution:

(a) For loose sand

$D_e = 0.10(80) + 2 = 10\,\text{ft}\ (3.05\,\text{m})$

(b) For dense sand

$D_e = 0.10(80) + 2 = 10\,\text{ft}\ (3.05\,\text{m})$

Observe there is no difference in foundation setting depths for the loose and dense sand conditions.

RUS method

This method has been suggested for wood pole structures in RUS Bulletin 200. According to this method:

$$H = \frac{S_e D_e^{3.75}}{(L - 2 - 0.662 D_e)} \tag{4.22}$$

where:
H = horizontal force in pounds applied 2 ft (0.61 m) from the pole top
D_e = embedment depth of pole in feet
L = total length of the pole in feet
S_e = Soil Constant, 140 for good soils, 70 for average soils, 35 for poor soils

Good soils: Very dense, well graded sand and gravel, hard clay, dense, well graded, fine and coarse sand.
Average soils: Firm clay, firm sand and gravel, compact sandy loam.
Poor soils: Soft clay, poorly compacted sands (loose, coarse, or fine sand), wet clays and soft clayey silt

The RUS Bulletin provides graphs for calculating the embedment depths using the above method with the following observations.

- The rule of thumb of "10 percent + 2 ft" is adequate for most wood pole structures in good soil and not subjected to heavy loadings.
- For Class 2 and larger class poles and poles of heights less than 60 ft, pole embedment depths should be increased 2 ft or more in poor soil (single pole structures).
- For Class 2 and larger class poles and poles of heights less than 40 ft, pole embedment depths should be increased 1–2 ft in average soil (single pole structures).
- For H-Frame wood structures, "10 percent + 2 ft" seems to be adequate for lateral strengths. Embedment depths are often controlled by pullout resistance.

This method is an improvement over 10% + 2′ rule because it takes into account to some extent the strength of the soil and loads acting on the pole. However, it does not consider interaction between the backfill and the in-situ soil.

Broms' method

From considerations of soil-pole interaction, a directly embedded pole is equivalent to a rigid pile carrying a large bending moment but relatively small vertical and lateral loads at the top. However, the effects of the backfill annulus between the pole and the surrounding soil must be considered in the analysis. For the case where the annulus backfill has shear strength and deformation characteristics equal to those of the surrounding in-situ soil, the performance of the foundation will be relatively unaffected by the annulus. Theoretical models developed for short rigid pile foundations can be used to predict the behavior of this foundation. However, in most situations, the backfill may be either stronger or weaker compared to the surrounding in-situ soil. In such a case, theoretical models for rigid piles are modified by using the strength and stiffness differences between the backfill and surrounding soil.

As discussed above, the basic difference between conventional rigid pile and directly embedded pole foundation is that the backfill is absent in rigid pile applications. Therefore to apply rigid pile theories to directly embedded pole foundations, certain assumptions can be made to simplify the design calculations as recommended in one of the CEA research reports (Haldar et al., 1997) on direct embedment foundations. Some of the assumptions are as follows:

Case (i): If the diameter of the backfill hole is very narrow and the backfill is stronger than or equivalent to surrounding in-situ soil, it can be assumed that the lateral soil failure occurs in the surrounding in-situ soil. For this condition, the thickness of the backfill annulus should be typically less than 0.5 to 1 foot (0.15 m to 0.30 m). The ultimate capacity can then be calculated using the pole diameter and the surrounding in-situ soil conditions.

Case (ii): If the diameter of the backfill hole is narrow and the backfill is much stronger than the in-situ soil (eg: concrete backfill), it can be assumed that the pole and backfill act as a single unit and failure occurs in surrounding in-situ soil. The ultimate capacity can be calculated using the diameter of the backfill hole as equivalent to the pole diameter and surrounding in-situ soil conditions.

Case (iii): If the diameter of the backfill hole is very wide, it can be assumed that the failure occurs in the backfill and effect of in-situ soil can be neglected. If the in-situ soils are granular type with high percentage of boulders and cobblestones, a backhoe type of equipment is used for excavations and the backfill annulus size may be much larger. This type of failure mechanism is suggested for a backfill hole size of 4 to 5 times the pole diameter. The ultimate capacity can be estimated using the diameter of the pole and the backfill soil conditions.

The rigid pile method is simple and easy to use. However, it should be cautiously used since it fails to consider the true engineering properties of the backfill and surrounding soil as well as the moment capacity contributions from skin friction and end bearing resistance which are usually ignored in conventional rigid pile theories.

Broms' method is one of the widely used methods by the utility industry to predict ultimate lateral capacity because of its simplicity. Broms (1964) proposed a rigid pile method for determining embedment by assuming a soil pressure distribution as shown in Figure 4.16. For cohesionless soils, the maximum lateral earth pressure at the base of the pile is equal to three times the Rankine's ultimate passive pressure. According to this method, the ultimate lateral capacity, H_u is given by:

$$H_u = \frac{0.5\gamma D_e^3 k_p B}{(e + D_e)} \tag{4.23}$$

where:
γ = unit weight of the soil
D_e = the embedded length of the pole
k_p = Rankine's earth pressure coefficient = $\tan^2(45+\frac{\phi}{2})$
ϕ = angle of internal friction of the soil
e = eccentricity of horizontal load
B = diameter of the pole

Example 4.5 A transmission pole structure 80 feet (24.38 m) long is to be installed in homogeneous soil deposits of (a) loose sand ($\phi = 30°$) and (b) dense sand ($\phi = 40°$) using direct embedment foundation. Determine the depth of embedment per Broms' rigid pile method. Water table is about 40 ft (12.19 m) below the ground level. The thickness of the backfill annulus = 0.75 ft (0.23 m). Use the data provided in Table 4.19 for the analysis.

The average diameter of the pole below the ground level is (a) 1.5 ft (0.46 m), (b) 2.5 ft (0.76 m). If the depth of foundation (D_e) is determined to be 10 ft (3.05 m) based on Rule of Thumb, determine if the design is sufficient.

The moment at the ground level is 250 kip-ft (339 kN-m) under extreme wind load case. A minimum factor of safety of 1.25 is required against lateral soil failure per internal utility standard. The resultant wind load (due to loads on wire and load on pole itself) acts at 10 ft (3.05 m) from the top of the pole.

Solution:

Since the hole is narrow and the backfill strength is higher than surrounding soil, conservatively apply case (i) discussed above for rigid piles, and assume that the failure occurs in the surrounding in-situ soil. The pole diameter governs the calculation. The problem is now simplified to that of a rigid pile under lateral load.

e = distance between the ground line and the resultanthorizontal load
= total length of the pole – embedment depth – distance from top of the pole to resultant force
= (80 – 10 – 10) = 60 ft (18.29 m)

Table 4.19 Design Data for Example E4.5.

		Surrounding In-Situ Soil	
Property	*Backfill (Crushed Stone)*	*Loose Sand*	*Dense Sand*
Unit Weight in lb/ft^3 (kN/m^3)	135 (21.21)	100 (15.71)	120 (18.85)
Drained Friction Angle ϕ in deg.	45	30	40
k_p = Rankine's Coefficient	5.82	3.0	4.6

According to Broms' method, the ultimate lateral capacity, H_u is given by,

$$H_u = \frac{0.5\gamma D_e^3 K_p B}{(e + D_e)}$$

Case (1) Loose Sand, $B = 1.5$ ft (0.46 m)

γ = Unit Weight of in-situ soil = 100 pcf (15.72 kN/m^3)
D_e = Depth of embedment of pole = 10 ft (3.05 m)
k_p = Rankine's earth pressure coefficient $= \tan^2(45 + \frac{\phi}{2}) = \tan^2(45 + \frac{30^o}{2}) = 3.0$
B = diameter of the pole = 1.5 ft (0.46 m)

Substituting above values,

$$H_u = \frac{0.5(100)(10^3)(3.0)(1.5)}{(60 + 10)} = 3214.3 \text{ lb} = 3.21 \text{ kips (14.3 kN)}$$

Moment Capacity $= M_u = H_u e = (3.21)(60) = 192.6$ kip-ft (261.15 kN-m)

Case (2) Loose Sand, $B = 2.5$ ft (0.76 m)

$\gamma = 100$ pcf (15.72 kN/m^3)
$D_e = 10$ ft (3.05 m)
$K_p = 3.0$
$B = 2.5$ ft (0.76 m)

Substituting above values,

$$H_u = \frac{0.5(100)(10^3)(3.0)(2.5)}{(60 + 10)} = 5357 \text{ lb} = 5.36 \text{ kips (23.84 kN)}$$

Moment Capacity $= M_u = H_u e = (5.36)(60) = 321.6$ kip-ft (436.07 kN-m)

Case (3) Dense Sand, $B = 1.5$ ft (0.46 m)

$\gamma = 120$ pcf (18.86 kN/m^3)
$D_e = 10$ ft (3.05 m)
$K_p = \tan^2(45 + \frac{\phi}{2}) = \tan^2(45 + \frac{40^o}{2}) = 4.60$
$B = 1.5$ ft (0.46 m)

Substituting above values,

$$H_u = \frac{0.5(120)(10^3)(4.6)(1.5)}{(60 + 10)} = 5914.3 \text{ lb} = 5.91 \text{ kips (26.29 kN)}$$

Moment Capacity $= M_u = H_u e = (5.91)(60) = 354.60$ kip-ft (480.81 kN-m)

Case (4) Dense Sand, $B = 2.5$ ft (0.76 m)

$\gamma = 120$ pcf (18.86 kN/m^3)
$D_e = 10$ ft (3.05 m)
$K_p = 4.60$
$B = 2.5$ ft (0.76 m)

Substituting above values,

$$H_u = \frac{0.5(120)(10^3)(4.6)(2.5)}{(60+10)} = 9857.1\,\text{lb} = 9.86\text{ kips }(43.86\,\text{kN})$$

Moment Capacity $= M_u = H_u e = (9.86)(60) = 591.6$ kip-ft (802.17 kN-m)

Table 4.20 summarizes the results of the calculations. In this table, the factor of safety is the ratio of the moment capacity of the foundation and the applied ground line moment.

The table shows that the Rule of Thumb method may not always provide safe designs since it does not take into account the variations in soil, changes in pole diameter and loads on the pole.

Example 4.6 Re-design the pole for Case (1) of Example E4.5 in which factor of safety is found to be less than unity. The minimum factor of safety required is 1.25 against lateral soil failure.

Solution:

Of all the soil and pole parameters in Broms' equation, an increase in embedment depth of pole significantly improves foundation capacity. So, let the embedment depth of foundation be increased to 12 ft (3.66 m).

$e = (80 - 12 - 10) = 58$ ft (17.68 m)

Case (1) Loose Sand, $B = 1.5$ ft (0.46 m)

$\gamma =$ Unit Weight of in-situ soil $= 100$ pcf (15.72 kN/m^3)
$D_e =$ Depth of embedment of pole $= 12$ ft (3.66 m)
$k_p =$ Rankine's earth pressure coefficient $= \tan^2\left(45 + \frac{\phi}{2}\right) = \tan^2\left(45 + \frac{30^\circ}{2}\right) = 3.0$
$B =$ diameter of the pole $= 1.5$ ft (0.46 m)

Table 4.20 Calculation Summary.

Case	*Moment Capacity of Foundation In kip-ft (kN-m)*	*Applied Ground Line Moment in kip-ft (kN-m)*	*Factor of Safety (FS)*	*Comments*
1. Loose Sand, B = 1.5 ft (0.46 m)	192.6 (261.13)	250.0 (338.96)	0.77	Not OK (FS < 1.25)
2. Loose Sand, B = 2.5 ft (0.76 m)	321.6 (436.03)	250.0 (338.96)	1.30	OK (FS > 1.25)
3. Dense Sand, B = 1.5 ft (0.46 m)	354.6 (480.77)	250.0 (338.96)	1.41	OK (FS > 1.25)
4. Dense Sand, B = 2.5 ft (0.76 m)	591.6 (802.10)	250.0 (338.96)	2.37	OK (FS > 1.25)

Substituting above values,

$$H_u = \frac{0.5(100)(12^3)(3.0)(1.5)}{(58+12)} = 5554.29\,\text{lb} = 5.55\,\text{kips}\ (24.71\,\text{kN})$$

Moment Capacity $= M_u = H_u e = (5.55)(58) = 321.9$ kip-ft (436.48 kN-m)
Factor of safety $= 321.9/250.0 = 1.29 > 1.25$, Design is OK

Hansen's method

Hansen's (1961) method is used to calculate the ultimate lateral and moment capacity of rigid piles and is described in Section 4.4.1.2. The main advantage of this method is that it can be applied to layered soils in addition to uniform soils. Based on Hansen's theory, RUS Bulletin 205 gives a series of graphs for determining the direct embedment depths for tangent poles in nine (9) different uniform soils. The parameters associated with the nine charts are given in Table 4.21 along with the charts (Figures 4.20 to 4.28).

The required setting depth for a pole can be determined from the ground line diameter and the ultimate moment. In these calculations, the effect of backfill is neglected. Important facts related to these charts are listed below.

- The diameters of poles range from 1 ft (0.305 m) to 4 ft (1.219 m).
- Ultimate moments considered at groundline range from 0 to 3500 kip-ft (4745.76 kN-m).
- Recommended range for embedment depth/diameter ratio range from 3 to 10
- For multi-layered soils, the predominant soil type should be considered for selecting the embedment depth.
- For horizontal loads greater than 40 kips (177.94 kN-m) or for stratified soils, use of actual equations is recommended.

Example 4.7 A transmission pole structure 80 feet long (24.38 m) is to be installed in homogeneous soil deposit of loose dry sand using a direct embedment foundation.

Table 4.21 Parameters Considered in the Design Charts.

Chart	*Soil/Type & Description*	*Unit weight in lbs/ft³ (kN/m³)*	*s_u for clay in kip/ft² (kPa)*	*ϕ for sand in degrees*	*Typical Blow Count Values, SPT 'N'*
1	Dense Dry Sand	140 (21.99)	0	41	>30
2	Dense Submerged Sand	85 (13.36)	0	41	>30
3	Medium Dry Sand	120 (18.85)	0	33	10–30
4	Medium Submerged Sand	65 (10.21)	0	33	10–30
5	Loose Dry Sand	95 (14.93)	0	28	0–10
6	Loose Submerged Sand	55 (8.64)	0	28	0–10
7	Stiff Saturated Clay	140 (21.99)	2.0 (95.76)	0	>8
8	Medium Saturated Clay	120 (18.85)	0.75 (35.91)	0	4–8
9	Soft Saturated Clay	100 (15.71)	0.25 (11.97)	0	0–4

(Courtesy: RUS/USDA.)

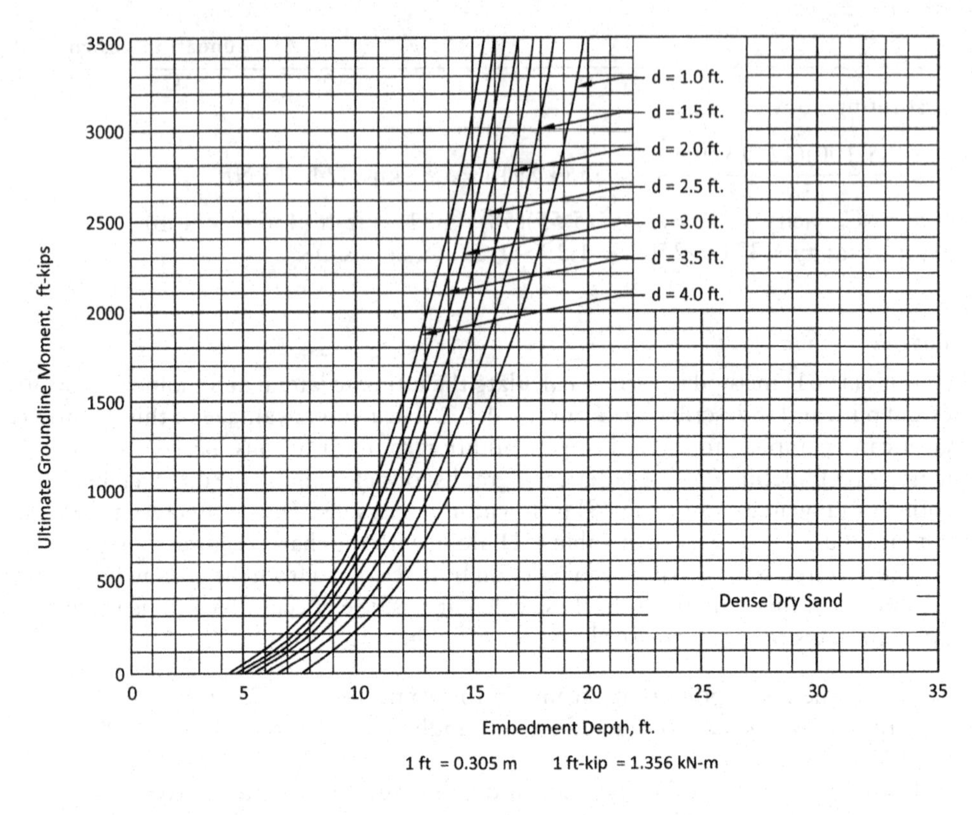

Figure 4.20 Embedment Depth for Dense Dry Sand (Source: RUS/USDA).

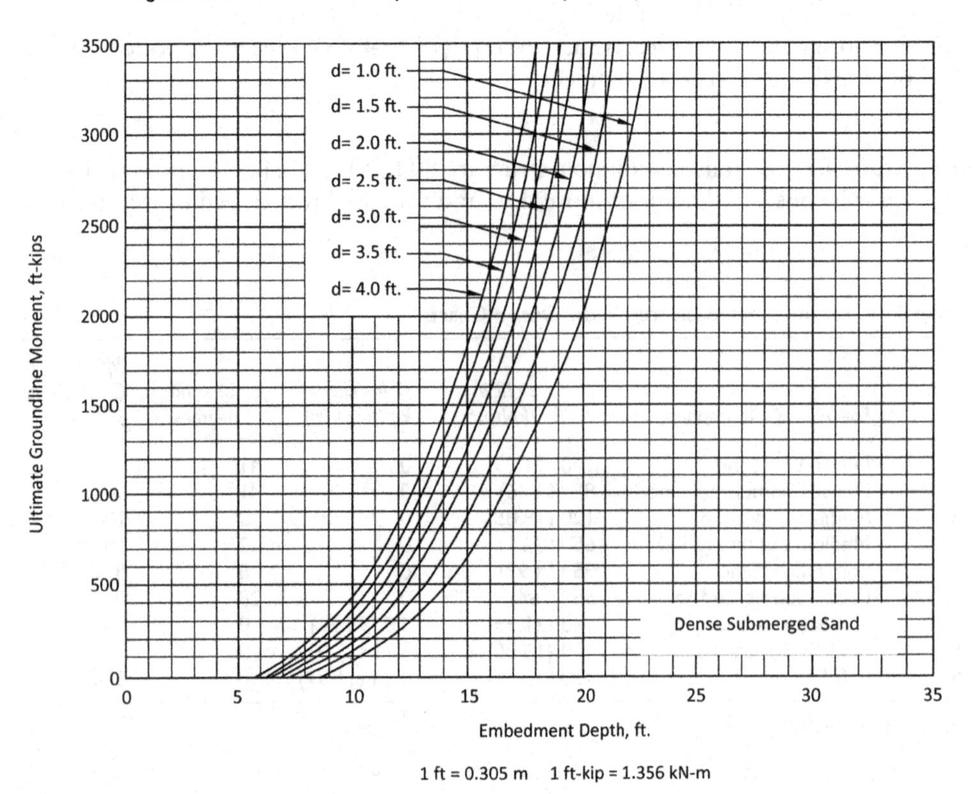

Figure 4.21 Embedment Depth for Dense Submerged Sand (Source: RUS/USDA).

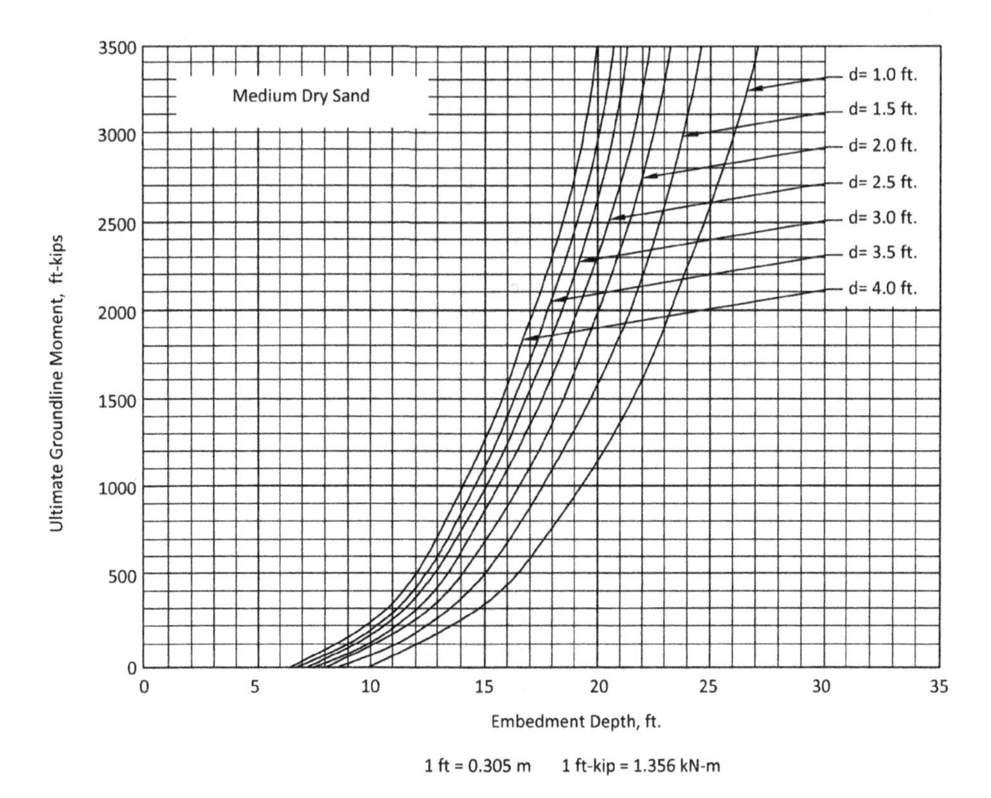

Figure 4.22 Embedment Depth for Medium Dry Sand (Source: RUS/USDA).

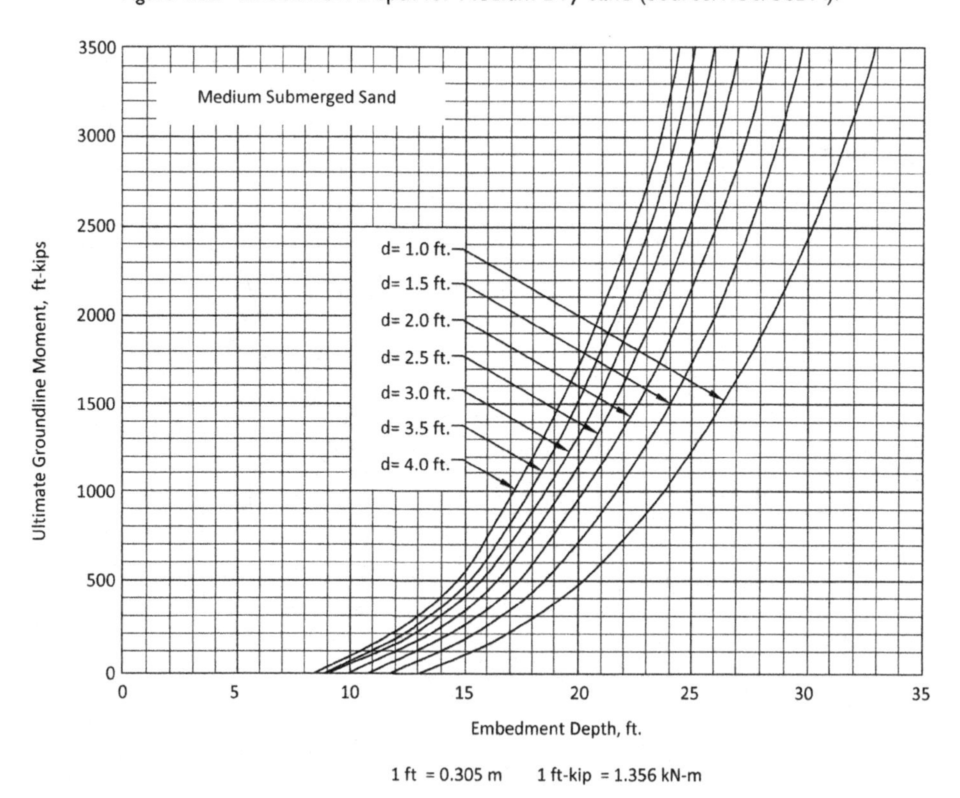

Figure 4.23 Embedment Depth for Medium Submerged Sand (Source: RUS/USDA).

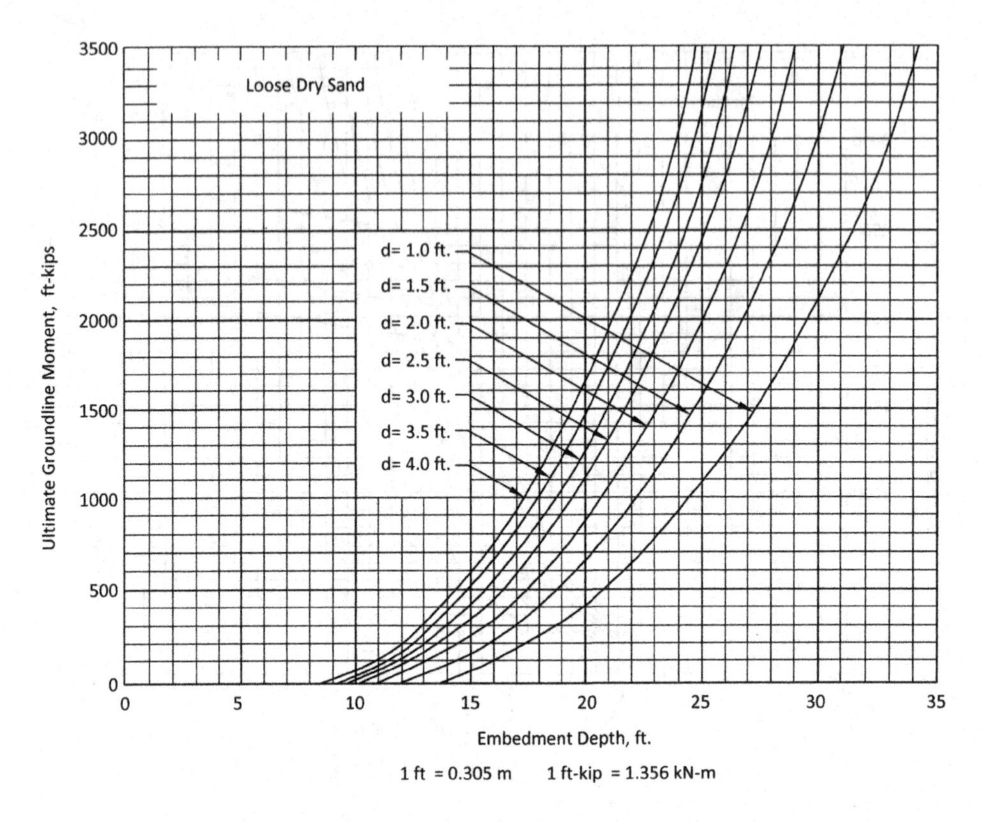

Figure 4.24 Embedment Depth for Loose Dry Sand (Source: RUS/USDA).

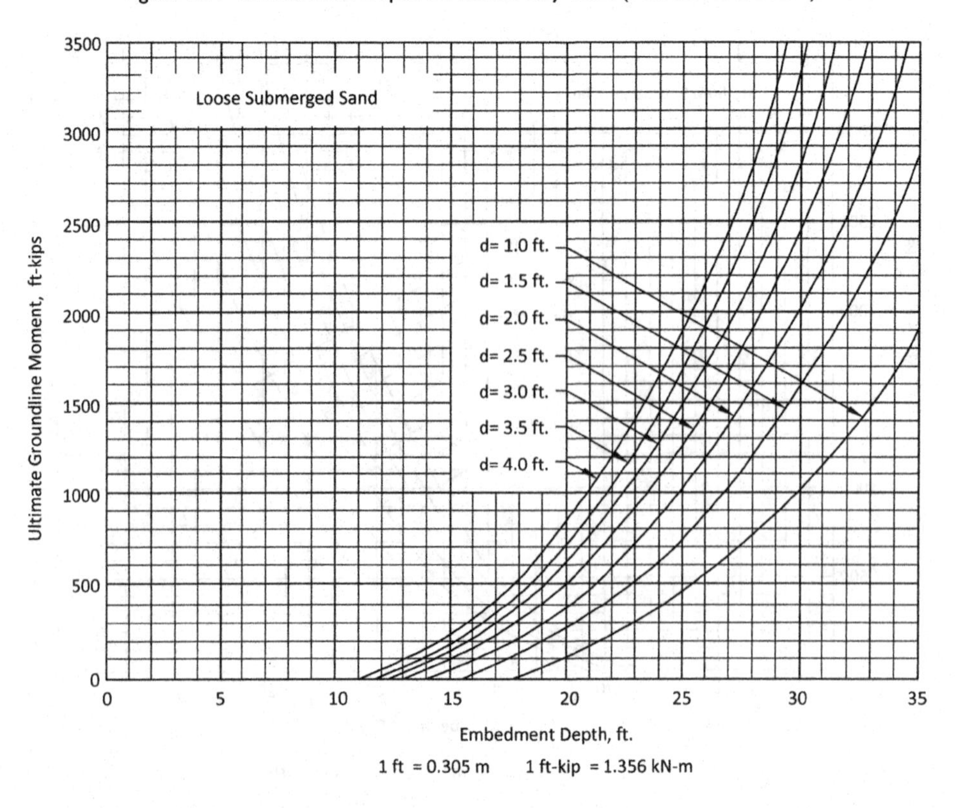

Figure 4.25 Embedment Depth for Loose Submerged Sand (Source: RUS/USDA).

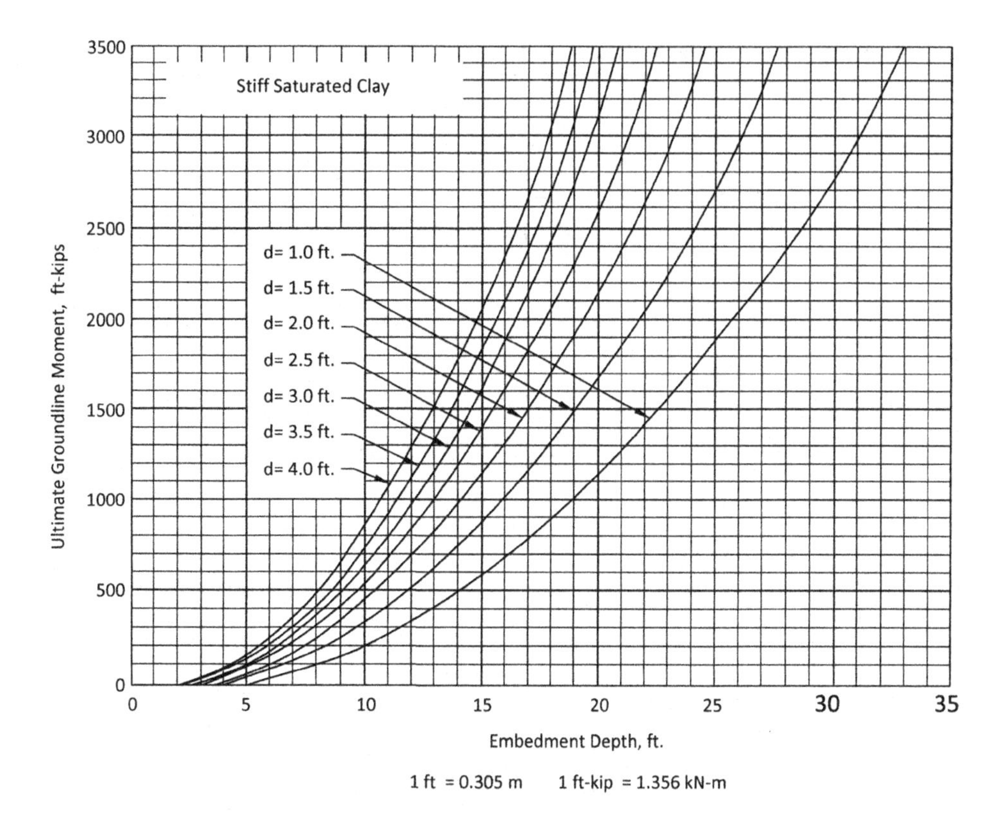

Figure 4.26 Embedment Depth for Stiff Saturated Clay (Source: RUS/USDA).

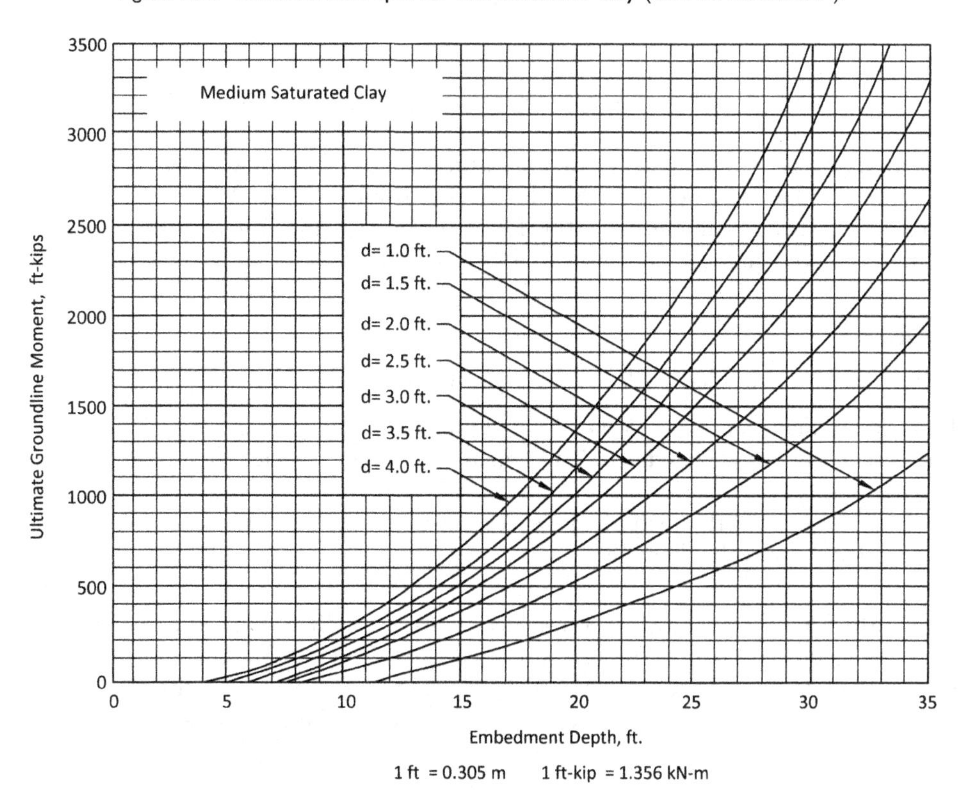

Figure 4.27 Embedment Depth for Medium Saturated Clay (Source: RUS/USDA).

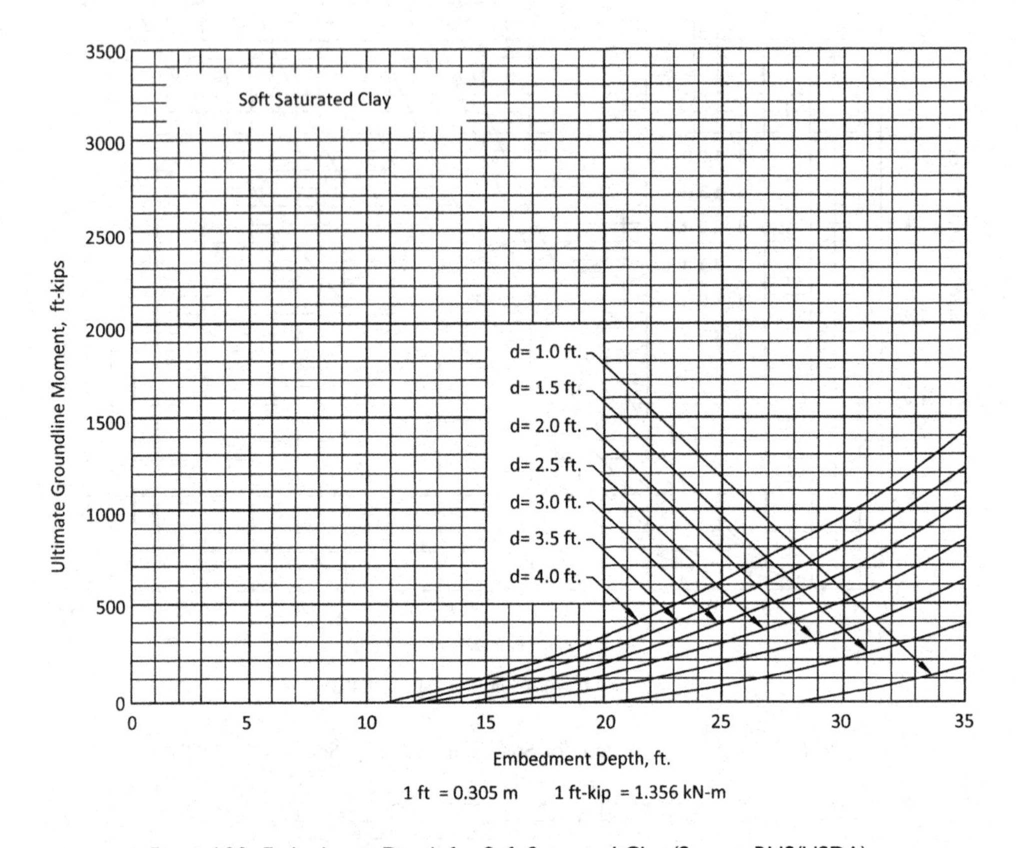

Figure 4.28 Embedment Depth for Soft Saturated Clay (Source: RUS/USDA).

Calculate the embedment depth of pole using the charts provided by RUS Bulletin 205. The average diameter of the pole below the ground is 2.5ft (0.76 m). The required moment capacity of foundation is 1000 kip-ft (1355.9 kN-m). What will be embedment depth if the pole diameter increases to 4 ft (1.22 m)?

Solution:

Using the chart provided for loose dry sand (Figure 4.24):

Embedment depth = 19.5 ft (5.94 m) when the pole diameter is 2.5 ft (0.76 m).
Embedment depth = 17.3 ft (5.27 m) when the pole diameter is increased to 4 ft (1.22 m).
The embedment depths in both the cases are more than the minimum embedment depth of 3 times the diameter.

Example 4.8 A transmission pole structure 80 ft (24.38 m) long is to be installed in a homogeneous soil deposit of stiff saturated clay using a direct embedment foundation. Calculate embedment depth of the pole foundation using the charts from RUS Bulletin 205. The average diameter of the pole below the ground is 2.0 ft (0.61 m).

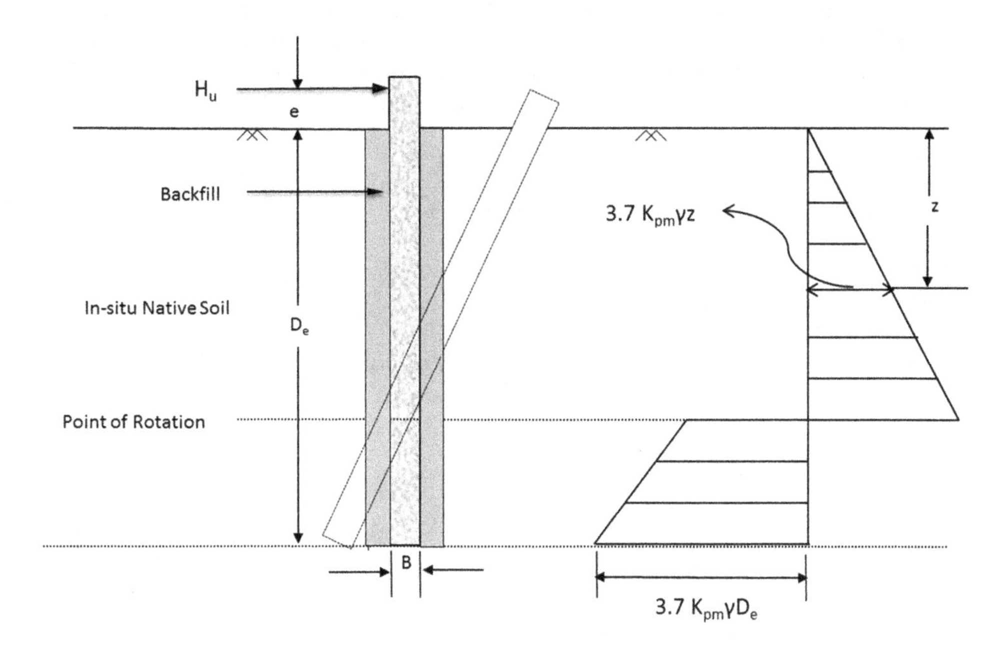

Figure 4.29 Pressure Distribution Under Lateral Load (with permission from CEATI International, Inc).

The required moment capacity of foundation is 1500 kip-ft (2033.90 kN-m). What will be embedment depth required if the pole diameter increases to 4 ft (1.22 m)?

Solution:

Using the chart provided for stiff saturated clay (Figure 4.26):

Embedment depth = 17 ft (5.18 m) when the embedded pole diameter is 2 ft (0.61 m).
Embedment depth = 13 ft (3.96 m) when the embedded pole diameter is increased to 4 ft (1.22 m).
The embedment depths in both the cases are more than the minimum embedment depth of 3 times the diameter.

CEA method

This method by Haldar et al. (1997) is developed from the first principles of earth pressure theory and considers both backfill and in-situ soil characteristics to estimate the capacity of directly buried pole foundations in cohesionless soils. This theory was a part of a major research project of the Canadian Electricity Association (CEA). The passive earth pressure theory is extended to rigid piles using shape factors and the pressure distribution shape recommended by Petrasovits and Award (1972). The pressure distribution of this theory is shown in the Figure 4.29. This theory is not valid when the backfill is very loose compared to the surrounding in-situ soil. The soil resistance at any depth z is given by:

$$p_z = 3.7 K_{pm} \gamma Z \qquad (4.24)$$

In the above equation, K_{pm} is the combined passive coefficient for backfill and the surrounding soil and considers friction angles of the backfill and the in-situ soil, and the thickness of the backfill. The average of unit weights of the backfill and the surrounding in-situ soils will be used as γ in the equation above.

$$K_{pm} = 2n\tan\left(45 + \frac{\phi_1}{2}\right) - n^2 + \frac{\tan^2(45 + \frac{\phi_2}{2})}{\tan^2(45 + \frac{\phi_1}{2})}\left[\tan\left(45 + \frac{\phi_1}{2}\right) - n\right]^2 \quad (4.25)$$

where:
ϕ_1 = friction angle of backfill
ϕ_2 = friction angle of in-situ native soil
n = thickness of annular space/embedment depth of pole
According to this method, the ultimate lateral capacity, H_u is given by:

$$H_u = \frac{3.7K_{pm}\gamma BD_e^2(2R^2 - 1)}{2} \quad (4.26)$$

$$M_u = H_u e \quad (4.27)$$

where:
K_{pm} = combined passive coefficient for backfill and surrounding soils
e = eccentricity of loading
D_e = embedment depth of pole
γ = average of unit weights of backfill and surrounding in-situ soils
R = ratio of depth of point rotation to total embedment depth of pole
The value R is 0.707 for pure bending moment and 0.794 for pure horizontal load. For a given combined moment and horizontal load, R can be calculated using the equation below.

$$(2R^2 - 1) = \frac{(1 - 2R^3)2D_e}{3e} \quad (4.28)$$

Example 4.9 A transmission pole structure 80 ft (24.38 m) long is to be installed in a homogeneous soil deposit of loose sand with a unit weight γ of 100 lb/ft^3 (15.72 kN/m^3) and $\phi = 30°$ using a direct embedment foundation. Calculate the ultimate moment capacity of the embedded pole foundation using the CEA method. The average diameter of the pole is 2.5 ft (0.76 m) and it is embedded to a depth of 12 ft (3.66 m). Water table is about 40 ft (12.19 m) below the ground level. The thickness of the backfill annulus is 0.75 ft (0.23 m). The backfill is crushed rock with a unit weight of $\gamma = 135$ lb/ft^3(21.22 kN/m^3) and $\phi = 45°$. The resultant of the load acts at 60 ft (18.29 m) above the groundline.

Solution:

The value of R is determined using the following equation.

$$(2R^2 - 1) = \frac{(1 - 2R^3)2D_e}{3e}$$

$D_e = 12$ ft (3.66 m)
$e = 60$ ft (18.29 m)
Calculated $R = 0.719$

$$K_{pm} = 2n \tan\left(45 + \frac{\phi_1}{2}\right) - n^2 + \frac{\tan^2(45 + \frac{\phi_2}{2})}{\tan^2(45 + \frac{\phi_1}{2})}\left[\tan\left(45 + \frac{\phi_1}{2}\right) - n\right]^2$$

n = thickness of annular space/embedment depth of pole = 0.75/12 = 0.0625
ϕ_1 = friction angle of backfill = 45°
ϕ_2 = friction angle of in-situ native soil = 30°

$$K_{pm} = 2(0.0625)\tan\left(45 + \frac{45}{2}\right) - (0.0625)^2 + \frac{\tan^2(45 + \frac{30}{2})}{\tan^2(45 + \frac{45}{2})}\left[\tan\left(45 + \frac{45}{2}\right) - 0.0625\right]^2 = 3.14$$

$$\gamma = \frac{(135 + 100)}{2} = 117.5 \text{ lb/ft}^3 \ (18.47 \text{ kN/m}^3)$$

$$H_u = \frac{3.7 K_{pm} \gamma B D_e^2 (2R^2 - 1)}{2}$$

$$H_u = \frac{3.7(3.14)(117.5)(2.5)(12)^2(2(0.719)^2 - 1)}{2}$$

$$= 8335.3 \text{ lb} = 8.335 \text{ kips } (37.08 \text{ kN})$$

$$M_u = H_u e = 8.335(60) = 500.1 \text{ kip-ft } (678.10 \text{ kN-m})$$

MFAD™ method

The details of the MFAD™ program for drilled shafts are discussed in Section 4.4.1.2. Bragg et al. (1988) modified the four-spring subgrade modulus model to account for the backfill in directly buried pole foundations as well as the effect of relatively smaller base area of directly buried poles compared to concrete drilled shafts. Several prototype tests have been conducted to validate the theoretical model developed.

4.4.2.3 Rock-socketed foundations

Rock socketed pole foundations are installed fully or partially in rock. DiGioia and Rojas-Gonzalez (1994) reported a testing program on rock socketed pole foundations. Several prototype drilled shaft and direct embedment foundations were tested in different subsurface conditions to provide an initial database of tests that could be used to verify and/or modify existing geotechnical design models. Based on the results of the full scale load tests, a provisional design guideline was proposed.

4.4.2.4 Special techniques to enhance capacity

There are several ways to enhance the capacity of a direct embedment pole foundation. A few methods are listed below.

Use of Special Backfills: Crushed stone backfill is commonly recommended. Special backfills such as lean concrete mix or flowable backfill made of fly ash increase the capacity significantly (Haldar et al., 2000).

Use of a Baseplate: Addition of a baseplate can help increase capacity. Baseplate usage is more common for pole structures to increase bearing capacity under compression load; however investigations (Prasad & Chari, 1996; Haldar et al., 2000) have shown that a baseplate also increases the moment capacity.

Use of Culverts: Culvert foundations are sometimes used to increase the stability of directly buried pole foundations. The pole is inserted in to a hollow metallic pipe and the annulus between the pole and metallic pipe is filled with crushed rock/gravel/concrete backfill.

Other techniques include the use of railroad ties or cribbing devices to brace a direct buried pole. Concrete thrust blocks are also occasionally used.

4.4.2.5 Deflections and rotations

A discussion on acceptable limits of deflections and rotations of drilled shaft foundations in presented in Section 4.4.1.4. Typically, the allowable deflection and rotation limits for direct embedment pole foundations are equal or greater than those of drilled shafts. Some of the theoretical models developed for predicting deflection/rotation behavior of drilled shaft foundations can be used for direct embedment foundations. However, these models need to consider the impact of backfill in addition to the in-situ soils. MFAD™ software considers strength and stiffness of both the backfill and in-situ soils to compute deflections and rotations. The CEA method by Haldar et al. (1997) used Bhushan et al. (1981)'s model for predicting deflection and rotation behavior of direct embedment foundations.

RUS Bulletin 205 used Davisson and Prakash (1963) model for predicting deflection and rotation of direct embedment pole foundations. The Bulletin gives formulae for calculating rotations and deflections at ground line for poles embedded in sands and clays; however, the equations are applicable only for loads which are 1/3 to 1/2 of ultimate loads (i.e.) within a serviceability range. At higher load levels, the relationship between load and deflection becomes non-linear. Broms (1964) indicated that the rigid pile deflections at groundline at the ultimate load are in the range of 20% of diameter of the pile.

The formulae for computing ground line deflections and rotations of embedded poles – both steel and concrete – in sand and clay are as follows:

For clay soils

$$y_g = \frac{\left\{(2.15P)\frac{D_r}{D_e}\right\}}{KD_e\left\{(1.87)\frac{D_r}{D_e} - 1\right\}} \tag{4.29}$$

$$\frac{D_r}{D_e} = \frac{\left\{\frac{M}{PD_e} + 0.683\right\}}{\left\{\frac{1.87M}{PD_e} + 1\right\}} \tag{4.30}$$

Table 4.22 Horizontal Subgrade Modulus for Granular Soils.

Soil Type	*n_h Values in Dry Soil in lb/in^3 (kN/m^3)*	*n_h Values in Submerged Soil in lb/in^3 (kN/m^3)*
Loose	9.4 (2551.60)	5.3 (1438.67)
Medium	28.0 (7600.52)	19.0 (5157.49)
Dense	75.0 (20358.53)	45.0 (12215.11)

(Source: RUS/USDA.)

Table 4.23 Modulus of Lateral Subgrade Reaction for Clays.

Cohesion in lb/ft^2 (kN/m^2)	*K values in psi (kN/m^2)*
1000–2000 (47.87–95.74)	700 (4826.50)
2000–4000 (95.74–191.48)	1400 (9653.00)
>4000 (>191.48)	2800 (19306.00)

(Source: RUS/USDA.)

For granular soils

$$y_g = \frac{\left\{(3.0P)\frac{D_r}{D_e}\right\}}{n_h D_e^2 \left\{\frac{1.5D_r}{D_e} - 1\right\}} \tag{4.31}$$

$$\frac{D_r}{D_e} = \frac{\left\{\frac{M}{PD_e} + 0.75\right\}}{\left\{\frac{1.5M}{PD_e} + 1\right\}} \tag{4.32}$$

where:
D_e = total depth of embedment
D_r = total depth of point of rotation
M = moment at groundline
P = shear force at groundline
y_g = deflection at groundline
n_h = coefficient of modulus variation (see Table 4.22)
K = modulus of lateral subgrade reaction (see Table 4.23)

4.4.3 Spread foundations under compression/moment

There are several situations where foundations of transmission structures are subjected to compressive loads that are large enough to control foundation design. The ability of a soil to sustain a foundation compression load without undergoing shear failure is called bearing capacity of the soil. Spread foundations such as concrete footings and grillage foundations of lattice transmission towers carry significant compressive loads. These foundations typically fall under the category of shallow foundations. Relevant theories for analysis and design are discussed in this section.

The foundations of H-Frame structures with X-braces and guyed single poles also carry compressive loads, but, these foundations typically fall under the category of deep foundations. Compressive loads do not generally govern the design of these foundations. In guyed pole structures, axial loads are significant since the pole carries the vertical components of the guy forces. In single concrete poles, compressive loads are significant due to the self-weight of the pole. To prevent these poles from sinking into the ground, the butt of the pole should have sufficient cross sectional area. Therefore, it is a common practice to increase the bearing area of pole butt by using concrete pads, crushed rock bed or pole anchors and base plates.

4.4.3.1 Bearing capacity theories

Three types of bearing failures are considered in soil analysis.

- General shear failure
- Local shear failure
- Punching shear failure

General shear failure takes place in soils that are relatively incompressible and reasonably strong. Punching shear failure takes place in very loose sand and in a thin layer of strong soil underneath by a very weak soil. Local shear failure is an intermediate case. Bearing capacity failure types in cohesionless and cohesive soils are provided in Tables 4.24 and 4.25, respectively.

Several theories have been developed over time to calculate the bearing capacity of shallow foundations. Some of the most widely used theories have been proposed by Terzaghi (1943), Meyerhof (1963), Hansen (1970) and Vesic (1973).

Terzaghi's theory

Terzaghi's theory (1943) is developed for continuous footings. Therefore it is simplified to a two-dimensional problem. Based on model test data, the theory was extended to

Table 4.24 Bearing Failure Capacity Modes in Cohesionless Soils.

Type of Bearing Capacity Failure	*Relative Density in %*
General	>65%
Local	30% to 65%
Punching	<30%

Table 4.25 Bearing Failure Capacity Modes in Cohesive Soils.

Type of Bearing Capacity Failure	*Consistency*
General	Very Stiff to Hard
Local	Medium to Stiff
Punching	Soft to Very Soft

square and round foundations using empirical coefficients. This theory includes the following major assumptions:

- Depth of the foundation is less than width ($D \leq B$)
- Foundation base is rough. Sliding does not occur at the base of the foundation
- Soil beneath the foundation is a homogeneous, semi-infinite mass
- Soil shear strength above base of the foundation is ignored and replaced with the surcharge load
- General shear failure is assumed. Factors are added to account for local shear failure

The failure mechanism assumed by Terzaghi is shown in Figure 4.30. Ultimate bearing capacity (q_u) for a strip footing under general shear failure is given by:

$$q_u = cN_c + qN_q + \frac{1}{2}\gamma BN_\gamma \tag{4.33}$$

where:
c = cohesion of soil
γ = unit weight of soil
$q = \gamma D$
D = depth of foundation
N_c, N_q, N_γ = bearing capacity factors (function of ϕ)

$$N_c = \cot\phi\left(\frac{a^2}{2\cos^2(45 + \frac{\phi}{2})} - 1\right) \tag{4.34}$$

$$N_q = \frac{a^2}{2\cos^2(45 + \frac{\phi}{2})} \tag{4.35}$$

$$N_\gamma = \frac{1}{2}\tan\phi\left(\frac{K_{p\gamma}}{\cos^2\phi} - 1\right) \tag{4.36}$$

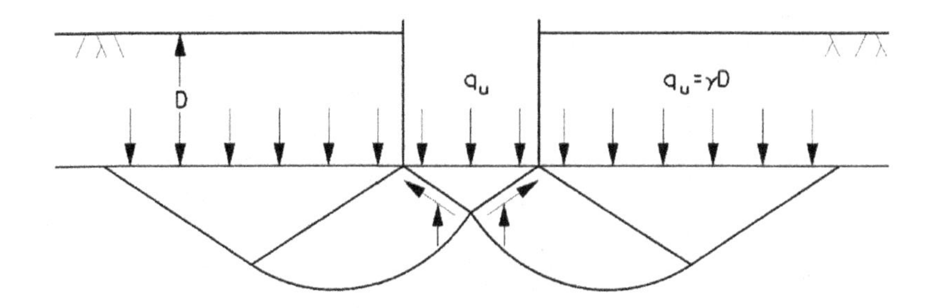

Figure 4.30 Bearing Capacity Failure.

where $K_{p\gamma}$ = coefficient of passive pressure and

$$a = e^{\left(\frac{3\pi}{4} - \frac{\phi}{2}\right)\tan\phi} \tag{4.37}$$

In the bearing capacity equation above, the contribution of the cohesive shear strength of the soil located below the foundation is calculated by the first term cN_c. If the soil below the foundation base is cohesionless, $c = 0$; then the contribution of the first term is zero.

The overburden pressure at the foundation base is quantified by the second term, qN_q. The q value is equal to γD, where γ is the total unit weight of soil located above the bottom of the footing. The second term indicates that the deeper the foundation, the greater is the ultimate bearing capacity due to the surcharge.

The third term $\frac{1}{2}\gamma BN_\gamma$ accounts for the frictional shear strength of the soil located below the base of the footing. The γ in this term represents the unit weight of soil located below the footing within a depth "B". The values of bearing capacity factors are summarized in Table 4.26 for different ϕ values in five degree intervals.

Effect of shape:

When square or circular footings are used, Terzaghi's equations are modified as below:

For square foundations with size $B \times B$,

$$q_u = 1.3cN_c + qN_q + 0.4\gamma BN_\gamma \tag{4.38}$$

For circular foundations with diameter B,

$$q_u = 1.3cN_c + qN_q + 0.3\gamma BN_\gamma \tag{4.39}$$

Corrections for Local Shear Failure

The above equations are applicable to relatively incompressible soil types where general shear failure is the predominant failure mode. In case of loose sands, a considerable amount of footing movement takes place prior to mobilization of full resistance. This

Table 4.26 Terzaghi's Bearing Capacity Factors.

ϕ *in degrees*	N_c	N_q	N_γ
0	5.7	1.0	0.0
5	7.3	1.6	0.5
10	9.6	2.7	1.2
15	12.9	4.4	2.5
20	17.7	7.4	5.0
25	25.1	12.7	9.7
30	37.2	22.5	19.7
35	57.8	41.4	42.4
40	95.7	81.3	100.4
45	172.3	173.3	297.5
50	347.5	415.1	1153.2

mode of failure is known as local shear failure. Terzaghi (1943) suggested following empirical correction factors to the soil parameters in loose sands.

$$c_{corrected} = \left(\frac{2}{3}\right)c \tag{4.40}$$

$$\phi_{corrected} = \tan^{-1}\left(\frac{2}{3}\tan\phi\right) \tag{4.41}$$

The $c_{corrected}$ and $\phi_{corrected}$ values shall be used instead of c and ϕ in the above equations when local shear failure predominates.

Effect of water table on bearing capacity

Figure 4.31 shows different locations of the water table. The unit weight of the soil needs to be corrected to correspond to the conditions. The submerged unit weight of soil, γ_{sub} is given by the following equation:

$$\gamma_{sub} = \gamma_{sat} - \gamma_w \tag{4.42}$$

where:
γ_{sat} = saturated unit weight of soil and
γ_w = unit weight of water = 62.4 lb/ft^3

In Equation 4.33 for the ultimate bearing capacity, γ is not factored in the first term; however, the second and third terms are impacted by the decrease in unit weight or in other words, decrease in the effective stress.

Condition (a): $0 \leq D_{wt} \leq D$
Water table depth is D_{wt} below the ground line as shown in the Figure 4.31. In this situation, both the second and third terms needs to be modified.

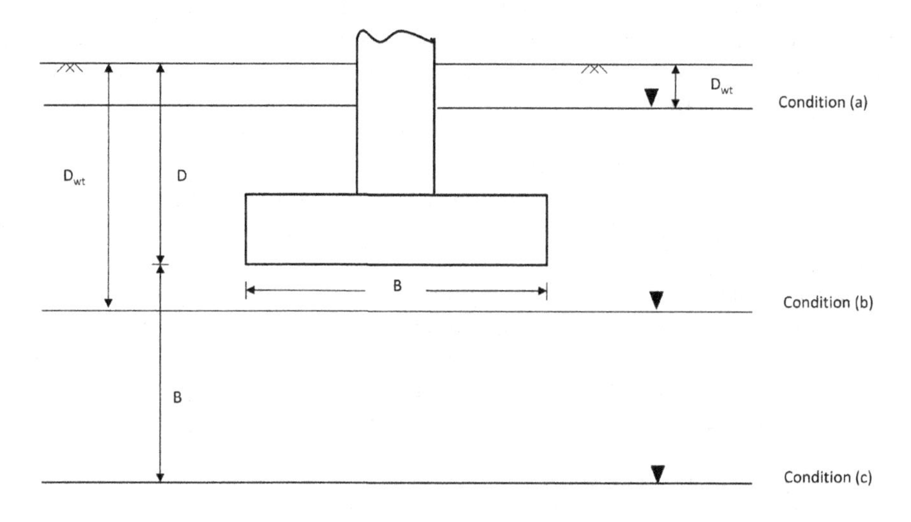

Figure 4.31 Water Table Conditions for Bearing Capacity.

In the second term, the value q has to be calculated using the following equation:

$$q = \gamma D_{wt} + (\gamma_{sat} - \gamma_w)(D - D_{wt}) \tag{4.43}$$

In the third term, the γ should be replaced by γ_{sub}.

Condition (b): $D \leq D_{wt} \leq D + B$

Water table depth is D_{wt} below the ground line as shown in the Figure 4.31. In this situation, only the third term needs to be modified.

In the third term, the value γ should be replaced by γ_{mod}

$$\gamma_{mod} = \gamma_{sub} + (\gamma - \gamma_{sub}) \frac{(D_{wt} - D)}{(B)} \tag{4.44}$$

Condition (c): $D_{wt} > D + B$

In this case, there is no need to correct any of the terms in bearing capacity equation.

Ultimate bearing capacity in saturated clays

In transmission line design, extreme wind loads and extreme ice loads are classified as transient loads because they are applied for relatively short periods. The magnitude of the loads resulting from extreme events is much higher than the everyday loads acting on the structure foundations. Transmission tower footings located in saturated clays create positive excess pore pressures in the soil medium when they are subjected to transient compressive loads. Due to low permeability, the excess pore pressures do not dissipate for a while. Thus, undrained conditions develop when transient loads are applied relatively rapidly to fine-grained soils such as clays. The most probable time for a bearing capacity failure is immediately after these extreme loads are applied. The shear strength is normally expressed by undrained shear strength (s_u) and $\phi = 0$ under these conditions.

Drained conditions also develop in fine-grained soils such as clays under long-term sustained loading conditions with $c = 0$. However, for lattice tower foundations, the long-term sustained loadings for a typical tangent lattice tower consist primarily of the self-weight of the tower and wires. These loads are not as significant as the loads associated with the extreme load cases; therefore, typically, undrained conditions control foundation design in saturated clays.

For footings supported by saturated sands and gravels, excess pore pressures are very small and dissipate rapidly. Thus, drained conditions develop under most loadings in coarse-grained soils such as sands.

In case of saturated clays under undrained conditions ($\phi = 0$), the ultimate bearing capacity Equation 4.33 for general shear failure can be simplified as follows.

(a) $c = s_u$ where is s_u is undrained shear strength
(b) From Table 4.26, $N_c = 5.7$, $N_q = 1.0$, $N_\gamma = 0$

For strip footings with a width of B,

$$q_u = 5.7 s_u + q \tag{4.45}$$

For square foundations with size $B \times B$,

$$q_u = 7.41 s_u + q \tag{4.46}$$

and for circular foundations with diameter B,

$$q_u = 7.41 s_u + q \tag{4.47}$$

Example 4.10 A lattice tower is supported on a square concrete footing. The footing is installed in a *sandy soil* at a depth of 6 ft (1.83 m). The base dimensions of the foundation are 7 ft × 7 ft (2.13 m × 2.13 m). The unit weight of soil shall be assumed as 120 lb/ft^3 (18.87 kN/m^3). The friction angle of sandy soil is 35 degrees ($c = 0$). Calculate the ultimate bearing capacity and ultimate compressive load that can be carried by the foundation using Terzaghi's theory for general shear failure. The water table is 14 ft (4.27 m) below the ground level.

Solution:

$q_u = 1.3cN_c + qN_q + 0.4\gamma BN_\gamma$
$B = 7$ ft (2.13 m)
$D = 6$ ft (1.83 m)
$D_{wt} = 14$ ft (4.27 m)
$D + B = 6 + 7 = 13$ ft (3.96 m)
$D_{wt} > D + B$, therefore water table has no impact on bearing capacity and is ignored.
$c = 0$
$q = \gamma D = (120)(6) = 720$ lb/ft^2 (34.48 kN/m^2)
For $\phi = 35$ degrees, from Table 4.26:
$N_c = 57.8$
$N_q = 41.4$
$N_\gamma = 42.4$
$q_u = (1.3)(0)(57.8) + (720)(41.4) + (0.4)(120)(7)(42.4)$
$= 0 + 29808 + 14246.4 = 44054.4$ lb/ft^2 (2109.38 kN/m^2)

Ultimate compressive capacity of the foundation
$= (q_u)B^2 = (44054.4)(7^2) = 2158665.6$ lbs $= 2158.67$ kips (9602.63 kN)

Example 4.11 For the data given in Example E4.10, recalculate the ultimate bearing capacity and ultimate compressive load that can be carried by the foundation using Terzaghi's theory for general shear failure with the following changes:

- Water table is at the ground line
- Assume saturated unit weight of soil = 120 lb/ft^3 (18.87 kN/m^3)

Solution:

Condition (a): $0 \leq D_{wt} \leq D$ is applicable in this problem.
In the second term, the value q has to be calculated using the following equation.

$$q = \gamma D_{wt} + (\gamma_{sat} - \gamma_w)(D - D_{wt})$$
$$D_{wt} = 0$$
$$q = 0 + (120 - 62.4)(6) = 345.6 \text{ lb/ft}^2 \ (16.55 \text{ kN/m}^2)$$

In the third term, the γ should be replaced by γ_{sub}

$$\gamma_{sub} = 120 - 62.4 = 57.6\,\text{lb/ft}^3\ (9.06\,\text{kN/m}^3)$$

$$q_u = (1.3)(0)(57.8) + (345.6)(41.4) + (0.4)(57.6)(7)(42.4)$$

$$= 0 + 14307.8 + 6838.3 = 21146\,\text{lb/ft}^2\ (1012.5\,\text{kN/m}^2)$$

Ultimate compressive load capacity of the foundation
$= (q_u)B^2 = (21146)(7^2) = 1036154\,\text{lb} = 1036.2\,\text{kips}\ (4609.4\,\text{kN})$

Note: The impact of water table is very significant in cohesionless soils. It can reduce ultimate bearing capacity by as much as 50%.

Example 4.12 A lattice tower is supported on a square concrete footing. The footing is installed in a *saturated clayey soil* at a depth of 6 ft (1.83 m). The base dimensions of the foundation are 7 ft × 7 ft (2.13 m × 2.13 m). The unit weight of soil may be assumed as 110 lb/ft^3 (17.29 kN/m^3). The undrained shear strength of soil is 1500 psf (71.82 kN/m^3). Calculate the ultimate bearing capacity and ultimate compressive load that can be carried by the foundation using Terzaghi's theory for general shear failure. The water table is at 14 ft below the ground level.

Solution:

In saturated clays, for square foundations with size $B \times B$, Equation 4.46 can be used.

$$q_u = 7.41 s_u + q$$

$$q = \gamma D = (120)(6) = 720\,\text{lb/ft}^2\ (34.48\,\text{kN/m}^2)$$

$$q_u = (7.41)(1500) + 720 = 11835\,\text{lb/ft}^2\ (556.67\,\text{kN/m}^2)$$

Ultimate compressive load capacity of the foundation
$= (q_u)\ B^2 = (11835)(7^2) = 579915\,\text{lb} = 579.92\,\text{kips}\ (2579.69\,\text{kN})$

Meyerhof's theory

Terzaghi's theory does not address rectangular footings which are also used for transmission towers. Meyerhof's (1963) theory can be used to calculate the bearing capacity of rectangular footings. The theory assumes logarithmic failure surface ending at the ground surface, and accounts for the additional soil resistance above the footing base, which was not factored in Terzaghi's theory. Meyerhof's theory is more comprehensive in nature and provides shape factors (s_c, s_q, s_γ), depth factors (d_c, d_q, d_γ) and inclined load factors (i_c, i_q, i_γ).

$$q_u = cN_c s_c d_c i_c + qN_q s_q d_q i_q + \frac{1}{2}\gamma B N_\gamma s_\gamma d_\gamma i_\gamma \tag{4.48}$$

where:
q = effective stress at the bottom level of the foundation

Table 4.27 Meyerhof's Bearing Capacity Factors.

ϕ *in degrees*	N_c	N_q	N_γ
0	5.14	1.0	0.0
5	6.5	1.6	0.1
10	8.3	2.5	0.4
15	11.0	3.9	1.1
20	14.8	6.4	2.9
25	20.7	10.7	6.8
30	30.1	18.4	15.7
35	46.1	33.3	37.1
40	75.3	64.2	93.7
45	133.9	134.9	262.7
50	266.9	319.0	873.7

Source: George Geoffrey Meyerhof; Some Recent Research on the Bearing Capacity of Foundations; *Canadian Geotechnical Journal*, 1963, 1(1): 16–26.

N_c, N_q, N_γ = bearing capacity factors (function of ϕ) – (see equations below and Table 4.27)

$$N_q = e^{\pi \tan\phi} N_\phi \tag{4.49}$$

$$N_c = (N_q - 1)\cot\phi \tag{4.50}$$

$$N_\gamma = (N_q - 1)\tan(1.4\phi) \tag{4.51}$$

$$N_\phi = \tan^2\left(45 + \frac{\phi}{2}\right) \tag{4.52}$$

Shape Factors
For $\phi = 0$

$$s_c = 1 + 0.2\left(\frac{B}{L}\right) \tag{4.53a}$$

$$s_q = 1 \tag{4.53b}$$

$$s_\gamma = 1 \tag{4.53c}$$

For $\phi \geq 10°$

$$s_c = 1 + 0.2\left(\frac{B}{L}\right)N_\phi \tag{4.54a}$$

$$s_q = s_\gamma = 1 + 0.1\left(\frac{B}{L}\right)N_\phi \tag{4.54b}$$

Depth Factors
For $\phi \geq 0$

$$d_c = 1 + 0.2\left(\frac{D}{B}\right) \tag{4.55a}$$

$$d_q = 1 \tag{4.55b}$$

$$d_\gamma = 1 \tag{4.55c}$$

For $\phi \geq 10°$

$$d_c = 1 + 0.2\left(\frac{D}{B}\right)\sqrt{N_\phi} \tag{4.56a}$$

$$d_q = d_\gamma = 1 + 0.1\left(\frac{D}{B}\right)\sqrt{N_\phi} \tag{4.56b}$$

Inclination Factors

$$i_c = i_q = \left(1 - \frac{\alpha}{90}\right)^2 \tag{4.57a}$$

$$i_\gamma = \left(1 - \frac{\alpha}{\phi}\right)^2 \tag{4.57b}$$

$$i_\gamma = 0 \quad \text{for } \phi = 0 \tag{4.57c}$$

where:
B = width of square/rectangular foundation or diameter of circular foundation
L = length of rectangular foundation
α = inclination of the load on the foundation from vertical axis in degrees

$$N_\phi = \tan^2\left(45 + \frac{\phi}{2}\right) \tag{4.58}$$

Example 4.13 A transmission tower is supported on a rectangular concrete footing and is installed in a sandy soil at a depth of 6.5 ft (1.98 m). The base dimensions of the foundation are 5 ft × 8 ft (1.52 m × 2.44 m). The unit weight of soil is 120 lb/ft^3 (18.87 kN/m^3). The friction angle of soil is 35 degrees. Calculate the ultimate bearing capacity and ultimate compressive load that can be carried by the foundation using Meyerhof's theory. Assume that the load is vertical. The water table is located a deeper depth and its effect may be neglected.

Solution:

With $c=0$ in Equation 4.48:

$$q_u = qN_q s_q d_q i_q + \frac{1}{2}\gamma B N_\gamma s_\gamma d_\gamma i_\gamma$$

$B = 5$ ft (1.52 m)
$L = 8$ ft (2.44 m)
$D = 6.5$ ft (1.98 m)
$q = \gamma D = (120)(6.5) = 780$ lb/ft^2 (37.35 kN/m^2)

For $\phi = 35$ degrees, from Table 4.27,

$N_q = 33.3$
$N_\gamma = 37.1$

Shape factors

$$s_q = s_\gamma = 1 + 0.1\left(\frac{B}{L}\right) N_\phi$$

$$\frac{B}{L} = \frac{5}{8} = 0.625$$

$$N_\phi = \tan^2\left(45 + \frac{35}{2}\right) = 3.69$$

$$s_q = s_\gamma = 1 + 0.1\,(0.625)(3.69) = 1.23$$

Depth Factors
For $\phi \geq 10°$

$$d_q = d_\gamma = 1 + 0.1\left(\frac{D}{B}\right)\sqrt{N_\phi}$$

$$\frac{D}{B} = \frac{6.5}{5} = 1.3$$

$$d_q = d_\gamma = 1 + 0.1(1.3)\sqrt{3.69} = 1.25$$

Inclination Factors

$$i_q = \left(1 - \frac{\alpha}{90}\right)^2$$

$$i_\gamma = \left(1 - \frac{\alpha}{\phi}\right)^2$$

For vertical load, $\alpha = 0$, therefore the inclination factors are unity.

$$q_u = qN_q s_q d_q i_q + \frac{1}{2}\gamma B N_\gamma s_\gamma d_\gamma i_\gamma$$

$$q_u = 780(33.3)(1.23)(1.25)(1) + 0.5(120)(5)(37.1)(1.23)(1.25)(1)$$
$$= 39935 + 17112 = 57047\,\text{lb/ft}^2\ (2731.48\,\text{kN/m}^2)$$

Ultimate compressive load that can be carried by the foundation
$= (q_u)(L \times B) = (57047)(8 \times 5) = 2281880\,\text{lb} = 2281.88\,\text{kips}\ (10150.71\,\text{kN})$.

4.4.3.2 *Foundations subjected to eccentric loads*

Figure 4.32 shows a footing is subjected to vertical concentric load that is applied at the center of gravity (CG) of the footing. In this case the distribution of pressure on the soil is uniform. When the base of the foundation is subjected to a moment loads in addition to a vertical load, the distribution of pressure over base of the foundation will

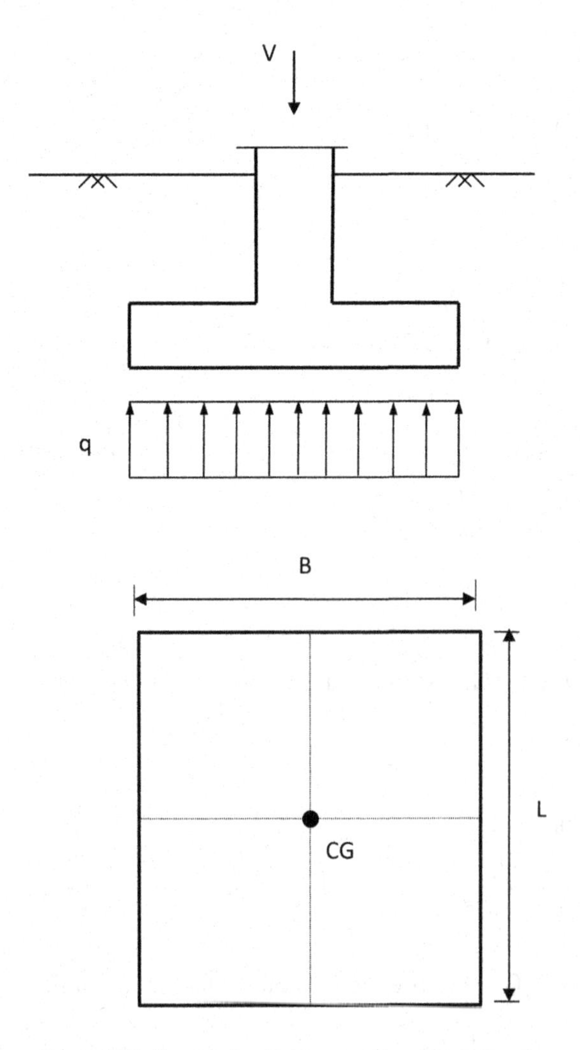

Figure 4.32 Foundation Subject to Concentric Load.

not be uniform. The bearing pressure on one side of the footing is higher than on the other side. The maximum pressure is denoted by q_{max} and the minimum pressure by q_{min}. Assuming the foundation is rigid the distribution of soil pressure varies linearly as shown in Figure 4.33.

$$q_{max} = \frac{V}{BL} + \frac{6M}{B^2L} \tag{4.59}$$

$$q_{min} = \frac{V}{BL} - \frac{6M}{B^2L} \tag{4.60}$$

where:
V = total vertical load
M = total moment

The moment can be represented by a vertical load V offset at a certain distance 'e' from the center of gravity of the footing. This offset is known as eccentricity. The eccentricity of load 'e' is given by the following equation.

$$e = \frac{M}{V} \tag{4.61}$$

Equations 4.59 and 4.60 above can be written in terms of 'e' as follows:

$$q_{max} = \frac{V}{BL} + \frac{6M}{B^2L} = \frac{V}{BL}\left(1 + 6\frac{M}{BV}\right) = \frac{V}{BL}\left(1 + 6\frac{e}{B}\right) \tag{4.62}$$

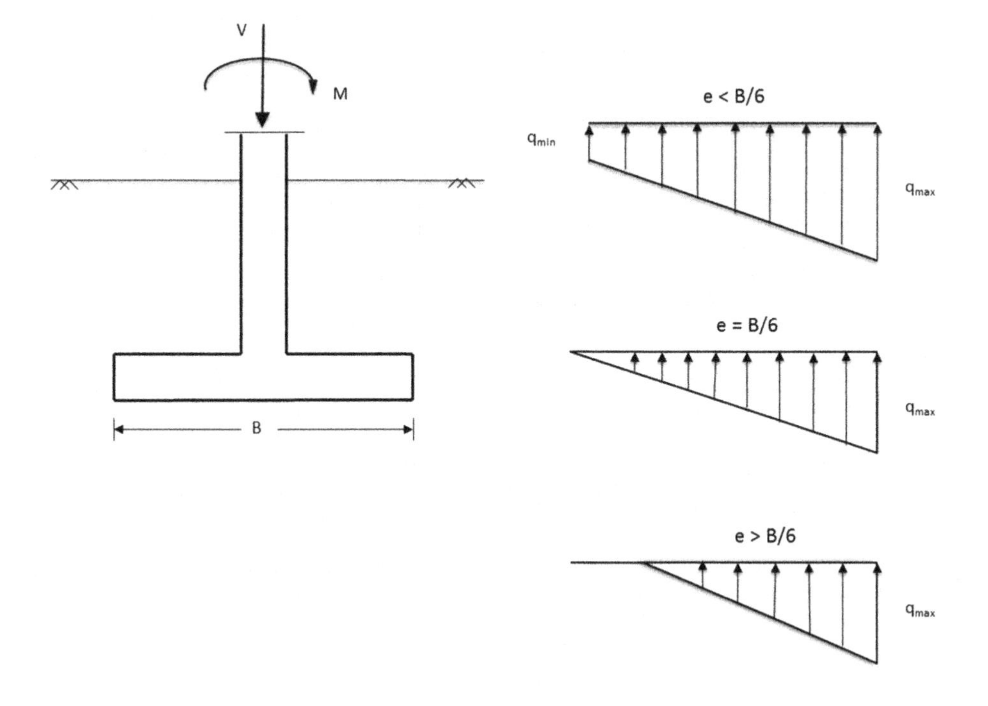

Figure 4.33 Foundation Subject to Eccentric Load.

$$q_{\min} = \frac{V}{BL} - \frac{6M}{B^2L} = \frac{V}{BL}\left(1 - 6\frac{M}{BV}\right) = \frac{V}{BL}\left(1 - 6\frac{e}{B}\right) \tag{4.63}$$

In the above equations, when $e = B/6$, $q_{\min}$ will become zero. For any value of 'e' higher than $B/6$, $q_{\min}$ will become negative. As soil cannot take any tension, the footing will separate from the soil underneath. In this situation, the value of $q_{\max}$ is

$$q_{\max} = \frac{4V}{3L(B - 2e)} \tag{4.64}$$

It is therefore a good practice to keep the eccentricity of load within the middle third of the footing dimension to avoid separation of footing from the soil underneath. This ensures that the entire plan area of the footing base remains in compression.

The footing should be designed so that the maximum pressure ($q_{\max}$) does not exceed the computed bearing capacity. Meyerhof (1953) suggested an effective area method for calculating the ultimate bearing capacity of eccentrically-loaded foundations. This is a simplified approach for eccentrically loaded footings. The effective dimensions of the foundation should be used in calculations. For a footing subjected to two moments in 'x' and 'y' directions shown in Figure 4.34a, the effective width B' and effective length L' are given in the equations below.

$$B' = B - 2e_B \tag{4.65}$$

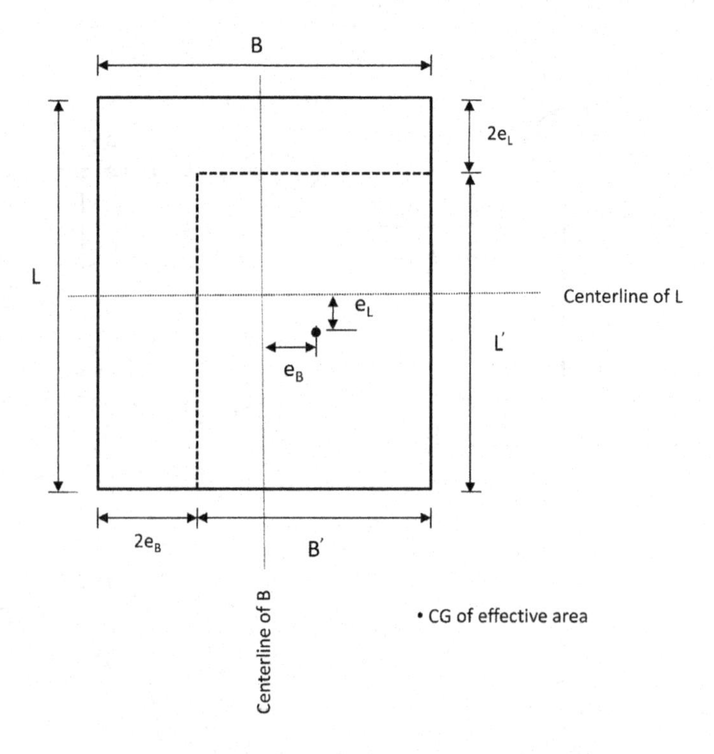

Figure 4.34a Equivalent Area Concept – Biaxial Bending.

$$L' = L - 2e_L \tag{4.66}$$

$$e_B = \frac{M_B}{V} \tag{4.67}$$

$$e_L = \frac{M_L}{V} \tag{4.68}$$

Equation 4.48 can be used with the following changes:

$$q'_u = cN_c s_c d_c i_c + qN_q s_q d_q i_q + \frac{1}{2}\gamma B' N_\gamma s_\gamma d_\gamma i_\gamma \tag{4.69}$$

In the above equation, B' is the lowest of B' and L'. The shape factor calculation should use the reduced dimensions of the foundation.

The total ultimate load capacity the foundation is calculated by multiplying q'_u with effective area ($B' \times L'$). It can be noticed from the above equations that eccentricity of loading decreases the bearing capacity of a foundation.

Grillage bases of lattice towers are usually subjected to moments about either axis of the footing base in addition to the axial load. In Figure 4.34b, the equivalent area concept for a foundation subjected to moment in one direction is shown.

Example 4.14 A transmission tower grillage base is installed in a saturated clayey soil at a depth of 8 ft (2.44 m). The base dimensions of the foundation are B = 7 ft (2.13 m) and L = 10 ft (3.05 m). The unit weight of soil is assumed as 115 lb/ft^3 (18.08 kN/m^3). The undrained shear strength ($\phi = 0$) of soil is 3000 psf (143.64 kN/m^2).

Calculate the ultimate bearing capacity and ultimate compressive load capacity of the foundation using Meyerhof's theory and the following additional data:

The backfill is crushed rock with a unit weight of 120 lb/ft^3 (18.87 kN/m^3).
The forces acting at the geometric centroid of grillage base are given below.
Vertical force including the foundation weight at the grillage base = V = 350 kips (1556.94 kN)
Moment at the grillage base acting along the width = M_B = 50 kip-ft (67.80 kN-m)
Moment at the grillage base along the length = M_L = 60 kip-ft (81.36 kN-m)
Load inclination acting along the main leg member = α = 30 degrees

Solution:

$$e_B = \frac{M_B}{V} = \frac{50}{350} = 0.143\text{ ft } (0.044\text{ m}) < \frac{B}{6} = \frac{7}{6} = 1.167\text{ ft } (0.36\text{ m})$$

$$e_L = \frac{dM_L}{V} = \frac{60}{350} = 0.171\text{ ft } (0.052\text{ m}) < \frac{L}{6} = \frac{10}{6} = 1.667\text{ ft } (0.51\text{ m})$$

No tension develops on both sides of the grillage foundation base.

$$B' = B - 2e_B = 7 - 2\,(0.143) = 6.714\text{ ft } (2.05\text{ m})$$

$$L' = L - 2e_L = 10 - 2\,(0.171) = 9.658\text{ ft } (2.94\text{ m})$$

$$q'_u = cN_c s_c d_c i_c + qN_q s_q d_q i_q + \frac{1}{2}\gamma B' N_\gamma s_\gamma d_\gamma i_\gamma$$

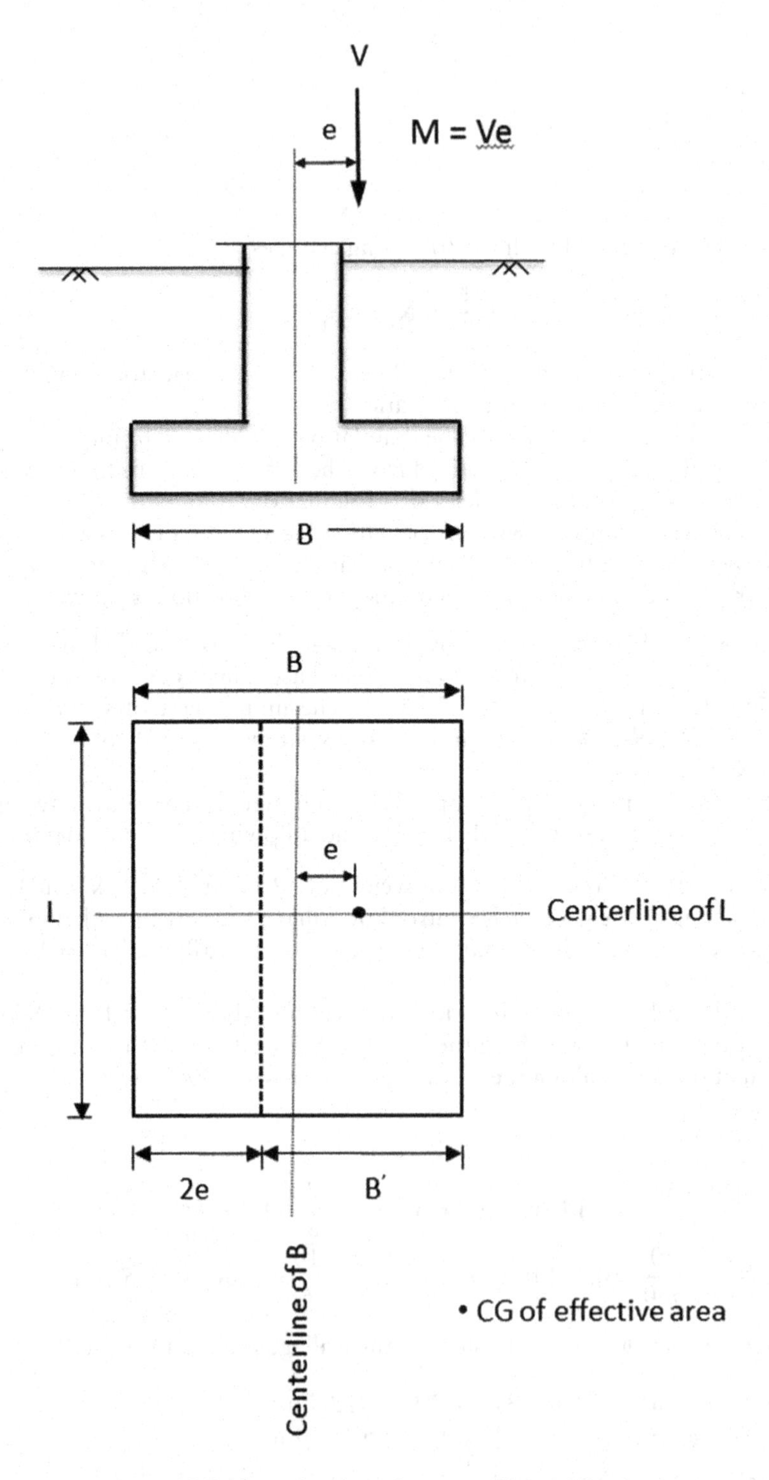

Figure 4.34b Equivalent Area Concept – Uniaxial Bending.

Undrained shear strength of soil, $s_u = 3000$ psf (143.64 kN/m^2)
For ($\phi = 0$), $N_c = 5.14$, $N_q = 1.0$, $N_\gamma = 0$ from Table 4.27.
Substituting bearing capacity factors in Equation 4.48

$$q'_u = cN_cs_cd_ci_c + qs_qd_qi_q$$

$$c = s_u = 3000 \text{ psf } (143.64 \text{ kN/m}^2)$$
$$q = \gamma D = (120)(8) = 960 \text{ lb/ft}^2 \ (45.97 \text{ kN/m}^2)$$

For $\phi = 0$

$$s_c = 1 + 0.2\left(\frac{B'}{L'}\right) = 1 + 0.2\left(\frac{6.714}{9.658}\right) = 1.139$$

$$s_q = 1$$

$$d_c = 1 + 0.2\left(\frac{D}{B'}\right) = 1 + 0.2\left(\frac{8}{6.714}\right) = 1.238$$

$$d_q = 1$$

$$i_c = i_q = \left(1 - \frac{\alpha}{90}\right)^2 = \left(1 - \frac{30}{90}\right)^2 = 0.444$$

$$q'_u = cN_cs_cd_ci_c + qs_qd_qi_q$$

$$q'_u = (3000)\,(5.14)\,(1.139)\,(1.238)\,(0.444) + (960)(1)(1)(0.444)$$
$$= 9654.098 + 426.24$$
$$= 10080.34 \text{ lbs/ft}^2 (482.66 \text{ kN/m}^2)$$

Ultimate compressive load that can be carried by the foundation
$= (q_u)(L' \times B') = (10080.34)(9.658 \times 6.714)$
$= 653647.672$ lb $= 653.65$ kips (2907.7 kN).

4.4.3.3 *Practical aspects*

Soil strength parameters for bearing capacity calculations

As discussed in section 4.4.3.1, selecting the type of drained conditions is very critical in determining the bearing capacity of soils. For cohesive soils such as clays, undrained soil conditions are considered while for cohesionless soils such as sands and gravels, drained soil conditions are used. Since silts are partially drained, undrained soil conditions are used as a conservative approach.

Soils are not always found saturated during field investigations. However, it is generally a good practice to conservatively assume that any dry soil will be saturated at least once during the life of the structure. Saturated soil samples are used for testing in the laboratory for the same reason. Bearing capacity is very sensitive to soil type and the location of the water table. Rather than the actual water table location found during investigations, the highest potential water table location (e.g.: flood plain levels near rivers) should be used in design.

Table 4.28 Presumptive Allowable Bearing Capacities in ksf (kPa).

Soil Descriptive	*Chicago 1966*	*Atlanta, 1950*	*Uniform Building Code 1964*
Clay, very soft	0.5 (23.94)	2.0 (95.76)	1.5 (71.82)
Clay, soft	1.5 (71.82)	2.0 (95.76)	1.5 (71.82)
Clay, ordinary	2.5 (119.70)	4.0 (191.52)	
Clay, medium stiff	3.5 (167.58)		2.5 (119.70)
Clay, stiff	4.5 (215.46)	4.0 (191.52)	
Clay, hard	6.0 (287.28)		8.0 (383.04)
Sand, compact and clean	5.0 (239.40)		
Sand, compact and silty	3.0 (143.64)		
Inorganic silt, compact	2.5 (119.70)		1.5 (71.82)
Sand, loose and fine			1.5 (71.82)
Sand, loose and coarse, or sand-gravel mixture, or compact and fine			2.5 (119.70)
Gravel, loose, and compact coarse sand		8.0 (383.04)	8.0 (383.04)
Sand-gravel, compact		12.0 (574.56)	8.0 (383.04)
Hardpan, cemented sand, cemented gravel	12.0 (574.56)	20.0 (957.61)	
Soft rock Sedimentary layered rock (hard shale, sandstone, siltstone)		30.0 (1436.41)	
Bedrock	200.0 (9576.05)	200.0 (9576.05)	

(Source: RUS/USDA.)

Presumptive bearing capacity values

Generally, soil data is needed for calculating bearing capacity values. However, there may be situations in which it is necessary to perform a transmission structure foundation design when soil data is either not known or is not available. RUS Bulletin 200 provides guidance for checking the bearing resistance of guyed pole bases in the absence of detailed soil information.

A summary of presumptive allowable bearing capacity values is provided in Table 4.28. These values are obtained from local building codes of Chicago, Atlanta and the Uniform Building Code. The allowable bearing capacity values recommended by these codes are very conservative since they consider the effects of differential settlement in buildings. It must be noted that the impact of settlement is not as critical as in a building when a pole settles in soil. Additionally, the Bulletin provides guidance on approximate *ranges* of ultimate bearing capacity values for different types of soils as shown in Table 4.29. These ultimate values are three times the allowable values.

4.4.4 Spread foundations under uplift

Spread foundations are constructed by excavating a trench and then backfilling the hole with excavated soil or barrowed backfill.

Table 4.29 Suggested Ranges of Presumptive Ultimate Bearing Capacities in psf (kPa).

Specific Description (Dry)	*Ultimate Bearing Capacity*
Clay, soft	2000–6000 (95.76–287.28)
Clay, ordinary	6000–9000 (287.28–430.92)
Clay, stiff	12000 (574.56)
Clay, hard	15000 (718.20)
Sand, loose	4500 (215.46)
Sand, compact and silty	9000 (430.92)
Sand, compact and clean	15000 (718.20)
Hardpan	40000 (1915.21)
General Description (Dry)	
Poor Soil	1500–4000 (71.82–191.52)
Average Soil	5000–9000 (239.40–430.92)
Good soil	12000–18000 (574.56–861.84)

(Source: RUS/USDA.)

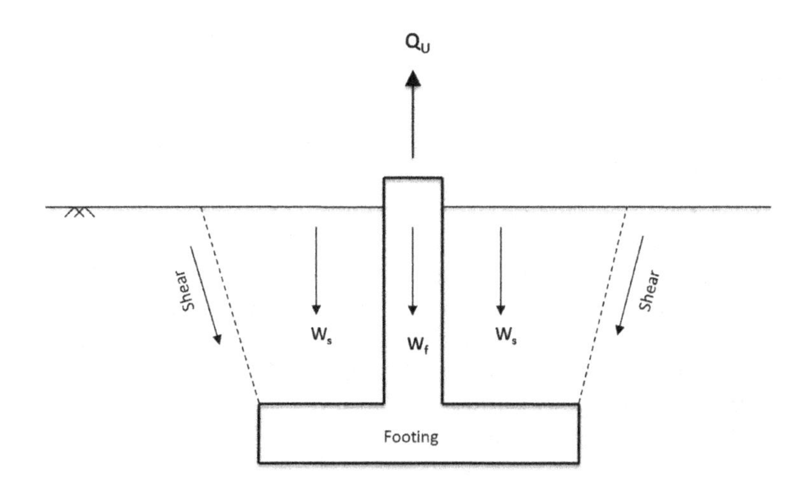

Figure 4.35 Shallow Foundation Under Uplift.

As shown in the Figure 4.35, when a shallow foundation is subjected to uplift load, the failure surface extends to the ground. The components contributing to uplift resistance are the weight of the foundation including soil weight (where applicable) and the shearing resistance of soil along the failure surface.

Various theories are available for calculating the uplift capacity of shallow spread foundations. The main difference among these theories is the assumed shape of the failure surface and how different components contributing to the capacity are accounted.

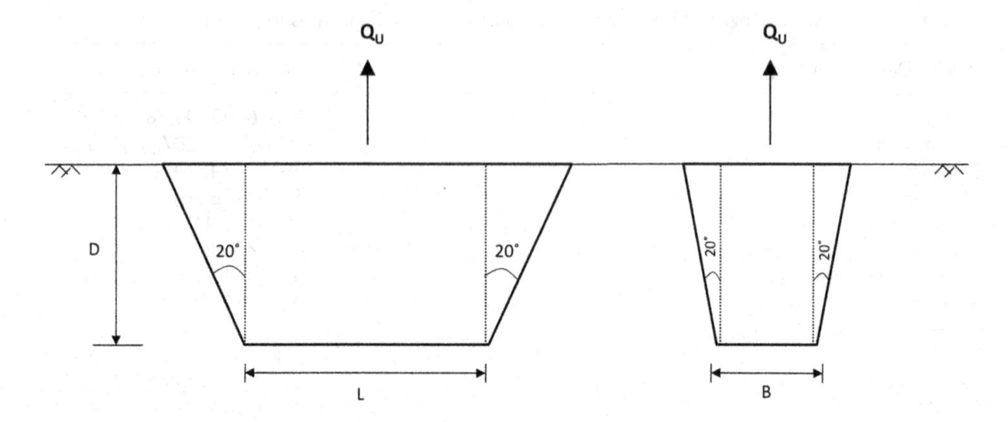

Figure 4.36 20 deg Cone Method.

4.4.4.1 Theoretical models

The following theories are discussed here:

- The 20 degree Cone Method
- Meyerhof and Adams Theory
- Shearing or Friction Method

a. Twenty Degree Cone Method

This simple method assumes a 20 degree failure plane. As shown in Figure 4.36, the uplift capacity of the foundation is based on the weight of the soil in the assumed 20 degree failure zone. Shear forces along the failure plane are ignored when calculating the uplift capacity. Soil strength parameters are, consequently, not required for computing uplift capacities.

The uplift capacity can be computed as below

$$Q_u = W_s + W_f \tag{4.70}$$

where:
W_s = weight of soil in the 20 degree cone of earth
W_f = weight of foundation

Design Standard No. 10 (1965) of the US Department of the Interior Bureau of Reclamation provides equations for calculating uplift capacities for different cone angles. A minimum factor of safety of 1.5 is recommended against the ultimate uplift capacities calculated from equations above.

If a rectangular foundation with plan dimensions $L \times B$ (B = width, L = length) is embedded to a depth D, the soil weight, W_s is given by the following equation for the 20 degree cone angle.

$$W_s = 13.88D^3 + 100BLD + 36.4D^2(B + L) \tag{4.71}$$

The equation above assumes that the unit weight of soil is 100 lbs/ft^3 (15.72 kN/m^3). The dimensions of foundation B, L and D are in feet. If the unit weight of soil is different, the value of W_s calculated using the Equation 4.71 must be scaled by a ratio of actual unit weight/100.

For a square foundation, the width and length are equal. By substituting B for L, the soil weight, W_s reduces to the following equation.

$$W_s = 13.88D^3 + 100B^2D + 72.8BD^2 \tag{4.72}$$

Other design practices assume cone angles between zero and thirty degrees. A cone angle of zero provides a very conservative design. The weight of soil in any cone angle greater than zero indirectly accounts for the soil strength contribution.

Example 4.15 A transmission tower grillage foundation is installed in a sandy soil at a depth of 5 ft (1.52 m). The base dimensions of the grillage are 5 ft × 5 ft (1.52 m × 1.52 m). The unit weight of soil is 110 lb/ft^3 (17.29 kN/m^3). The weight of the grillage foundation is 3500 lb (15.57 kN). Calculate the ultimate uplift capacity of grillage foundation using the 20 degree cone method. The water table is at 30 ft (9.14 m) below the ground level and its effect can be neglected.

Solution:

$$\begin{aligned} W_s &= 13.88D^3 + 100BLD + 36.4D^2(B+L) \\ &= 13.88(5)^3 + 100(5)(5)(5) + 36.4(5)^2(5+5) \\ &= 1735 + 12500 + 9100 = 23335 \text{ lb } (103.8 \text{ kN}) \end{aligned}$$

Correction for unit weight = (110/100)(23335) = 25669 lb = 25.67 kips (114.19 kN)

$$\begin{aligned} Q_u &= W_s + W_f \\ &= 25.67 + 3.50 = 29.17 \text{ kips } (129.76 \text{ kN}) \end{aligned}$$

b. Meyerhof and Adams Theory

Meyerhof and Adams (1968) theory for uplift capacity considers the following:

- Curved failure surface
- Circular, rectangular and square footings
- Shallow and deep footings

Circular footings

Shallow Footings ($D \leq H$)

$$Q_u = \pi BDc + s_f\left(\frac{\pi}{2}\right)B\gamma D^2K_u\tan\phi + W_f + W_s \tag{4.73}$$

where:
B = diameter of foundation
D = depth of foundation

H = limiting depth which separates shallow foundations to deep foundations (see Table 4.30)
S_f = Shape factor
K_u = nominal uplift coefficient of earth pressure on the vertical rupture surface
W_f = weight of foundation
W_s = weight of soil contained in a cylinder of length D

Deep Footings ($D > H$)

$$Q_u = \pi BHc + s_f\left(\frac{\pi}{2}\right) B\gamma(2D - H)HK_u \tan\phi + W_f + W_s \quad (4.74)$$

W_s = weight of soil contained in a cylinder of length H

In the above equations, the value of K_u is around 0.95 for soil friction angles between 30 to 48 degrees.

The shape factor is approximately given by the following equation.

$$S_f = 1 + m\frac{D}{B} \quad (4.75)$$

For deep footings, the value of S_f is limited by the maximum value given below

$$S_{f\text{Max}} = 1 + m\frac{H}{B} \quad (4.76)$$

The values of liming H/B, m and $S_{f\text{Max}}$ for different friction angles are summarized in Table 4.30 below.

Rectangular footings

For rectangular footings of width B and length L, the uplift capacity is given by the following equations.

Shallow Footings ($D \leq H$),

$$Q_u = 2(B + L)Dc + \gamma D^2(2S_fB + L - B)K_u \tan\phi + W_f + W_s \quad (4.77)$$

Deep Footings ($D > H$),

$$Q_u = 2(B + L)Hc + \gamma H(2D - H)(2S_fB + L - B)K_u \tan\phi + W_f + W_s \quad (4.78)$$

Table 4.30 Design Parameters for Meyerhof-Adams Theory.

ϕ *in degrees*	20	25	30	35	40	45	48
Limiting H/B	2.5	3.0	4.0	5.0	7.0	9.0	11.0
m	0.05	0.10	0.15	0.25	0.35	0.50	0.60
$S_{f\text{Max}}$	1.12	1.30	1.60	2.25	3.45	5.50	7.60

Source: G. G. Meyerhof, J. I. Adams; The Ultimate Uplift Capacity of Foundations; Canadian Geotechnical Journal, 1968, 5(4): 225–244.

Square footings

For square footings with base dimensions of $B \times B$, use B in place of L in the above equations.

For cohesionless soils, the value of $c = 0$ in the equations above. Based on tests, Meyerhof and Adams (1968) suggested using the equations above for long-term capacity of shallow foundations in clays. Drained soil parameters must be determined from laboratory testing and appropriate soil strength parameters c and ϕ should be used in calculations.

There is an upper limit to the capacities calculated above. The uplift capacity should not be more than the bearing capacity of soil above the foundation, computed with the contribution from N_γ neglected.

Example 4.16 A concrete circular footing has a diameter of 5 ft (1.52 m). The base of the footing is located at a depth of 6 ft (1.83 m) in a cohesionless soil with a $\phi = 30$ degrees. The unit weight of soil is 110 pcf (17.29 kN/m^3). Calculate the ultimate uplift capacity using Meyerhof and Adam's theory. The weight of the footing base is 6 kips (26.69 kN).

Solution:

$D = 6$ ft (1.83 m)
$B = 5$ ft (1.52 m)
$W_f = 6$ kips (26.69 kN)
$D/B = 6/5 = 1.2$

This implies a shallow footing because the maximum limit for D/B (or H/B) is 4 for $\phi = 30$ degrees from Table 4.30. The first term of Equation 4.73 is neglected because $c = 0$. The value of $m = 0.15$ for $\phi = 30$ degrees and $K_u = 0.95$.

$$S_f = 1 + m\frac{D}{B}$$

$$= 1 + 0.15(1.2) = 1.18$$

$W_s = \frac{\pi}{4}(5)^2(6)(110) = 12959$ lb (57.65 kN)

Substituting the values,

$$Q_u = s_f\left(\frac{\pi}{2}\right)B\gamma D^2 K_u \tan\phi + W_f + W_s$$

$$= 1.18(\pi/2)(5)(110)(6^2)(0.95)\tan 30° + 6000 + 12959$$

$$= 20129 + 6000 + 12959 = 39088 \text{ lb} \quad (173.88 \text{ kN})$$

Table 4.31 Variation of N_u with *D/B*.

D/B	1	2	3	4
N_u	2	4	6	8

Source: G. G. Meyerhof, J. I. Adams; The Ultimate Uplift Capacity of Foundations; *Canadian Geotechnical Journal*, 1968, 5(4): 225–244.

Empirical equations for short-term loading

The equations listed below can be used to calculate short-term capacity of shallow footings in saturated cohesive soils under undrained conditions ($\phi = 0$) with an undrained shear strength of s_u. Here, A_f is the area of the footing in plan view.

$$Q_u = A_f(s_u N_u) + W_f + W_s \tag{4.79}$$

The value of N_u is given in Table 4.31 and its maximum value is limited to 9.

In case of saturated clayey soils, the soil above the foundation will be compressed under the applied load, but at the same time, the soil below the foundation will be relieved of some stress. Because of this pore water pressure below the foundation decreases and at the same time, pore water pressure above the foundation will increase. This differential pore pressure will create a suction force and it increases the uplift capacity of foundation. However, this force is usually neglected in the design on a conservative side.

c. Shearing or Friction Method

As shown in Figure 4.37, the failure surface is assumed to extend vertically upward to the ground surface for shallow foundations. The ultimate uplift capacity is equal to sum of the shearing or friction along the failure surface (T_s), the weight of soil inside the failure surface (W_s) and the weight of foundation (W_f).

For a footing with a depth of D, the shear resistance along the failure surface (T_s) can be calculated as shown below:

For cohesionless soils

$$T_s = (\text{area of failure surface})\,\frac{1}{2}\gamma D K \tan\phi \tag{4.80}$$

Circular footings ($B = \text{diameter}$)
Substituting surface area of cylinder $= (\pi B)D$ in Equation 4.80:

$$T_s = (\pi BD)\frac{1}{2}\gamma DK \tan\phi = \left(\frac{\pi}{2}\right)\gamma BD^2 K \tan\phi \tag{4.81}$$

Rectangular footings ($B = \text{Width}$ and $L = \text{Length}$)
Substituting surface area of failure surface $= 2\,(B + L)D$:

$$T_s = 2(B + L)D\frac{1}{2}\gamma DK \tan\phi = (B + L)\gamma D^2 K \tan\phi \tag{4.82}$$

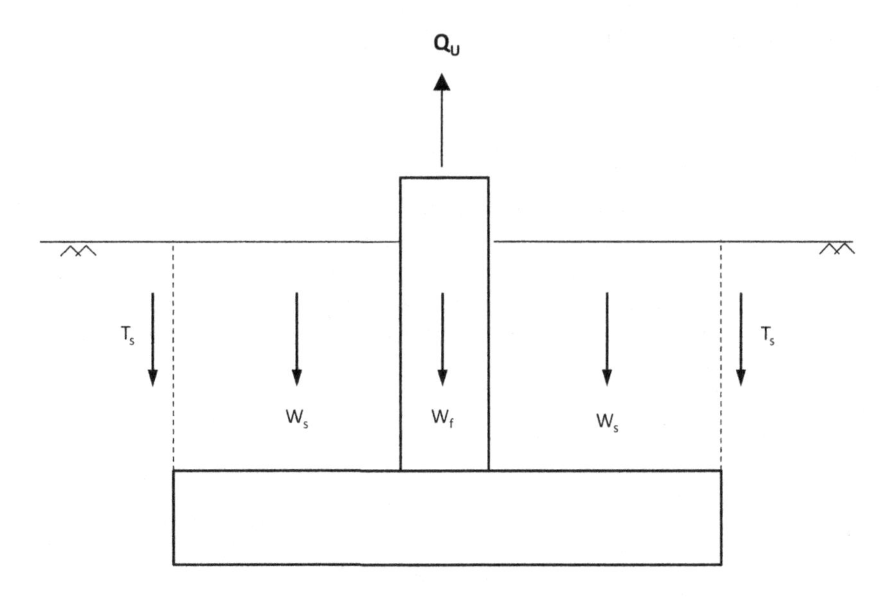

Figure 4.37 Shearing or Friction Method.

Square footings ($B = \text{width}$)
Substituting surface area of failure surface $= 4BD$:

$$T_s = 4BD\frac{1}{2}\gamma DK \tan\phi = 2B\gamma D^2 K \tan\phi \tag{4.83}$$

The value of K is typically considered to be equal to coefficient of lateral earth pressure at rest. In the absence of test data, it can be conservatively estimated as $K = K_0 = 1 - \sin\phi$.

Ireland (1963) suggested a value of $K = 0.5$ for a cohesionless soil with a friction angle of 30 degrees.

For saturated cohesive soils under undrained conditions ($\phi = 0$) with an undrained shear strength of s_u, the equations are as shown below.

$$T_s = (\text{area of failure surface})s_u \tag{4.84}$$

Circular footings

Substituting surface area of cylinder $= (\pi B)$ D in Equation 4.84:

$$T_s = (\pi BD)s_u \tag{4.85}$$

Rectangular footings

Substituting surface area of failure surface $= 2\ (B + L)D$:

$$T_s = 2(B + L)Ds_u \tag{4.86}$$

Square footings
Substituting surface area of failure surface $= 4BD$:

$$T_s = 4BDs_u \tag{4.87}$$

4.4.4.2 *Practical considerations*

Concrete spread footings and grillage foundations are constructed by excavating the native soil, constructing/placing the foundation and then backfilling the excavation with compacted native soil or borrow backfill. The interaction between the backfill and surrounding in-situ native soil is very complex and the uplift capacity depends on backfill compaction, and the size and shape of the excavation. The failure surface is more complex; failure surfaces can be cone or wedge or combination of cylinder/rectangle and cone/wedges. Extensive testing and follow up analyses have led to very sophisticated uplift capacity models (Kulhawy, 1985; Stas & Kulhawy, 1984; Kulhawy et al., 1983) for shallow foundations.

1. The theories described above provide a good estimate of uplift capacity only when the backfill is in a medium to dense condition.
2. In case of excavations such as those shown in Figure 4.38a, the soil properties of both native soil and backfill must be considered in the analysis and the lesser of the two should be used for design under drained loading.
3. In case of excavations such as those shown in Figure 4.38b, the properties of backfill soil must be considered for design under drained loading.
4. The evaluation of K in the theories above is more involved. Some in-situ techniques such as pressuremeters can be used for accurate predictions. For a long transmission line, a few full scale pilot tests can be performed to calibrate the theoretical models for more reliable/economical foundations.

4.4.4.3 *Foundation movements*

Movement of foundations under applied loads can impact tension in the conductors and ground wires. In practice, the structure-foundation interaction is typically ignored in line design. Complete fixity is assumed at the base. The movement of foundations in guyed structures may slacken guy wires and impact the stability of the structure. In case of lattice towers, the foundations should be idealized with appropriate spring stiffness reflecting the load-displacement behavior and the analysis should include the effect of soil-structure interaction. Excessive differential movements may create additional loads within the structure which can impact structural stability. Based on limited analysis of four representative structures, the EPRI Report EL-2870 (Kulhawy et al., 1983) indicates that differential foundation movements even as small as 1 in (25 mm) may prove critical.

The current practice in the industry considers ultimate capacity in design, but, neglects the effect of foundation movement. Trautmann and Kulhawy (1988) proposed the following generalized load-displacement relationship for estimating the vertical displacement of a foundation under uplift load.

$$\frac{Q}{Q_u} = \frac{\frac{z}{D}}{0.013 + 0.67\left(\frac{z}{D}\right)} \tag{4.88}$$

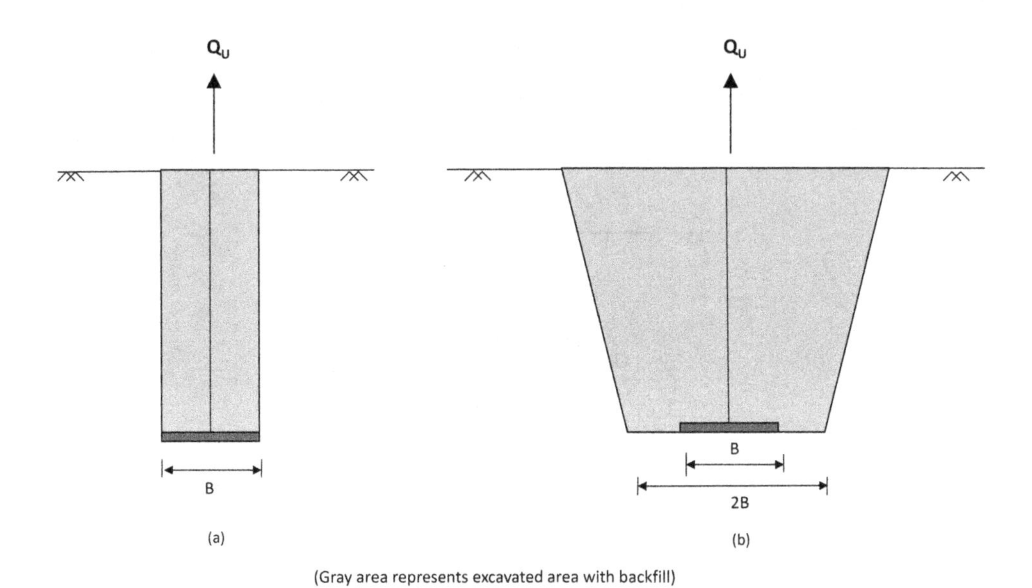

Figure 4.38 Practical Considerations in Uplift Capacity Calculations.

where:
Q_u = ultimate uplift capacity
Q = applied uplift load
z = foundation displacement
D = foundation depth

The above relationship reflects full-scale pullout tests on 75 individual foundations. The foundations tested included grillage and spread footings in granular and cohesive soils. The equation predicts the displacement with 95% confidence level. Prasad and Haldar (2001) also suggested following equation for grillage foundations backfilled with well-compacted granular materials.

$$\frac{Q}{Q_u} = \frac{\frac{z}{D}}{0.0056 + 0.8279\left(\frac{z}{D}\right)} \tag{4.89}$$

4.4.5 Drilled shafts under uplift

Drilled shaft foundations used for lattice towers and H-Frame structures with X-braces are typically controlled by uplift load. In this section, the traditional cylindrical shear model in cohesive and cohesionless soils is discussed. The failure surface is assumed to be along the shaft-soil interface.

As shown in Figure 4.39, the uplift capacity of drilled shaft is given by the following equation:

$$Q_u = Q_s + W_f \tag{4.90}$$

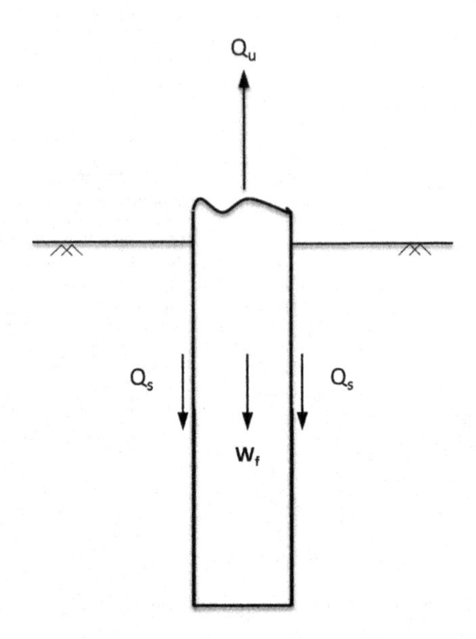

Figure 4.39 Concrete Drilled Shaft Foundation Under Uplift.

where:
Q_s = shaft side resistance
W_f = weight of concrete drilled shaft

Uplift Capacity in Cohesionless Soils

The Q_s over an elemental length ΔD of shaft with a diameter B is given by:

$$Q_s = \sum (\pi B)\Delta D f \tag{4.91}$$

where f = unit skin resistance

Unit skin resistance in cohesionless soils is given by:

$$f = K\sigma' \tan\delta \tag{4.92}$$

where:
K = coefficient of horizontal soil stress
σ' = effective vertical stress at the depth under consideration
δ = soil-foundation material friction angle

Both IEEE 691 (2001) and EPRI EL-2870 (Kulhawy et al., 1983) provide guidance for K and δ values. For drilled shafts, the value of K equals 0.666 to 1.0 times the at-rest coefficient of horizontal soil stress (K_0). A value of 0.666 is generally used for slurry construction and 1.0 is used for dry construction. The value of $K_0 = 1 - \sin\phi$ is often commonly used; however, it is too conservative for soil layers located near the surface. For drilled shaft construction, the value of δ equals ϕ for dry construction. For smooth concrete (slurry) construction, the value of δ ranges from 0.8ϕ to ϕ. For sand/steel surfaces, δ varies from 0.7ϕ to 0.9ϕ (rough steel) and 0.5ϕ to 0.7ϕ (smooth steel). For sand/timber construction, the value of δ is between 0.8ϕ to 0.9ϕ.

Example 4.17 Determine the ultimate uplift capacity of a drilled concrete shaft installed in a two-layered granular soil formation for the data given below. Assume unit weight of concrete = 150 lb/ft^3 (23.58 kN/m^3) and slurry construction.

Drilled shaft diameter = B = 5 ft (1.52 m)
Drilled shaft depth = D = 20 ft (6.10 m)

Layer 1: Loose sand – 0 ft to 6 ft (0 m to 1.83 m)
Unit weight of soil = 100 lb/ft^3 (15.72 kN/m^3)

Friction angle of soil = $\phi = 28°$
Layer 2: Medium dense sand – 6 ft to 20 ft (1.83 m to 6.10 m)
Unit weight of soil = 120 lb/ft^3 (18.87 kN/m^3)
Friction angle of soil = 35°
Water table is located at the interface of Layer 1 and Layer 2 (i.e.) at a depth of 6 ft (1.83 m) from ground level.

Solution:

Layer 1

$$f = K\sigma' \tan \delta$$

For slurry construction,

$$K = \left(\frac{2}{3}\right)K_0 = \left(\frac{2}{3}\right)(1 - \sin\phi) = \left(\frac{2}{3}\right)(1 - \sin 28°) = 0.354$$

Effective vertical stress at ground surface = 0

Effective vertical stress at the bottom of Layer 1 = (100)(6) = 600 lb/ft^2 (28.73 kN/m^2)

$$\delta = 0.8\phi = 0.8(28°) = 22.4°$$

The unit skin friction (f) is given by,

$$f = K\sigma' \tan \delta$$

$$= (0.354)\left(\frac{0 + 600}{2}\right)(\tan 22.4°)$$

$$= 43.77\,\text{lb/ft}^2\ (2.10\,\text{kN/m}^2)$$

Q_{s1} = shaft side resistance for Layer 1 (0 – 6 ft, $\Delta D = 6$ ft)

$$Q_{s1} = (\pi B)\Delta D f$$

$$= (\pi)(5)(6)(43.77)$$

$$= 4125.22\,\text{lb}\ (18.35\,\text{kN/m}^2)$$

Layer 2

$$f = K\sigma' \tan\delta$$

For slurry construction,

$$K = \left(\frac{2}{3}\right)K_0 = \left(\frac{2}{3}\right)(1 - \sin\phi) = \left(\frac{2}{3}\right)(1 - \sin 35°) = 0.284$$

Effective vertical stress at the top of layer 2 = (100)(6) = 600 lb/ft^2 (28.73 kN/m^2)

Effective vertical stress at the bottom of layer 2 = (100)(6) + (120 − 62.4)(20 − 6) = 1406.4 lb/ft^2 (67.34 kN/m^2)

$$\delta = 0.8\phi = 0.8(35°) = 28°$$

The unit skin friction (f) is given by,

$$\begin{aligned} f &= K\sigma' \tan\delta \\ &= (0.284)\left(\frac{600 + 1406.4}{2}\right)(\tan 28°) \\ &= 151.49\ \text{lb/ft}^2\ (7.25\ \text{kN/m}^2) \end{aligned}$$

Q_{s2} = shaft side resistance for layer 2 (6 − 20 ft, $\Delta D = 14$ ft)

$$\begin{aligned} Q_{s2} &= (\pi B)\Delta D f \\ &= (\pi)(5)(14)(151.49) \\ &= 33314.39\ \text{lb}\ (148.20\ \text{kN}) \end{aligned}$$

$$\begin{aligned} Q_s &= \sum (\pi B)\Delta D f \\ &= Q_{s1} + Q_{s2} \\ &= 4125.22 + 33314.39 \\ &= 37439.61\ \text{lb} = 37.44\ \text{kips}\ (166.55\ \text{kN}) \end{aligned}$$

$W_f = \left(\frac{\pi}{4}\right)(5)^2(6)(150) + \left(\frac{\pi}{4}\right)(5)^2(14)(150 - 62.4) = 41751.8$ lb = 41.75 kips (185.73 kN)

The ultimate uplift capacity, Q_u is

$$\begin{aligned} Q_u &= Q_s + W_f \\ &= 37.44 + 41.75 = 79.19\ \text{kips}\ (352.27\ \text{kN}) \end{aligned}$$

Uplift capacity in cohesive soils

The method described here is called the Alpha (α) Method. The α is known as the empirical adhesion factor and is typically determined from the lab and field tests. It takes into account the shear resistance between the foundation material and surrounding soil. The unit friction resistance, f is given by the following equation:

$$f = \alpha s_u \tag{4.93}$$

where s_u is the undrained shear strength of the soil.

Das and Seeley (1982) recommended the following equations for α:
For cast-in-situ concrete piles which include drilled shaft foundations, α is calculated as:

For $s_u \leq 80\,\text{kN/m}^2$ (1.67 ksf)

$$\alpha = 0.9 - 0.00625 s_u \tag{4.94a}$$

For $s_u > 80\,\text{kN/m}^2$ (1.67 ksf)

$$\alpha = 0.4 \tag{4.94b}$$

For pipe piles, the equations for α are given below and are used in the uplift calculations in Section 4.4.5.1.

For $s_u \leq 27\,\text{kN/m}^2$ (0.56 ksf)

$$\alpha = 0.715 - 0.0191 s_u \tag{4.95a}$$

For $s_u > 27\,\text{kN/m}^2$ (0.56 ksf)

$$\alpha = 0.2 \tag{4.95b}$$

In the above equations for determining the value of α, s_u should be in kPa or kN/m^2 ($1\,\text{kN/m}^2 = 20.885\,\text{lb/ft}^2$).

Example 4.18 Determine the ultimate uplift capacity of a drilled concrete shaft installed in a uniform stiff clay soil for the data given below. Assume unit weight of concrete = $150\,\text{lb/ft}^3$ ($23.58\,\text{kN/m}^3$). The water table depth is 25 ft (7.62 m) below the ground surface.

Drilled shaft diameter = B = 5 ft (1.52 m)
Drilled shaft depth = D = 20 ft (6.10 m)
Undrained shear strength = s_u = 1.5 ksf ($71.82\,\text{kN/m}^2$)

Solution:

Using the Equation 4.94a, the α is given by,

$$\begin{aligned}\alpha &= 0.9 - 0.00625s \\ &= 0.9 - 0.00625(71.82) = 0.45\end{aligned}$$

$$f = \alpha s_u$$

$$= (0.45)(1.5) = 0.675\ \text{ksf}\ (32.32\ \text{kN/m}^2)$$

$$Q_s = \sum (\pi B)\Delta D f$$

$$= (\pi)(5)(20)(0.675) = 212.06\ \text{kips}\ (943.33\ \text{kN})$$

$W_f = \left(\frac{\pi}{4}\right)(5)^2(20)(150) = 58904.9\ \text{lb} = 58.91\ \text{kips}\ (262.06\ \text{kN})$
The ultimate uplift capacity, Q_u is

$$Q_u = Q_s + W_f$$

$$= 212.06 + 58.91 = 270.97\ \text{kips}\ (1205.38\ \text{kN})$$

4.4.5.1 *Application to direct embedment poles*

Direct embedment foundations used for cross-braced H-Frame structures experience significant uplift loads. The traditional cylindrical shear model for straight sided drilled shafts discussed above can be extended to direct embedment pole foundations. However, the analysis is complicated due to the presence of two soil types: backfill and in-situ soil. The failure surface can occur at either the interface of pole/backfill interface or the backfill/in-situ soil interface.

The following steps shall be considered for conservative estimate of uplift capacity:

1. Using the pole diameter and *backfill* properties, calculate the ultimate uplift capacity according to the procedure described in Section 4.4.5.
2. Using the pole diameter and *in-situ soil* properties, calculate the ultimate uplift capacity according to the procedure described in Section 4.4.5.
3. The *smaller* value from 1 and 2 is the ultimate uplift capacity of direct embedment pole foundation.

Use the values of α recommended in Equations 4.95 for pipe piles in cohesive soils.

Skin friction values for preliminary designs

The RUS Bulletin 200 provides some guidance for ultimate skin friction values for various soils. These values are summarized in Table 4-32a. The RUS recommends using pole "bearing" shoes typically for double X-braced H-Frame structures. Per RUS, the increase in uplift capacity is 2.0 to 2.5 times in a dry hole with natural backfill. If native backfill is good, then the use of aggregate backfill and bearing shoes is usually not necessary. Table 4.32b gives skin friction values of some common soils. A Factor of Safety of 2.0 is usually recommended while calculating allowable capacities.

Table 4.32a Ultimate Skin Friction Values.

Conditions	*Ultimate Skin Friction in psf (kPa)*
Soils which are wet or soils which are susceptible to frequent wettings	Not greater than 100 (4.79)
Native soil is used as backfill and soil is not subject to frequent wettings	100 and 500 (4.79–23.94)
Aggregate backfill is used	250 and 1000 (11.97–47.87)

(Source: RUS/USDA.)

Table 4.32b Skin Friction Values of Various Soils.

Soil Type	*Ultimate Skin Friction in psf (kPa)*
Soft Clay and Silt	200–600 (9.58–28.73)
Sandy Silt	400–1000 (19.16–47.88)
Stiff Clay	800–2000 (38.32–95.76)
Very Stiff Clay	1000–4000 (19.16–191.52)
Loose Sand	250–700 (11.97–33.52)
Dense Sand	700–1400 (11.97–67.03)

(Source: Terzaghi and Peck, 1948.)

Table 4.33 Holding Strength of Typical Screw Anchors (Single Helix).

	Holding Strength in kips (kN)	
Anchor size in inch (cm)	*Poor soil*	*Average soil*
8 (20)	6.0 (27)	15 (67)
11-5/16 (29)	9.5 (42)	15 (67)

4.4.6 Anchors under pullout loads

Anchors have been discussed in detail in Section 4.3.6. In this section, models for predicting pullout capacity of two common types of anchors, helical anchors for use in soil and rock anchors for use in rock are briefly discussed.

4.4.6.1 Helical anchors

Helical anchors consist of a square steel shaft fitted with one or more helically deformed plates. One to four plates are used for most common applications. Typical capacities of single helical anchor with plate sizes 8 inches (20 cm) to 12 inches (86 cm) are shown in Table 4.33. The values provided are for illustration purposes only and they need to be verified with the manufacturer.

The design of helical anchors includes both field methods and theoretical models. Some of the field methods include measurement of twist of the shaft, installation torque and pull testing of anchors. Of these, the installation torque method is popular due to its simplicity and relative accuracy. One method proposed by Hoyt and Clemence (1989) is given below:

$$\text{Ultimate capacity in kips} = (K_T) \times \text{Installation Torque in kip-ft} \tag{4.96}$$

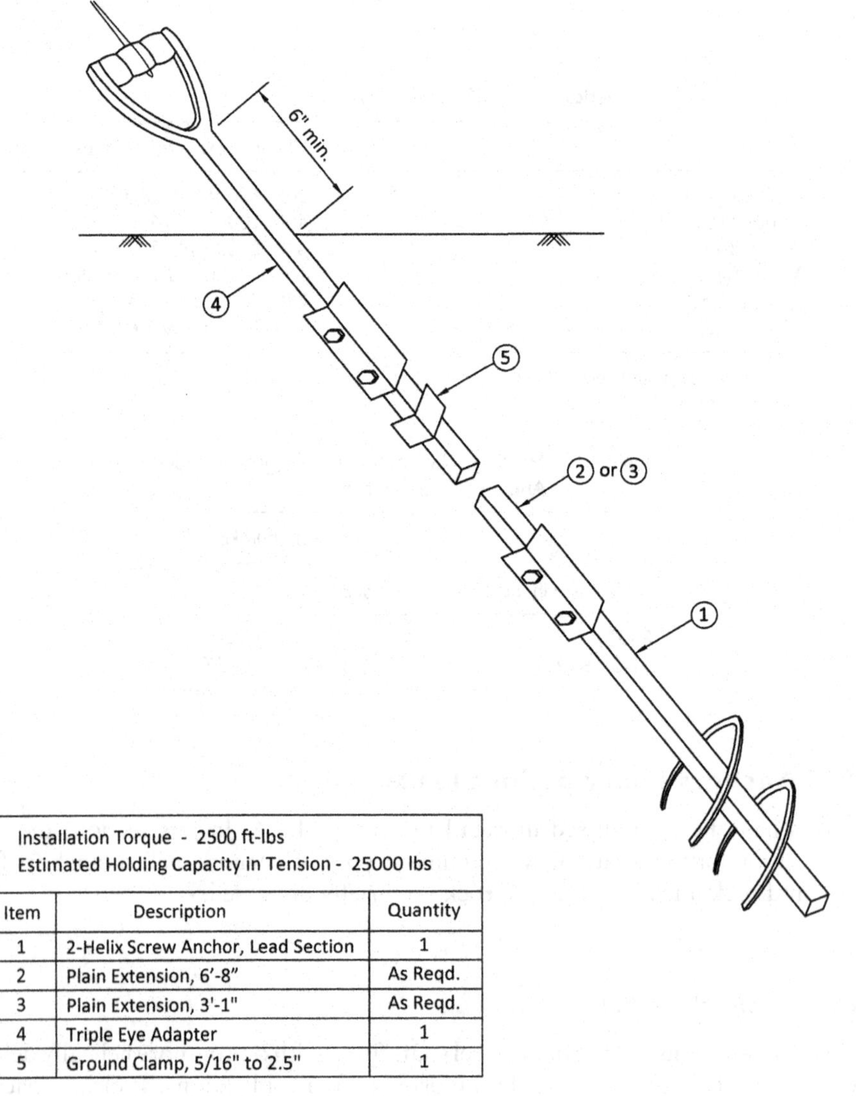

Installation Torque - 2500 ft-lbs Estimated Holding Capacity in Tension - 25000 lbs		
Item	**Description**	**Quantity**
1	2-Helix Screw Anchor, Lead Section	1
2	Plain Extension, 6'-8"	As Reqd.
3	Plain Extension, 3'-1"	As Reqd.
4	Triple Eye Adapter	1
5	Ground Clamp, 5/16" to 2.5"	1

1 ft = 30.5 cm 1 in = 25.4 mm 1 lb = 4.45 N

Figure 4.40 Twin Helix Screw Anchor Assembly.

The recommended value of $K_T = 10\,\text{ft}^{-1}$ for all square (or circular) – shaft anchors with shaft size less than 3.5 inches (89 mm). For example, as shown in Figure-4.40, the holding capacity of a 2-helix anchor which is installed to a torque of 2500 lb-ft (3389.83 N-m) is given by $10\,\text{ft}^{-1}$ (2500 ft-lbs) = 25000 lbs or 25 kips (111.21 kN). In this case, the square shaft size is 1.5 inches (38.1 mm) and the diameter of the top and bottom helical plates are 10 inches (25.4 cm) and 8 inches (20.3 cm), respectively. A power digger or anchor installing machine equipped with dial torque indicator or calibrated hydraulic pressure gauges are used for continuous measurement of installation torque. The installation torque must be maintained for the last few feet of installation, and at a minimum, for the distance between the top and bottom plate for multi-helix anchors. During installation, the anchor must be in line with guy wire. Any misalignment introduces additional bending moments and decrease the pullout capacity of helical anchor. Typically, the strength of an anchor is governed by the geotechnical capacity than the structural capacity.

Several theoretical models are available to estimate the pull out capacity of helical anchors. As shown in the Figure 4.41, uplift capacity of multi-helix, deep helical anchor (minimum depth/diameter ration of top plate is 5.0) in cohesive soils is given by the sum of following components:

- R_1, Resistance of soil along the truncated pyramid between the top and bottom helical plates
- R_2, Bearing resistance on top helical plate
- R_3, Adhesion/frictional resistance along the shaft above the top helical plate

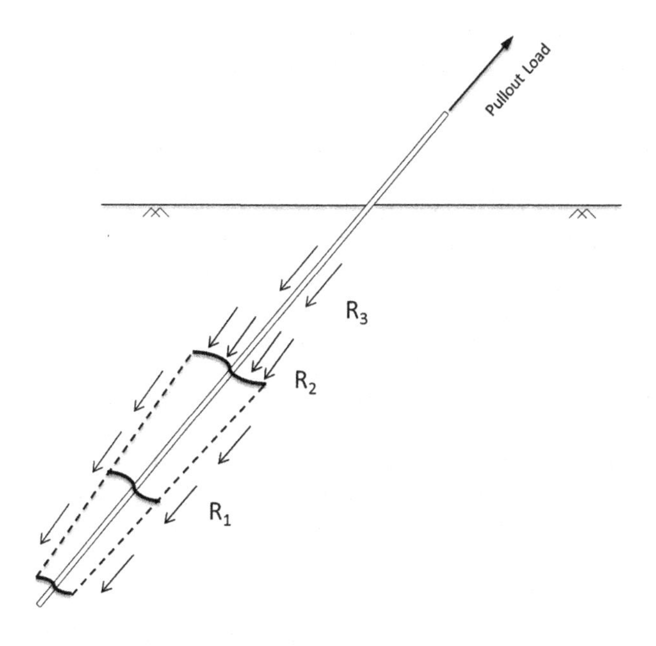

Figure 4.41 Failure Mechanisms for Helical Anchor in Tension.

The resistance of the soil along the truncated pyramid between the top and bottom helical plates is given by the product of surface area around the truncated pyramid and undrained shear strength (c_u) of soil. For cylindrical failure to occur between top and bottom helical plates, the spacing between consecutive helical plates should not be more than of 2 to 3 times the average diameter.

The bearing resistance on top helical plate is given by the area of the top plate times $9c_u$. The adhesion resistance along the shaft above the top helical plate is given by the surface area of the shaft times αc_u. The value of α (adhesion factor) is unity for soft clays and is 0.4 for stiff clays.

Example 4.19 A deep square shaft helical anchor with 3 plates is installed in homogeneous clayey soil with an undrained cohesion of 1.5 ksf (71.82 kN/m^2).

Average diameter of 3 helical plates = 10 inches (25.4 cm)
Distance between top and bottom plate = 40 inches (101.6 cm)
Diameter of top plate = 12 inches (30.48 cm)
Side of square shaft = 1 inch (25.4 mm)
Depth of top plate from GL = 10 ft (3.05 m)
Determine the ultimate pullout capacity of the anchor.

Solution:

Resistance of soil along truncated pyramid between the top and bottom helical plates = R_1

R_1 = (Surface area around the truncated pyramid)(undrained cohesion of soil)
$= \pi\left(\frac{10}{12}\right)\left(\frac{40}{12}\right)(1.5) = 13.1$ kips (58.27 kN)

Bearing resistance on top helical plate = R_2

R_2 = (Area of top plate)(9)(undrained cohesion of soil)
$= \left(\frac{\pi}{4}\right)\left(\frac{12}{12}\right)^2(9)(1.5) = 10.60$ kips (47.15 kN)

Adhesion resistance along the shaft above the top helical plate = R_3

R_3 = (surface area of the shaft)(α)(c_u)
$=(4)\left(\frac{1}{12}\right)(10)(0.7)(1.5) = 3.5$ kips (15.57 kN)

Assumed medium stiff clay with $\alpha = 0.7$ in the above equation.

Ultimate Pullout Capacity = $R_1 + R_2 + R_3 = 13.1 + 10.6 + 3.5 = 27.2$ kips (120.99 kN)

Figures 4.42a and 4.42b show typical ultimate holding capacities of various multi-helix anchors in clay and sand. The charts are based on a soil unit weight of 125 lbs/ft^3 (19.63 kN/m^3) and the top-most helix at 15 ft (4.57 m) below the ground. Ground water table is assumed to be below the tip of the anchor. The charts can only be used for preliminary sizing of anchors; actual subsurface and loading conditions must be verified before establishing the true strength of the anchor.

4.4.6.2 *Grouted rock anchors*

A typical grouted rock anchor is shown in Figure 4.43. In this case, ¾ inch (19.05 mm) anchor rod is used and the diameter of the grouted hole is 2 inches (50.8 mm).

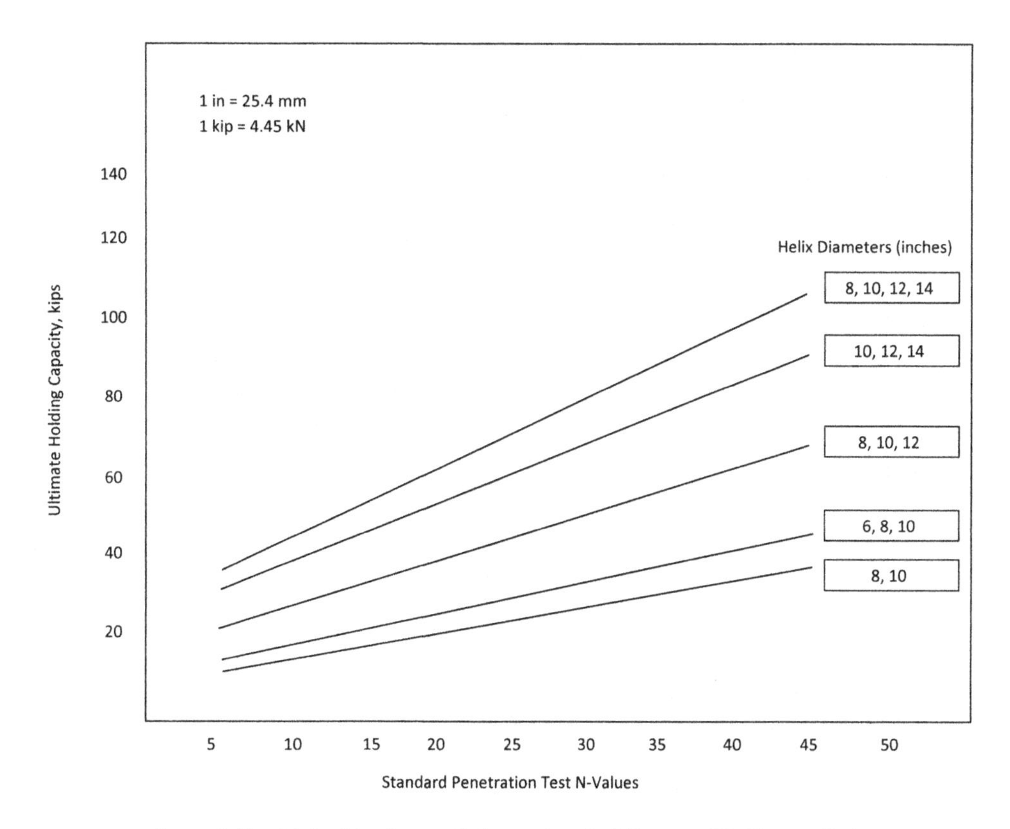

Figure 4.42a Helical Anchor in Cohesionless Soil (Source: AB Chance Co.).

When a grouted rock anchor is subjected to pullout, the following modes of failures can occur:

- Pullout failure of the rock mass
- Failure of the grout-rock bond
- Failure of the grout-steel bond
- Failure of steel tendon and associated hardware connections (steel wire/cable/bar)

In rock mass failure, the failure is assumed to be a cone with a certain apex angle at the anchorage extending to the top surface. The resistance to pullout is calculated as the sum of the weight of rock mass within the cone and the shear resistance along the failure plane. The designer should have a proper understanding of in-situ rock characteristics before applying this method. One such important property is RQD of rock mass.

The in-situ quality of rock is designated by Rock Quality Designation (RQD) and is measured using the recovered core samples. It is defined as the ratio of total length of intact hard and sound pieces of core of length greater than 4 inches (100 mm) to

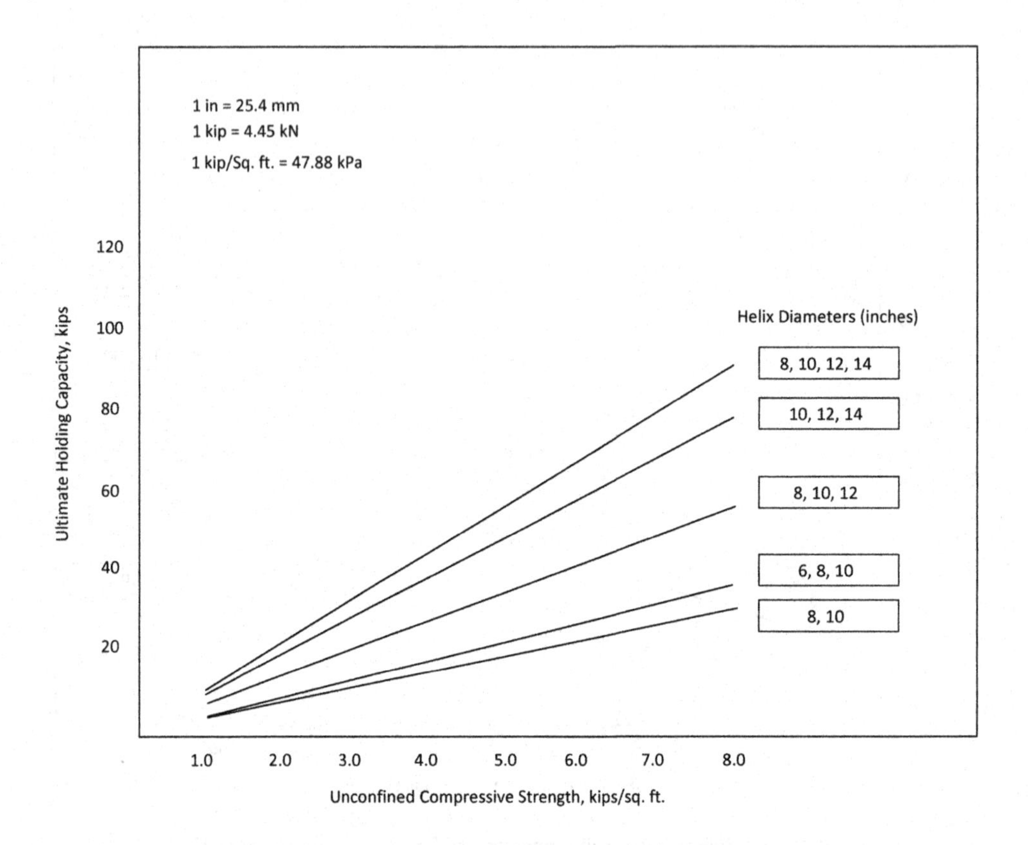

Figure 4.42b Helical Anchor in Cohesive Soil (Source: AB Chance Co.).

total length of drilling. The higher the RQD value, the better is the quality of rock. Typically, if RQD is close to 100%, the rock quality is excellent where as if the RQD is less than 25%, it is very poor rock (Peck et al., 1974).

Another important failure mode is the separation of grout-rock interface. The equation used to quantify the strength of the rock-grout connection is given below:

$$P_u = \pi B L T_s \tag{4.97}$$

where:
P_u = ultimate pullout capacity
B = diameter of the anchor shaft
L = length of anchor shaft
T_s = average ultimate bond stress along the shaft length between the rock and the grout plug

Typical values of T_s are provided in Table 4.34.

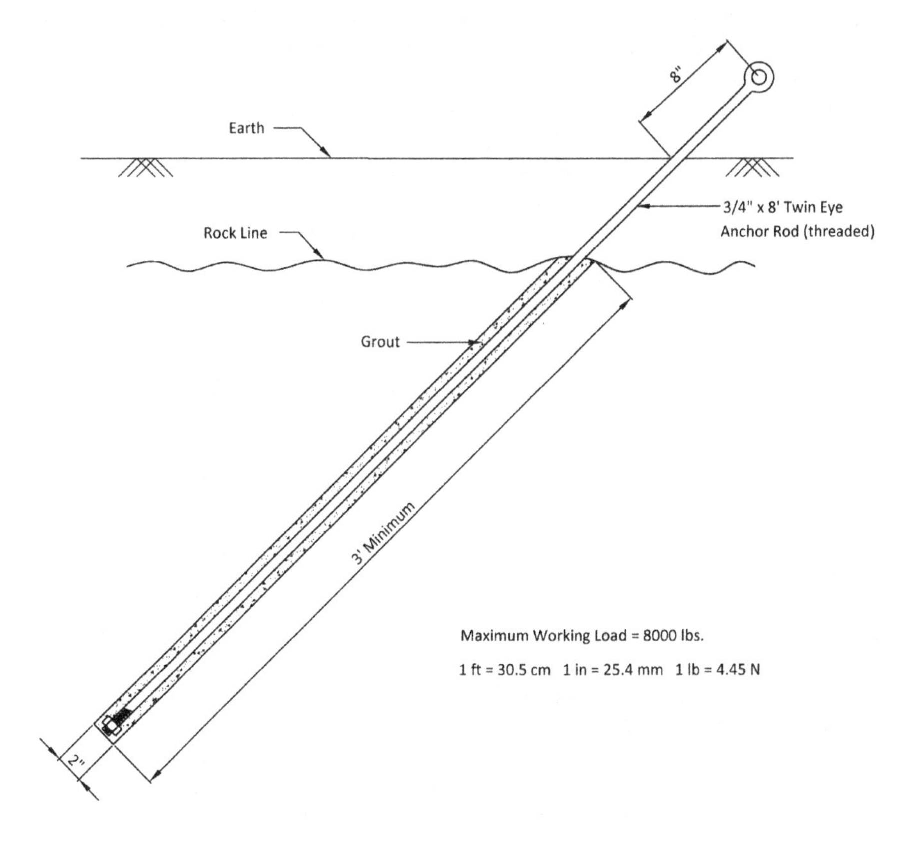

Figure 4.43 Typical Rock Anchor.

Table 4.34 Typical Ultimate Bond Stress between Grout Plug and Rock.

Rock Type	*Ultimate Bond Stress in psi (kPa)*
Granite and Basalt	250–450 (1700–3100)
Dolomite Limestone	200–300 (1400–2100)
Soft Limestone	150–220 (1000–1500)
Slates and Hard Shales	120–200 (800–1400)
Soft Shales	30–120 (210–800)
Sandstone	120–250 (800–1700)
Concrete	200–400 (1400–2800)

(with permission from ASCE).

In addition to the above, it is also necessary to check the grout-steel bond. It is necessary to install a deformed steel bar to at least the minimum development length. The steel tendon yield strength should be checked against the pull out load applied. All hardware used should have sufficient mechanical strength to resist the tensile forces.

Example 4.20 Assuming grout-rock interface failure as the controlling mode, calculate the ultimate pullout capacity of a grouted rock anchor in a sand stone formation with the following data:

Diameter of anchor rod = 2.257 inches (57.33 mm)
Diameter of grouted shaft = B = 5 inches (127 mm)
Average ultimate bond stress along the shaft between rock and grout plug = T_s = 120 psi (827.59 kN/m^2)
Length of the anchor shaft = 6 ft (1.83 m)

Solution:

$P_u = \pi BLT_s = \pi(5)(6)(12)(120) = 135717$ lbs = 135.7 kips (603.65 kN)

4.5 COMPUTER PROGRAMS

4.5.1 CAISSON

This program is a part of the PLS suite of software. CAISSON™ is a Microsoft Windows program for the design of concrete drilled shaft foundations and directly embedded pole foundations subjected to moment and shear loads. CAISSON™ uses ultimate strength design concepts for both geotechnical and structural design.

The program is based on Broms' Theory modified to layered soils. The software does not have the capabilities to estimate deflections and rotations of drilled shaft foundations or direct embedment foundations, but, only checks the ultimate lateral resistance of the soil for moment and shear loads. Its use is limited to unguyed single pole foundations. Despite these limitations, it produces fast designs, easy to use and have user friendly features.

Soil parameters can be input as either sand or clay for each soil layer. The typical output contains the required depth of the pier shaft and the suggested reinforcement, along with moment and shear distribution along the length of the pier and the bearing (axial) stress under the pier. In case of direct embedment foundations, depth of direct embedment pole is calculated.

4.5.2 FAD

FAD™ stands for Foundation Analysis and Design and it is part of EPRI's (Electric Power Research Institute) TLWorkstation™. FAD™ has essentially two modules: MFAD™ (Moment Foundation Analysis and Design) and CUFAD™ (Compression/Uplift Foundation Analysis and Design). MFAD™ is for design of foundations governed by moment loads and CUFAD™ is for the design of foundations subjected compressions and uplift.

As discussed in earlier section 4.4.1.2, MFAD™ software uses the semi-empirical theoretical model for ultimate capacity and non-linear load deflection response of drilled shaft foundations which was originally developed by Davidson (1981). Using this model, a computer program PADLL (Pier Analysis and Design for Lateral Loads) was originally developed. Later, this program was upgraded to MFAD™ which uses Hansen's (1961) theory for predicting ultimate lateral capacity of the foundation.

Design capabilities of MFAD™ include:

- Analysis and design of drilled shafts in soil and rock
- Concrete design for drilled shafts
- Analysis and design of direct embedment foundations in soils with backfills
- Design of both drilled shaft and direct embedment foundations to match or exceed limits of rotation and deflection required at the ground line (including total and non-recoverable components).

In addition to the strength parameters, this program requires the pressuremeter modulus for calculating rotations and deflections.

Recently, the MFAD™ program incorporated the LRFD design by including a strength factor for geotechnical design. In addition, LRFD design concepts are included in the HFAD™ (H-Frame Foundation Analysis and Design) and TFAD™ (Tower Foundation Analysis and Design). HFAD™ is developed for the analysis and design of H-Frame foundations and it can be used for drilled shaft foundations and direct embedment pole foundations. TFAD™ is developed for the analysis and design of Tower foundations. These three latest programs are also commercially available under the name of FAD Tools.

4.5.3 LPILE

This program is a part of the Ensoft, Inc. suite of software. LPILE™ is a powerful special-purpose program for analyzing and designing piles and drilled shafts. LPILE™ uses the non-linear *p-y* method to compute moment capacities. A variety of soil types such as soft clay, stiff clay, sand, silt, liquefied sand, rocks and elastic sub-grades can be defined in the input section. The soil-resistance (*p-y*) curves are generated internally in LPILE™ and adjusted for soil layering effects. The program calculates deflection, bending moment, shear force and soil response over the length of the pile. Although, the program was originally developed for offshore foundation design, it is used by US utility engineers for foundation designs.

4.5.4 SHAFT

SHAFT™ (2012) is a computer program used to evaluate the axial capacity and short-term load-settlement of drilled shafts or bored piles in different types of soil and rock. The models used in the SHAFT™ program are based on the Federal Highway Administration (FHWA) Manual. For models in soil, SHAFT™ produces a single graph showing the upper-bound, lower-bound and the average load versus settlement curves. It also computes side, end-bearing and total resistance for every foot of embedment and predicts load-settlement pattern of the pier top and bottom. There is also a provision to model shafts with a bell at the bottom. The latest version of this program incorporates LRFD concepts.

4.5.5 Heli-Cap

The Heli-Cap™ (2012) software was developed by Chance-Hubbell Power Systems to facilitate analysis and design of helical anchors in cohesive and granular soils. Both vertical and inclined anchors can be analyzed. Input is simple and consists of soil data including the soil type, SPT blow count, density, friction angle and/or cohesion of each layer of soil. The program computes capacities of anchors in both tension and compression along with the installation torque needed.

The HeliCap™ database contains a family of helical anchors ranging from 6 inches to 14 inches (15.2 cm to 35.6 cm) plate diameter with square rods ranging from 1½ inches to 2¼ inches (38 mm to 57 mm). Graphical output consists of a sectional profile of the anchor system showing each soil layer and the anchor.

PROBLEMS

P4.1 Rework Example E4.1 if the depth of drilled shaft is reduced to 16 ft (4.88 m). What is the ultimate moment capacity? What is factor of safety for the moment capacity of foundation?

P4.2 Rework Example E4.2 if the undrained shear strength of soil increased to 2 ksf (95.76 kN/m^2)?

P4.3 Rework Example E4.3 if the angle of internal friction of soil is reduced to 30 degrees.

P4.4 Rework Example E4.4 if the pole length increased to 100 feet (30.48 m)?

P4.5 A transmission pole structure 80 feet (24.38 m) long is to be installed in a homogeneous soil deposit of loose sand with $\gamma = 100$ lb/ft^3 (15.71 kN/m^2) and $\phi = 30°$ using a direct embedment foundation. Determine the depth of embedment per Broms' rigid pile method. Water table is about 40 ft (12.19 m) below the ground level. The thickness of the backfill annulus = 0.75 ft (0.23 m). The backfill is concrete.

The average diameter of the pole below the ground level is 1.5 ft (0.46 m). If the depth of foundation (D_e) is determined to be 10 ft (3.05 m) based on Rule of Thumb, determine design is safe.

The moment at the ground level is 250 kip-ft (339 kN-m) under extreme wind load case. A minimum factor of safety of 1.25 is required against lateral soil failure per a utility's internal standard. The resultant wind load (due to loads on wire and load on pole itself) acts at 10 ft (3.05 m) from the top of the pole.

P4.6 Rework P 4.5 if the average diameter of the pole below the ground is increased to 2.5 ft (0.76 m).

P4.7 Rework E4.7 if the soil type is loose submerged sand.

P4.8 Rework E4.8 if the soil type is medium saturated clay.

P4.9 Rework E4.9 if the average diameter of the pole increased to 3.0 ft (0.91 m).

P4.10 Rework E4.10 if the unit weight of the soil is increased to 125 lb/ft^3 (18.87 kN/m^3) and the friction angle of sandy soil is increased to 40 degrees ($c = 0$).

P4.11 Rework E4.11 if the unit weight of soil increased to 125 lb/ft^3 (18.87 kN/m^3) and the friction angle of sandy soil increased to 40 degrees ($c = 0$).

P4.12 Rework E4.12 if the unit weight of soil is increased to 120 lb/ft^3 (18.87 kN/m^3) and the undrained shear strength of soil is increased to 3000 psf (143.64 kN/m^2).

P4.13 Rework E4.13 if the friction angle of sandy soil is increased to 40 degrees ($c = 0$).

P4.14 Rework E4.14 if the undrained shear strength of soil is decreased to 1500 psf (71.82 kN/m^2).

P4.15 Rework E4.15 if the unit weight of soil is increased to 125 lb/ft^3 (18.87 kN/m^3).

P4.16 Rework E4.18 if the diameter of the drilled shaft is increased to 6 ft (1.83 m).

Chapter 5

Design deliverables

The design activity for overhead transmission lines comprises of several phases, each of which produces a specific output or a by-product. These are the Design Deliverables which constitute the contents of the project's permanent design record and involves contributions of many disciplines.

The following definitions apply to the contents of this Chapter per ASCE 48-11 and refer to the situation where design is performed by an external consultant hired by the Owner.

Owner – Utility owning the transmission line or their designated representative
Engineer – Party responsible for the design of the structure; may be an agent of the Owner or Fabricator
(Note that some utilities also have in-house divisions performing all required designs).

Deliverables are the responsibility of the Engineer and are submitted to the Owner at the start, during and at the end of a project. These deliverables cover every issue critical to the project: assumptions made, standards adopted, design and analysis types, clearances, load and weather cases, materials and hardware used, structural and foundation drawings etc. They also cover electrical issues as well, namely conductor selection, ampacity needs, insulation design, EMF studies etc. The format of deliverables can vary; Engineering consulting firms in the United States usually provide both print and electronic forms of design records to the Owner. During construction process, on-site construction contractors generally request full-size paper drawings which are easily readable.

This chapter discusses the various types of deliverables, their content and scope *with particular reference to the standards established by Rural Utilities Service (RUS) of United States Department of Agriculture (USDA)*. This will be followed by a brief discussion on the type of numerical calculations that comprise typical Engineering design of a transmission structure. Various types of drawings are considered next; these include Plan and Profile drawings, Structure Framing drawings, Assembly and Sub-Assembly drawings, Foundation drawings and As-Built records (after construction is completed). This is only an example of a template for deliverables and any utility can develop their own system on this basis by adding sections relevant to their particular project.

5.1 DESIGN REPORTS

Prior to any design, the criteria adopted for structural, electrical and geotechnical aspects of the line are communicated by the Engineer to the Owner for their assessment. These preliminary guidelines take into account the various material and design requirements mandated by the Owner. For example, most utilities indicate the type of conductors and ground wires (or optical ground wires) they preferred to be used on the line as well as structural material. Owners often stipulate that wood poles less than Class 2 shall not be used on their transmission structures or that all insulators shall be polymer insulators. Others indicate a particular manufacturer of hardware or site-specific weather or load conditions that must be checked.

For projects on transmission lines in rural areas owned by electric cooperatives, financed and coordinated with RUS/USDA, guidelines stipulated by RUS are followed.

5.1.1 Preliminary design data book

The Rural Utilities Service (RUS) of United States Department of Agriculture requires a preliminary design data book prepared for all transmission line projects constructed according to RUS standards. The contents of the book vary from project to project depending on voltage, location and design preferences. The book is prepared during the early planning stages of a project and contains the following:

(a) Route Map
(b) Design Summary
(c) Clearance Tables
(d) Right of Way Width
(e) Sag and Tension Data
(f) Structure Strength
(g) Conductor Separation
(h) Insulator Swing Limits
(i) Guying Calculations
(j) Anchor Checks
(k) Foundation Loads
(l) Vibration Dampers
(m) Framing and Assembly Drawings

5.1.1.1 Route map

This is a map showing the general location of the project, including starting and ending substations, districts or counties covered and important topographical and environmental features shown.

5.1.1.2 Design summary

Figure 5.1 shows a typical summary outline in RUS format taken from Bulletin 200 (2015). This is a two-page document summarizing the most important design features of the project, including line, wire, structure, insulator, environmental and meteorological data. The most important feature in this design summary is the span limits. These limits are useful in performing structure spotting (i.e.) determining the locations of tangent structures as a function of structure strength, conductor separation, under-build (if any) and galloping.

5.1.1.3 Clearance tables

Vertical and horizontal wire clearances applicable to the project are generally shown in tabular form. RUS requires the following clearances listed:

a. conductor clearances to ground
b. conductor clearances to other wires and objects
c. conductor clearance from structure (insulator swing)
d. vertical separation between wires of different circuits
e. clearance to crossing wires

Item 'c' will be discussed later in Subsection 5.1.1.8.

5.1.1.4 Right of way width

Equation 2.2 from Chapter 2 governs Right-of-Way width calculations. For a given set of data relevant to Figure 2.2, the required width is computed and compared with recommended corridor width for that voltage (Table 2.5).

5.1.1.5 Sag and tension data

This contains the sag and tension values for each wire used on the line. Figure 5.2a shows a typical sag table computed using Alcoa SAG10^{TM} program. Note that the values are compiled in terms of a given ruling span (RS) and for each weather case relevant to the project. Both initial and final wire tensions and sags are shown. Programs such as Southwire SAG10^{TM} as well as PLS-CADD^{TM} are also used for the purpose. The tables can be customized to include an extra column showing tensions as a percentage of wire breaking strength to enable checking compliance with established limits (Table 2.14).

Stringing charts are needed for on-site construction purposes. These charts show the sags and tensions for various incremental spans and temperatures, and are very useful for field Engineers to assess sags during installation and stringing of a new line. Figure 5.2b shows the stringing chart for the same conductor shown in Figure 5.2a.

Sag templates are often used to verify potential for uplift at a location on a hilly, uneven terrain. These templates are developed for the same vertical and horizontal scales as those of plan and profile drawings. They are useful in situations where uplift possibility needs to be quickly assessed without recourse to digital computers. A typical sag template is shown in Figure 5.3. This template is a combined curve for

TRANSMISSION LINE DESIGN DATA SUMMARY	I. GENERAL INFORMATION			
	BORROWER:			DATE:
	LINE IDENTIFICATION:			
	VOLTAGE		LENGTH	
	TRANSMISSION	UNDERBUILD	TRANSMISSION	UNDERBUILD
	______kV	______kV	______mi	______mi.
	TYPE OF TANGENT STRUCTURE:		BASE POLE: ______HT. ______CL	
	DESIGNED BY:			

II. CONDUCTOR DATA	TRANSMISSION	OHGW	UNDERBUILD	COMMON NEUTRAL
SIZE *(kcmil or in.)*				
STRANDING				
MATERIAL				
DIAMETER *(in)*				
WEIGHT *(lbs./ft.)*				
RATED STRENGTH *(lbs.)*				

III. DESIGN LOADS (Wires)		TRANSMISSION *(lbs./ft.)*	OHGW *(lbs./ft.)*	UNDERBUILD *(lbs./ft.)*	COMM.NEUTRAL *(lbs./ft.)*
NESC:______LOADING DISTRICT					
a. ICE: ______in.	Vertical.				
b. WIND ON ICED COND. ___psf	Transverse				
c. CONSTANT K______	Resultant + K				
HEAVY ICE*(NO WIND)* _____in.	Vertical.				
HIGH WIND*(NO ICE)* _____psf	Transverse				
EXTREME HIGH WIND/ICE					
ICE: ______in.	Vertical.				
WIND ON ICED COND. ___psf	Transverse				

IV. SAG & TENSION DATA

SPANS AVERAGE*(EST)* ______ft MAXIMUM*(EST)* ______ft.. RULING*(EST)* ______ft.

SOURCE OF SAG-TENSION DATA:	TRANSMISSION		OHGW		UNDERBUILD		COMM.NEUTRAL	
TENSIONS *(% RATED STRENGTH)*	INITIAL	FINAL	INITIAL	FINAL	INITIAL	FINAL	INITIAL	FINAL
NESC a. UNLOADED (0° 15° 30°) ____°F								
NESC b. LOADED (0° 15° 30°) ____°F								
MAXIMUM ICE 32 °F								
HIGH WIND (NO ICE) ____°F								
UNLOADED LOW TEMPERATURE ____°F								
SAGS *(FT)*								
NESC DISTRICT LOADED ____°F								
UNLOADED HIGH TEMP(120°FOR OHGW &U.B.)___°F								
MAXIMUM ICE 32°F								
LOADED 1/2" ICE, NO WIND 32°F								

V. CLEARANCES

MINIMUM CLEARANCES TO BE MAINTAINED AT:______

CLEARANCES IN FEET	RAILROADS	HIGHWAY	CULTIVATED FIELDS			ADDITIONAL . ALLOWANCE
TRANSMISSION						
UNDERBUILD						

VI. RIGHT OF WAY

WIDTH ______ FT. *(MIN.)* ______ FT. *(MAX.)*

Figure 5.1 Transmission Line Project Summary.

VII. CONDUCTOR MOTION DATA							
	HISTORY OF CONDUCTOR GALLOPING:						
	HISTORY OF AEOLIAN VIBRATION:						
	a. TYPE OF VIBRATION DAMPERS USED *(IF ANY)*						
	b. TYPE OF ARMOR RODS USED *(IF ANY)*						
VIII. INSULATION							
	NO. OF THUNDERSTORM DAYS/YR____________ELEV.ABOVE SEA LEVEL (MIN, MAX, ft.)________________						
	CONTAMINATION EXPECTED?___________MAX EST. FOOTING RESISTANCE_________Ω SHIELD ANGLE ________°						
	STRUCTURE TYPE	STRUCTURE DESIGNATION	NO. OF BELLS / POLYMER / PIN OR POST	60 HZ DRY FLASHOVER	INSULATOR SIZE (DIAMETER & LENGTH)	SML / M & E RATING / POST STRENGTH	OTHER
	TANGENT						
	ANGLE						
	STRAIN STRUC						
IX. INSULATOR SWING							
	CRITERIA: (1)________PSF ON BARE CONDUCTOR AT _________(6 psf MIN) FOR _______________in. CLEARANCE (2)________PSF HIGH WIND ON BARE CONDUCTOR AT _________° F FOR _____________in. CLEARANCE						
	ALLOWABLE SWING ANGLE:			ANGLE IN DEGREES			
		STRUCTURE. TYPE	NO. OF INSULATORS.	6 psf MIN. WIND(1)	HIGH WIND (2)	NO WIND	OTHER
X. ENVIRONMENTAL AND METEORLOGICAL DATA							
	TEMPERATURE: MIN_________° MAX._________° AVERAGE YEARL LOW ___________°			EXTREME 10 SEC. WIND GUSTS *(mph)*: 10 YR. _________ 50 YR._________ 100 YR_________			
	MAXIMUM HEIGHT OF SNOW ON THE GROUND UNDER THE CONDUCTOR*(ft.)*:			DESCRIBE TERRAIN AND CHARACTERISTICS OF SOIL			
	CORROSIVENESS OF ATMOSPHERE:						
XI. STRUCTURE DATA (FOR SINGLE POLES AND H-FRAMES)							
	POLE MATERIAL:________________________ ARM MATERIAL: Trans_____________________ Underbuild_____________________			TYPE OF FOUNDATION: TANGENT:__________ANGLE:__________ DEADEND_______________GUYED STRUCTURES__________			
	TANGENT STRUCTURE TYPE_______________ SUMMARY OF SPANS *(ft.)* FOR TANGENT STRUCT.			BASE POLE ____FT.____CL	OTHER HEIGHTS/CLASSES AND BRACING		
	LEVEL GROUND SPAN						
	MAX. HORIZON. SPAN LIMITED BY STRUCTURE STRENGTH						
	MAX. VERTICAL SPAN LIMITED BY STRUCTURE STRENGTH						
	MAX. HORIZONTAL SPAN LIMITED BY COND. SEPARATION						
	MAX. SPAN LIMITED BY UNDERBUILD						
	MAX. SPAN LIMITED BY GALLOPING						
	EMBEDMENT DEPTH FOR BASE POLE:______________________			PRESERVATIVE OF WOOD POLE (*TYPE &RETENT.*)____________________ CORROSION PROTECTION FOR STEEL POLES ______________________			
	GUYING: TYPE OF ANCHORS: ______________________ GUY SIZE AND R.B.S.:___________________________						
XII. LINE DESCRIPTION							
	TANGENTS______________%	LIGHT ANGLES ____________%		AVERAGE NUMBER OF LINE ANGLES PER mi. ______________			
	MEDIUM ANGLES__________%	DEADEND & HEAVY ANGLES____________%		MAXIMUM DISTANCE BETWEEN FULL DEADENDS (*mi.*)______________			

Figure 5.1 Transmission Line Project Summary. *(cont'd)*

```
Textbook Example - Sag and Tension Data

Conductor:  336.4 Kcmil 26/ 7 Stranding ACSR "LINNET"
Area = 0.3070 sq.in    Diameter = 0.720 in    Weight = 0.463 lb/ft.   RTS = 14100 lbs
Data from Chart No. 1-782
English Units
Limits and Outputs in Average Tensions.

Span = 800.0 Feet                Special Load Zone
Creep is NOT a Factor
```

Design Points					Final		Initial	
Temp	Ice	Wind	K	Weight	Sag	Tension	Sag	Tension
°F	in	psf	lb/ft	lb/ft	Ft	lb	Ft	lb
0.0	0.50	4.00	0.30	1.649	23.29	5684	22.06	6000*
0.0	0.00	6.00	0.00	0.586	17.91	2625	14.11	3330
32.0	1.00	0.00	0.00	2.602	27.59	7582	27.59	7582
32.0	0.50	0.00	0.00	1.222	22.85	4291	20.51	4778
60.0	0.00	18.70	0.00	1.214	24.00	4061	21.44	4541
60.0	0.00	6.00	0.00	0.586	21.06	2234	16.94	2775
-20.0	0.00	0.00	0.00	0.463	15.82	2345	11.70	3169
-10.0	0.00	0.00	0.00	0.463	16.41	2261	12.17	3046
0.0	0.00	0.00	0.00	0.463	16.99	2184	12.66	2928
30.0	0.00	0.00	0.00	0.463	18.70	1985	14.19	2614
60.0	0.00	0.00	0.00	0.463	20.34	1826	15.77	2353
90.0	0.00	0.00	0.00	0.463	21.53	1726	17.35	2139
120.0	0.00	0.00	0.00	0.463	22.31	1666	18.91	1963
167.0	0.00	0.00	0.00	0.463	23.52	1580	21.28	1746
212.0	0.00	0.00	0.00	0.463	24.66	1508	23.43	1586

* Design Condition

Figure 5.2a Typical Sag-Tension Chart.

both the conductor and the optical ground wire (OPGW). Uplift is usually checked for a very cold temperature. In the United States, this temperature is typically −20 deg. F; hence this curve is often called "cold curve". Utilities also employ the sag template for structure spotting purposes.

5.1.1.6 Structure strength

For transmission structures, structure strength is usually expressed in terms of *allowable spans*. For a given structure type, pole material and specific design criteria (such as weather cases, load and strength factors and wire configuration), various allowable spans can be computed using design spreadsheets, PLS-POLE™ or any structural finite element analysis program. Often specific criteria such as conductor separation and additional under-build wires (Distribution circuits, telephone and/or other communication cables) also affect the maximum possible spans.

Stringing Chart

Conductor: 336.4 Kcmil 26/ 7 Stranding ACSR "LINNET"

Ruling Span: 800.0 Feet Stringing Sag Table Using Initial Sag

Special Load Zone Max Tension = 7582 lb

Design: 6000 lb @ 0 Deg °F, 0.50 in Ice, 4.00 psf Wind, Initial

H Tens	2927	2815	2710	2612	2519	2432	2351	2274	2203	2136	2074	2015	1960
(lbs)	-----	-----	-----	-----	-----	-----	-----	-----	-----	-----	-----	-----	-----
Temp °F>	0	10	20	30	40	50	60	70	80	90	100	110	120
Sag	Ft-In	Ft-In	Ft-In	Ft-In	Ft-In	Ft-In	Ft-In	Ft-In	Ft-In	Ft-In	Ft-In	Ft-In	Ft-In
Span	-----	-----	-----	-----	-----	-----	-----	-----	-----	-----	-----	-----	-----
700.0	9- 9	10- 1	10- 6	10-11	11- 4	11- 8	12- 1	12- 6	12-11	13- 4	13- 9	14- 1	14- 6
720.0	10- 4	10- 8	11- 1	11- 6	11-11	12- 5	12-10	13- 3	13- 8	14- 1	14- 6	14-11	15- 4
740.0	10-10	11- 4	11- 9	12- 2	12- 8	13- 1	13- 6	14- 0	14- 5	14-11	15- 4	15- 9	16- 3
760.0	11- 6	11-11	12- 5	12-10	13- 4	13-10	14- 3	14- 9	15- 3	15- 8	16- 2	16- 8	17- 1
780.0	12- 1	12- 7	13- 0	13- 6	14- 0	14- 6	15- 0	15- 6	16- 0	16- 6	17- 0	17- 6	18- 0
800.0	12- 8	13- 2	13- 9	14- 3	14- 9	15- 3	15-10	16- 4	16-10	17- 5	17-11	18- 5	18-11
820.0	13- 4	13-10	14- 5	14-11	15- 6	16- 1	16- 7	17- 2	17- 9	18- 3	18-10	19- 4	19-11
840.0	14- 0	14- 7	15- 1	15- 8	16- 3	16-10	17- 5	18- 0	18- 7	19- 2	19- 9	20- 4	20-11
860.0	14- 8	15- 3	15-10	16- 5	17- 1	17- 8	18- 3	18-10	19- 6	20- 1	20- 8	21- 4	21-11
880.0	15- 4	16- 0	16- 7	17- 3	17-10	18- 6	19- 1	19- 9	20- 5	21- 0	21- 8	22- 4	22-11
900.0	16- 1	16- 8	17- 4	18- 0	18- 8	19- 4	20- 0	20- 8	21- 4	22- 0	22- 8	23- 4	24- 0

Figure 5.2b Typical Stringing Chart.

Knowledge of such span limits helps in determining structure locations along the alignment.

Figures 5.4 and 5.5 show typical allowable horizontal spans computed for 69 kV single poles and braced H-frames. These spans are determined for a governing load case and for a given set of wires, load and strength factors. Although the examples shown in the figures refer to a few NESC weather cases, other weather conditions may also govern, depending on the structure type, climate and location.

Poles in H-Frames are subject to shear, moment and axial loads. Transverse (wind) loads on the frame cause one leg to resist compression while the other goes into uplift in case of X-braced H-Frame structure. The uplift capacity of a pole is usually less than the compression capacity; therefore it is a limiting factor taken into consideration while determining allowable spans. Uplift capacity is also a function of depth of embedment in soil. But soil data is often not readily available; in such cases it is customary to assume reasonable values for skin friction to compute foundation uplift capacity (600 psf. in Figure 5.5).

Additionally, allowable spans for H-frames can also be limited by axial loads, flexure, brace location and strength and horizontal spacing between conductors.

5.1.1.7 Conductor separation

This contains the spacing between conductors and between conductors and overhead ground wires (OHGW). For systems designed per RUS standards, Bulletin 200 provides

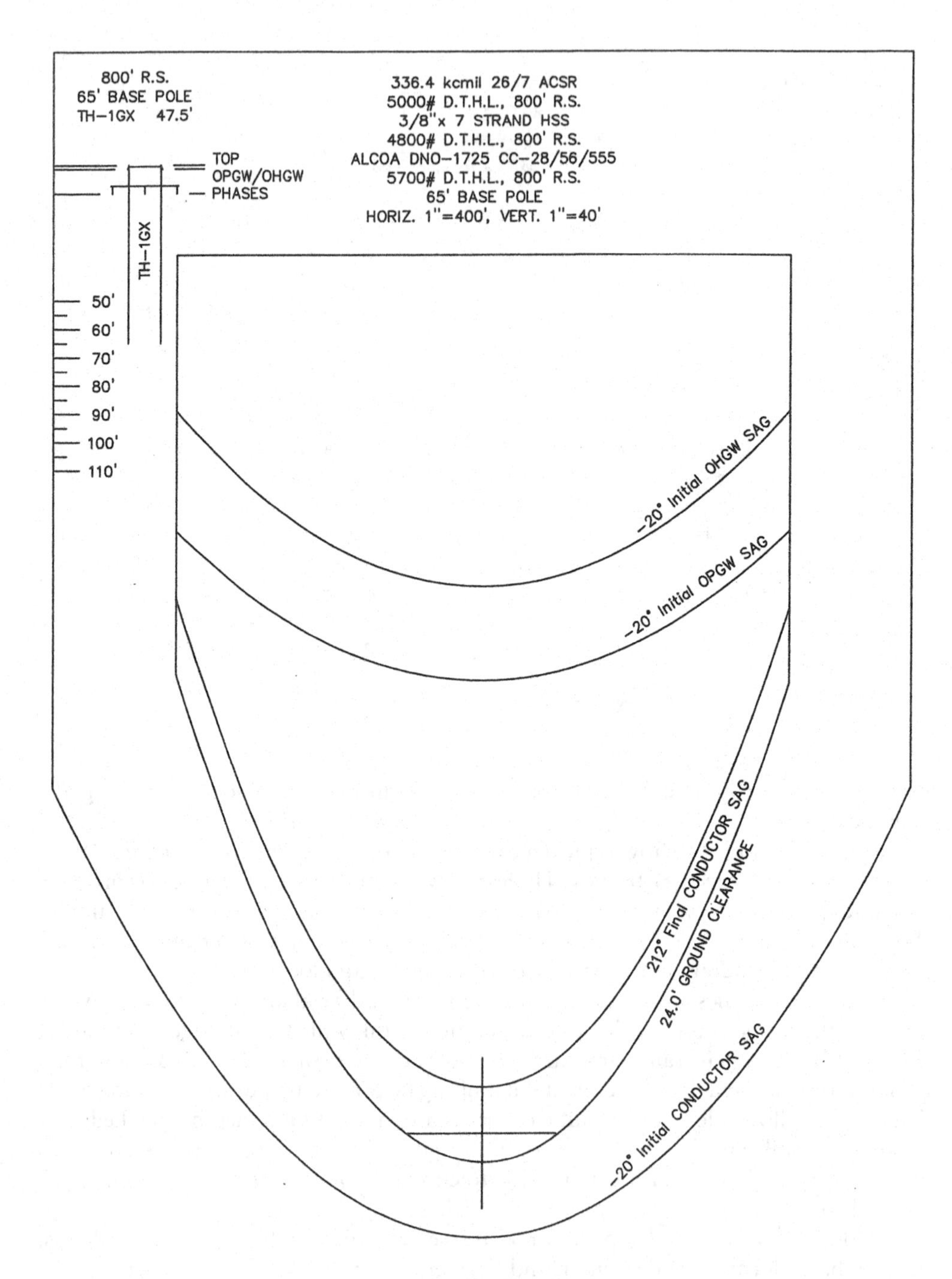

Figure 5.3 Sag Template.

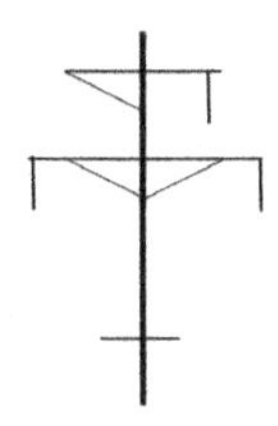

Pole Length (ft.)	Class 1 Poles	Class 2 Poles
45	601	498
50	570	458
55	549	444
60	534	435
65	524	428
70	517	424
75	513	409
80	510	408

1 in. = 25.4 mm, 1 ft. = 0.3048 m, 1 psf. = 47.89 Pa

Embedment – 10% of Pole Length + 2 ft.

Loading	½" ice, 4 psf. wind LF = 2.50 (wind), 1.50 (axial) – NESC Heavy LF = 1.00 for wind and axial - All other weather cases SF = 0.65 – NESC Heavy SF = 0.75 - Others
Conductor	3
OPGW	1
Poles	D-Fir or SYP (E = 1900 ksi, F_b = 8000 psi) F_b = maximum fiber stress for the given wood grade

Figure 5.4 Allowable Spans for Single Wood Poles.

specific guidelines with reference to required separation between OHGW and phase wires. Mid-span wire separation requirements are assuredly satisfied if the OHGW (or OPGW) sag is less than or equal to 80% of the conductor sag under the same weather conditions.

5.1.1.8 Insulator swing limits

Insulator swing under moderate and high winds affects the required clearance of a conductor from the structure surface (see Tables 2.9a and 2.9b for distances to be maintained from the structure under various weather situations). Figures 5.6a and 5.6b show an example of the geometrical aspects of these clearances and how swing

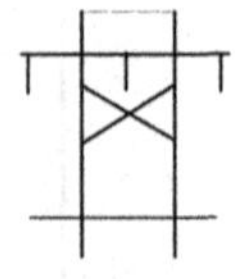

Height (ft)	Uplift Capacity * (lbs)	Location of X-Brace	Pole Class			
			H2	H1	1	2
45	13800	X-Brace @ 8'-8" from Pole Top	1265 f	1241 f	1225 f	1210 f
50	15600		1183 b	1177 b	1173 b	1165 b
55	17100		1106 b	1103 b	1097 b	1089 b
60	18900		1041 b	1036 b	1030 b	1006 p
65	20700		980 b	976 b	970 b	880 p
70	24000		926 b	922 b	916 b	780 p
75	25800		880 b	873 b	867 b	683 p
80	27900		832 b	829 b	787 p	617 p
85	30000		791 b	788 b	703 p	549 p

1 in. = 25.4 mm, 1 ft. = 0.3048 m, 1 psf. = 47.89 Pa

* corresponds to gravel backfill with Factor of Safety = 3.0
f_{sf} = 600 psf. (ultimate skin friction)

Embedment – 10% of Pole Length + 2 ft.

Conductor	3
OHGW	1
OPGW	1
Poles	D-Fir or SYP (E = 1900 ksi, F_b = 8000 psi) F_b = maximum fiber stress for the given wood grade
Loading	NESC Heavy Loading - 4 psf. wind, ½" ice, 0° F NESC Extreme Wind - 19.0 psf. (86 mph) wind, no ice, 60° F LF = 2.50 (wind), 1.50 (axial) – NESC Heavy LF = 1.00 for wind and axial - All other weather cases SF = 0.65 – NESC Heavy SF = 0.75 – Others

Controlling Item: b = brace, f = foundation, p = pole bending

Figure 5.5 Allowable Spans for H-Frames.

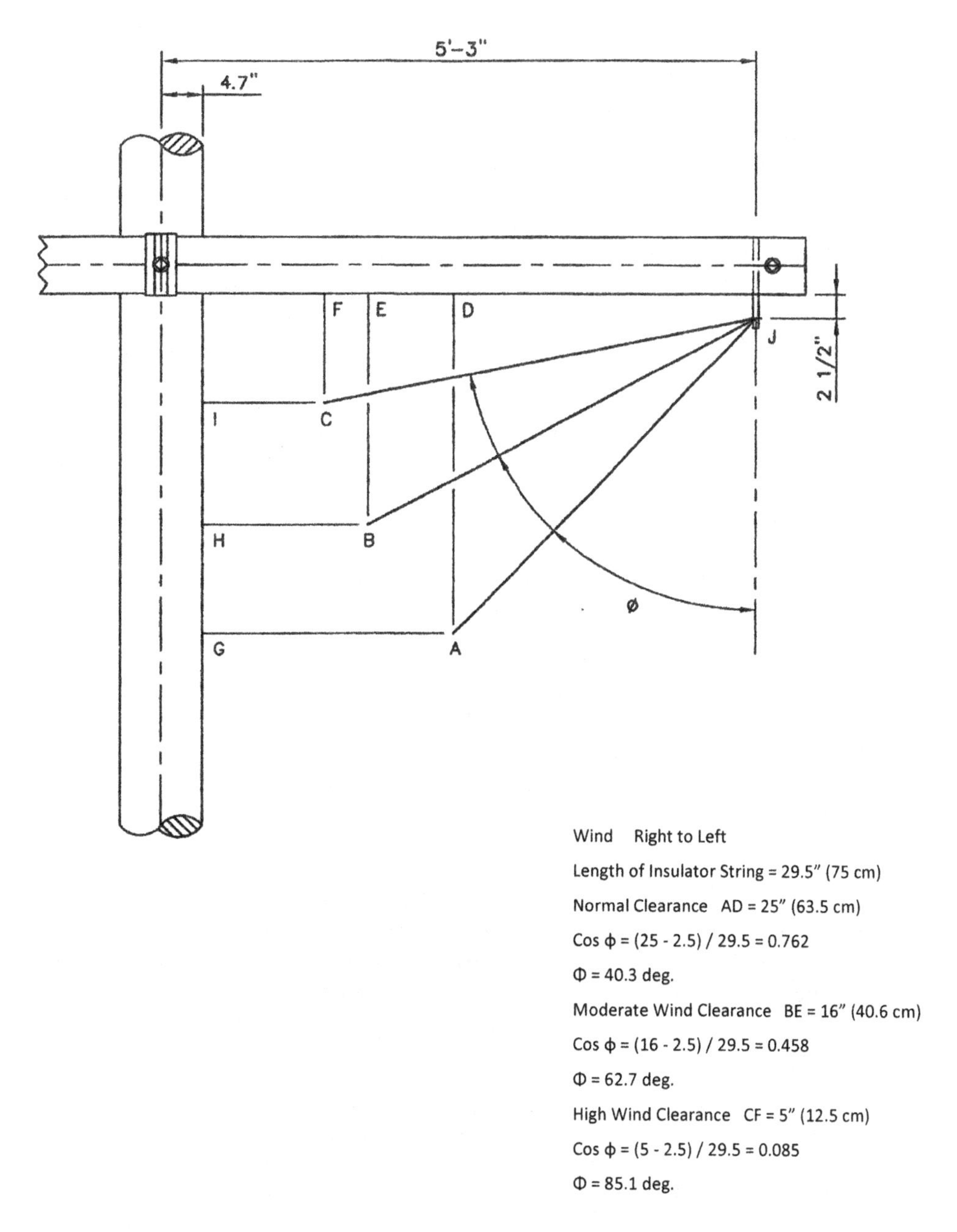

Figure 5.6a Insulator Swing – H-Frame.

limits are computed. The length of insulator strings is voltage-dependent; which in turn requires increased distance to be maintained from the structure surface. It is also important to consider swing in both the directions – wind from right and left – to determine effects of insulator swing.

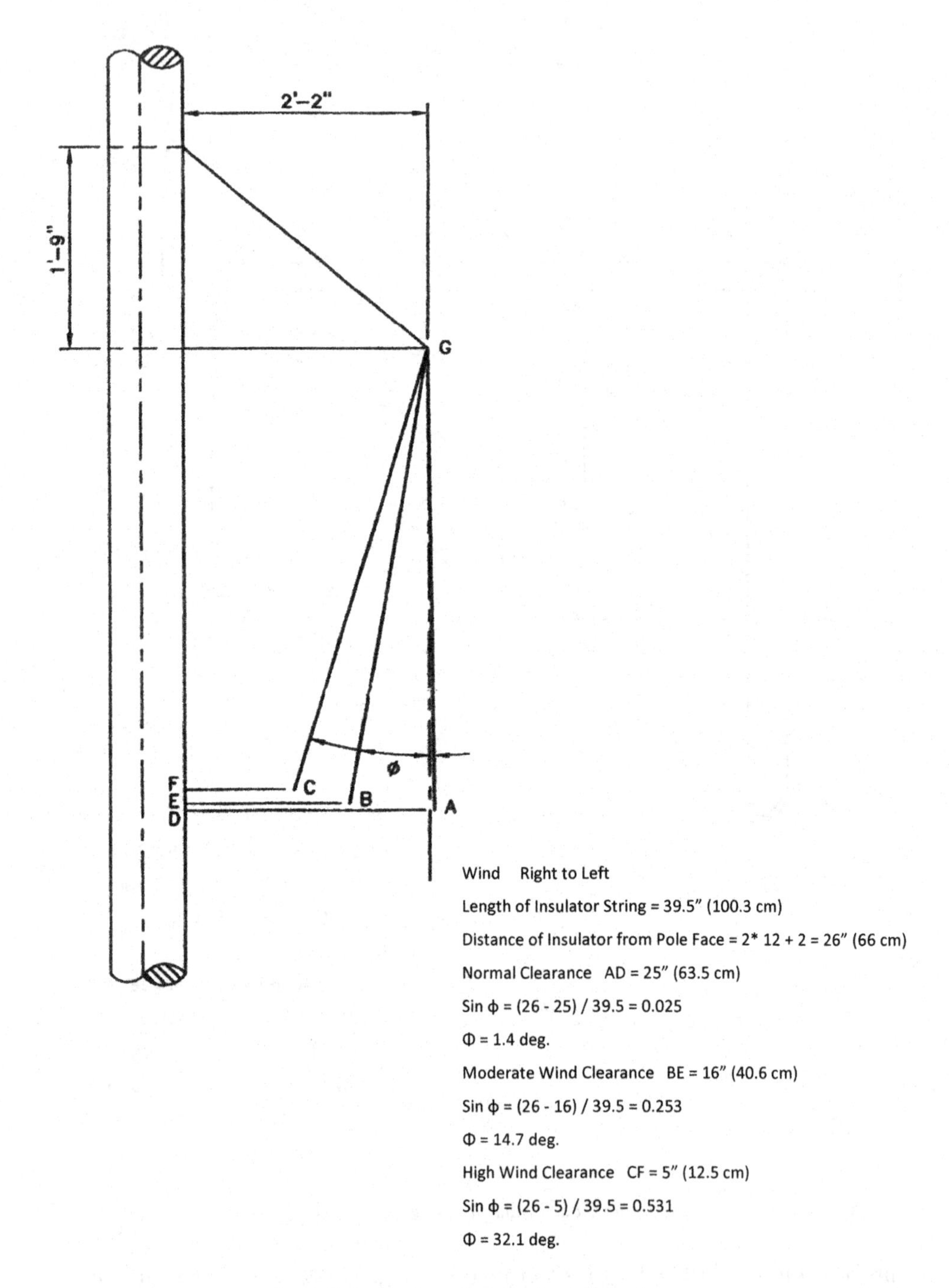

Figure 5.6b Insulator Swing – Angle Structure.

5.1.1.9 Guying calculations

Currently calculations involving guy wires and anchors are internally performed by powerful programs such as PLS-POLE™. However, in situations where such programs are not readily available, guying guides assist in such checks. These guides generally show standard guy-anchor systems and their design capacities for a given set of wire tensions. Figure 5.7 shows a typical guying guide for a double deadend pole, guyed in both the directions, full tension on one side and slack tension of the other. Note that the anchors and configurations of guy wires shown are RUS types.

5.1.1.10 Anchor checks

Anchors are usually checked in tandem with guy wires. For RUS-type anchors, their strengths are readily known (for example: TA-2L is rated at 16,000 lbs.) If helical types of anchors are used, then programs such as HeliCAP™ (2007) can be used to check for anchor holding capacity by inputting exact soil layer information.

5.1.1.11 Foundation loads

For guyed structures, vertical loads on pole resulting from guy wire tension components need to be computed and checked against allowable bearing capacity of soil at that location. These axial loads also include the weight of the pole, iced conductors and hardware with appropriate load factors. For large foundation loads, it may be necessary to install a concrete bearing pad below the pole butt to distribute bearing pressure on a larger area (Refer to Chapter 4 for details).

5.1.1.12 Vibration dampers

Vibration dampers are installed on conductors, shield wires and optical ground wires to assist in reducing wire damage due to Aeolian vibration. RUS Bulletin 200 recommends dampers if the initial unloaded design tension (see Section 2.3.3.2) for a bare wire (no ice, no wind) measured at 0° F (also called the Average Annual Minimum Temperature) exceeds 20% of the wire's rated breaking strength. The requirement (number and type) and location of dampers is usually determined by the manufacturer. Dogbone and other types of dampers are typical for conductors and overhead ground wires while spiral dampers are recommended for optical ground wires. The arrangement of dampers along the span depends on the span length, diameter, weight and tension in the wire.

However, one rule of thumb adopted by several utilities in the mid-western United States is to install one damper at each end of the span for all spans up to 600 ft. (183 m) and two dampers at each end for spans exceeding 600 ft. and up to 1200 ft. (366 m). For special situations, such as river crossings, wire manufacturers will provide custom damper recommendations based on the wire type, diameter, governing tension and span.

5.1.1.13 Framing and assembly drawings

The final feature of a typical preliminary design data book is the individual structure framing drawings and the associated detail drawings of all component assemblies

GUYING GUIDE

STRUCTURE TS-5GS / TS-5GH R.S. 800 FT.

CONDUCTOR TYPE 336.4 MCM 26/7 ACSR MAX. TENSION 6000#, W_H 0.5737, $2W_H$ 1.1474

O.P.G.W. ALCOA 0.555 5700#, 0.5183, 1.0366

O.H.G.W. 3/8" HSS MAX. TENSION 4800# W_H 0.4533 $2W_H$ 0.9066

GUY WIRE:

TYPE 7/16" EHSS, 65 % ULT. TENSION 13,520#, HOR. STRGTH. 1/1 LEAD 9560#, 2/1L 6046#

HOR. STRGTH. OF ANCHORS:

TA-3L- 1/1 LEAD 9192#, 2/1L 5813#, TA-5L- 1/1 LEAD 18,384#, 2/1L 11627#

LINE DEADEND

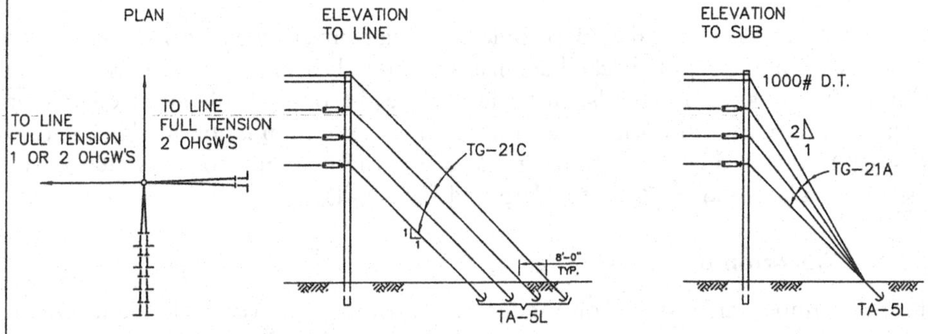

PLAN

TO LINE
FULL TENSION
1 OR 2 OHGW'S

TO LINE
FULL TENSION
2 OHGW'S

ELEVATION
TO LINE

TG-21C

1
1

8'-0"
TYP.

TA-5L

ELEVATION
TO SUB

1000# D.T.

2
1

TG-21A

TA-5L

REQUIRED EACH WAY

ONE OHGW	2 – OHGW's
4 – TG-21C	4 – TG-21C
4 – TA-5L	4 – TA-5L

REQUIRED

4 – TG-21A
1 – TA-5L

GUYING COMBINATION OHGW & CONDUCTORS

LINE DEADEND

OHGW & OPGW	10,500#	USE 1 – TG-21C & TA-5L (18,384#)
OPGW	5700#	USE 1 – TG-21C & TA-5L (18,384#)
EACH CONDUCTOR	6000#	USE 1 – TG-21C & TA-5L (18,384#)

SUB DEADEND

OHGW & OPGW 2000# D.T. USE 1 – TG-21A (6046#)

Figure 5.7 Guying Guide.

that comprise the system. Table 5.1 shows a typical construction drawing index for a 69 kV transmission line built per RUS guidelines and RUS standard framing assemblies of Bulletin 810 (1998). The set covers *all* structure types: tangent, light angle, medium angle, large angle and deadends (single poles and H-frames) as well as insulators and associated hardware, OHGW, foundations, grounding, cross arms, braces, guy and anchor assemblies etc. A given transmission line may use only selected components of this listing per design requirements and not all of them.

For all RUS-standard structures, these framing and assembly details can be found in Bulletins 810 and 811 for transmission structures and in Bulletins 803 (1998) and 804 (2005) for distribution structures along with the design manual for HV transmission lines, Bulletin 200. Drilling details for wood poles and cross arms can also be found in these Bulletins.

5.2 ENGINEERING CALCULATIONS

A transmission structure analysis involves both input design variables as well as interpretation of the computer output to assess structure response. This section discusses various forms of input and how output summaries can be examined to modify and strengthen the structure.

5.2.1 Structural analysis input

Input data covers a wide range of items from geometry to weather cases to hardware. These generally include:

a. Structure geometry
b. Load cases, load and strength factors, design wind loads parameters, criteria file etc.
c. Point loads due to wind and ice on wires only
d. Compatible hardware such as insulators, clamps, yoke plates, anchor shackles etc.
e. Direct Embedment depths
f. Foundation capacities
g. Baseplate geometry

The geometrical configuration of the structure is the primary input for any structural analysis. The basic geometry shows height of the structure above ground, component and insulator attachment points etc.

If the Engineer wants to quickly analyze a stand-alone structure, simple spreadsheets can be used to determine wire loads due to various weather conditions. Figures 5.8a, b and c show spreadsheets for calculating loads due to wires for a single tangent pole with both transmission and distribution circuits. Figure 5.8a gives basic system information such as wires, gust factors, wind speed calculations etc. while Figures 5.8b and 5.8c show calculation of loads for two specific load cases, namely,

Table 5.1 Typical Construction Drawing Index.

Drawing Number	*Description*
LA-1	Large Angle Three-Pole Deadend
TH-1G	Tangent H-Frame
TH-2G	Tangent H-Frame Double Cross Arm
TH-3G	Medium Angle Three-Pole
TH-4G	Large Angle Three-Pole
TH-5 or 5G	Large Angle Deadend Three-Pole
TH-7G	Two-Pole Double Deadend
TH-9G	Tangent H-Frame with Swing Brackets
TP-69G	Single Pole Tangent with Post Insulators
TS-1, TSS-1	Single Pole Tangent
TS-3G, TS-4G	Small and Medium Angle Single Poles
TS-5G	Vertical Double Deadend Pole
TM-1	Insulator Assembly with Suspension Clamps
TM-2	Insulator Assembly with Cushioned Suspension
TM-3P	Polymer Line Post
TM-4	OHGW Assemblies
TM-6	OHGW Support Assemblies
TM-9	Pole Grounding Assemblies
TM-9 S-1, S-2	Multiple Pole Grounding Assemblies
TM-16C	H-Frame Grounding Assemblies
TM-16R	Single Pole Grounding Assemblies
TM-55	Woodpecker Protection Assembly
TM-101, 102, 103	Foundation Units
TM-110A, 110B, 110C	X-Braces
TM-F6	Concrete Foundation Pads
TM-D-C	Vibration Damper – Conductor
TM-D-GW	Vibration Damper – Overhead Ground Wire
OPT-D, 2D	Vibration Damper – Optical Ground Wire
TMF-4*	OPGW Tangent Assemblies
TMF-112T, 112TD*	OPGW Deadend Assemblies
TM-124T, 124TD, 124TDS*	OPGW Line Splicing Assemblies
TM-N	Structure Numbering Assembly
TA-2, 3	Single Road Screw Anchor
TA-2P, 3P	Single Rod Plate Anchor
TA-2L, 3L	Single Rod Log Anchor
TA-4L, 5L	Double Rod Log Anchor
TG-1G	PVC Guy Guard
TG-1, 2, 3	Guying Assemblies
TG-11, 21, 31	Guying Assemblies
TG-35	Heavy Duty Guy Attachments
TG-25	Medium Duty Guy Attachments
TG-15	Light Duty Guy Attachments
TD-92	Guy Link Assemblies
TD-4, 5	Cross Arm to Pole Details
TD-9	Miscellaneous Details
TCD-20	Cross Arms
TCD-91	Cross Arms

*Assemblies adapted from TM-4 to OPGW (Reference: Bulletin 810, RUS/USDA).

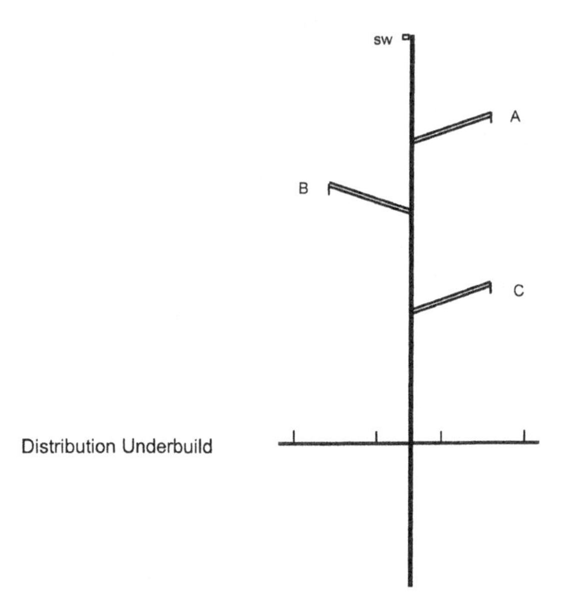

Conductor	477 ACSR	Hawk 26/7	3-phase, Single Circuit
Shield Wire	3/8" 7-str HS		Ruling Span = 300 ft (T), 300' (D)
Distribution	477 ACSR	Pelican 18/1	3-phase + 1 Neutral

NESC Formula for Wind Pressures for Extreme Wind Loading

q	$= 0.00256 * V^2 * K_Z * G_{RF} * C_d * I$		(psf)
	=	21.21	for Structure
	=	19.91	for Wire @ 50' to 80'
	=	18.70	for Wire @ 35' to 50'

Wind Velocity, V 90 mph

	Structure for 80' pole	Wire Hw = 50'-80'	Wire Hw = 35'-50'	
Kz	1.10	1.20	1.10	for 80' pole
GRF	0.93	0.80	0.82	Span = 300'
Cd	1.0	1.0	1.0	cylindrical pole
I	1.0	1.0	1.0	Importance factor

Figure 5.8a Basic Structure Data and Wind Pressure.

	Transmission	OHGW	Distribution
	Conductor 477 ACSR 26/7	3/8" HS Steel	Conductor 477 ACSR 18/1
S_{wd} Wind Span (ft)	300	300	300
S_{wt} Weight Span (ft)	300	300	300
$\theta/2$ (Line Angle/2)	0	0	0
d_w Diameter of Wire (in)	0.858	0.360	0.814
w_{bw} Bare Wire Weight (lbs/ft)	0.656	0.273	0.518
Tension Difference TD	0	0	0
T_w Sag Tension (lbs)	6825	2700	4130
Ultimate Strength (lbs)	19500	10800	11800

	NESC Heavy	
	t_{ri} Radial Ice (in)	0.5
	Wind Speed (mph)	40
	Pressure p_w (psf)	4
	Temperature (F)	0
Load Factors	Vertical LF_v	1.50
	Transverse Wind LF_t	2.50
	Wire Tension LF_{wt}	1.65

Conductor 477 ACSR 26/7

											USE
V =	wire weight with ice	S_{wt}	LF_v	no. of wires							
	1.501	300	1.50	1					=	675.30	680
T =	d_w with ice	$p_w/12$	S_{wd}	LF_t	2	$sin(\theta/2)$	T_w	LF_{wt}			
	1.858	0.33	300	2.50	2	0	6825	1.65	=	464.50	470

3/8" HS Steel

											USE
V =	wire weight with ice	S_{wt}	LF_v	no. of wires							
	0.808	300	1.50	1					=	363.56	370
T =	d_w with ice	$p_w/12$	S_{wd}	LF_t	2	$sin(\theta/2)$	T_w	LF_{wt}			
	1.36	0.33	300.0	2.50	2	0	2700	1.65	=	340.00	350

Conductor 477 ACSR 18/1

											USE
V =	wire weight with ice	S_{wt}	LF_v	no. of wires							
	1.335	300	1.50	1					=	600.89	610
T =	d_w with ice	$p_w/12$	S_{wd}	LF_t	2	$sin(\theta/2)$	T_w	LF_{wt}			
	1.81	0.33	300.0	2.50	2	0	4130	1.65	=	453.50	460

Figure 5.8b Typical Wire Loads by Spreadsheet – NESC Heavy.

	Transmission	OHGW	Distribution
	Conductor 477 ACSR 26/7	3/8" HS Steel	Conductor 477 ACSR 18/1
S_{wd} Wind Span (ft)	300	300	300
S_{wt} Weight Span (ft)	300	300	300
$\theta/2$ (Line Angle/2)	0	0	0
d_w Diameter of Wire (in)	0.858	0.360	0.814
w_{bw} Bare Wire Weight (lbs/ft)	0.656	0.273	0.518
Tension Difference TD	0	0	0
T_w Sag Tension (lbs)	6825	2700	4130
Ultimate Strength (lbs)	19500	10800	11800

	Extreme Wind	
	t_{ri} Radial Ice (in)	0
	Wind Speed (mph)	90
	Pressure p_w (psf)	19.90
	Temperature (F)	60
Load Factors	Vertical LF_v	1.00
	Transverse Wind LF_t	1.00
	Wire Tension LF_{wt}	1.00

Conductor 477 ACSR 26/7

V =	wire weight with ice	S_{wt}	LF_v	no. of wires							USE
	0.656	300	1.00	1					=	196.80	200
T =	d_w with ice	$p_w/12$	S_{wd}	LF_t	2	$\sin(\theta/2)$	T_w	LF_{wt}			
	0.858	1.66	300	1.00	2	0	6825	1.00	=	426.86	430

3/8" HS Steel

V =	wire weight with ice	S_{wt}	LF_v	no. of wires							USE
	0.273	300	1.00	1					=	81.90	90
T =	d_w with ice	$p_w/12$	S_{wd}	LF_t	2	$\sin(\theta/2)$	T_w	LF_{wt}			
	0.36	1.66	300.0	1.00	2	0	2700	1.00	=	179.10	180

Conductor 477 ACSR 18/1

V =	wire weight with ice	S_{wt}	LF_v	no. of wires							USE
	0.518	300	1.00	1					=	155.40	170
T =	d_w with ice	$p_w/12$	S_{wd}	LF_t	2	$\sin(\theta/2)$	T_w	LF_{wt}			
	0.81	1.66	300.0	1.00	2	0	4130	1.00	=	404.97	410

Figure 5.8c Typical Wire Loads by Spreadsheet – Extreme Wind.

NESC Heavy and Extreme Wind. Others can be developed in a similar manner. The equations used in the spreadsheet are from Section 3.3.3 of Chapter 3.

Tensile strength is usually a controlling factor for most line hardware. Therefore wire tensions (i.e.) sag and tension data is critical in ascertaining the maximum tension expected on an insulator and thereby on attachment hardware. In the case of suspension clamps used for conductors, the wire diameter and clamp strength is the governing factor in selecting an appropriate clamp. As discussed in Section 2.3.4, insulators are also rated by their strength relative to their configuration: cantilever, suspension, angle and deadend etc.

Computer programs such as PLS-POLE™ require pole embedment depths specified in input to calculate the pole height above ground. The default value is the rule-of-thumb depth (i.e.) 10% of pole length plus 2 ft. If various foundation capacities (compression, uplift) of an embedded pole are known, including them in the input facilitates an evaluation of their adequacy. For guy anchors, pull out strength of an anchor is generally input.

Spreadsheets can also be used to quickly determine the *approximate* geometry of a base plate and concrete pier for steel poles supported on drilled shafts. Figure 5.9 shows an *example* spreadsheet used for the purpose. Note that these values are intended only for quick guidance but can be useful in quickly modeling a base plate in PLS-POLE™ or checking a pier in CAISSON. (For more details on exact sizing of baseplates and drilled shafts, see Equation 4.19 of Chapter 4).

5.2.2 Structural analysis output

The objective of a transmission structure analysis is often to identify the areas of maximum and minimum usage and use the information to achieve a balanced design. Some utilities require usage levels of components included as a part of the Design Report. To this end, output from PLS-POLE™, TOWER™ and PLS-CADD™ provide level of usage of various components of a transmission structure. This assists the Engineer to identify and modify areas of overstress of these components. For tower angle members, the output contains usage in both tension as well as compression and connection usages in tension, shear and bearing. Steel poles are usually checked for stress interaction (i.e.) combined flexural, shear, torsion and axial stress. Any overstress is basically due to insufficient moment of inertia (pole diameter and/or thickness). If deflection limits are input, the usage summaries also contain warnings if deflections are excessive. Again, this assists the design Engineer to stiffen the pole by increasing the diameter or thickness or both. Similarly, for wood and concrete poles, use of a pole of a higher class can assist in preventing overstress.

In addition to structural elements (angle members, cross arms and pole shafts), both TOWER™ and PLS-POLE™ provide insulator usage data in their output. Suspension strings, clamps and strain insulators are checked for overstress. It is important for the design Engineer to specify the correct strength rating for insulators and the corresponding strength factor (SF).

For Engineered steel poles supported on concrete piers, PLS-POLE™ assists in determining the required base plate thickness or check adequacy of a given thickness.

Pole GL Diameter (in)	Anchor Bolt Circle * (in)	Base Plate Diameter (in)	Re-Bar Circle (in)	Pier Shaft Diameter (in)
12	18	24	26	32
18	24	30	32	38
24	30	36	38	44
30	36	42	44	50
36	42	48	50	56
42	48	54	56	62
48	54	60	62	68
54	60	66	68	74
60	66	72	74	80
66	72	78	80	86
72	78	84	86	92
78	84	90	92	98

* center-to-center

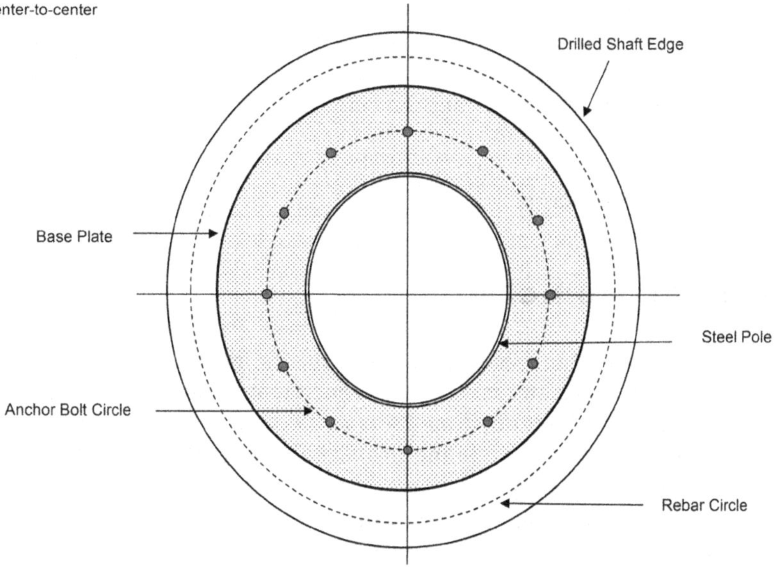

Figure 5.9 Example Template for Base Plate Geometry.

The output summaries also contain ground reactions needed for designing foundations. Reactions at the pole ground line or tower legs are generally listed for each load case. If guy anchors are present, then vertical and shear forces at anchorage points are shown. If foundation strength (uplift, compression or anchor) is input,

PLS-POLE™ or TOWER™ will provide relevant usage levels. These usage data can help to upgrade, modify or strengthen the required element to meet established design criteria.

Wire usages are generally determined in PLS-CADD™ with reference to tension limits imposed on various wires. This helps in ensuring that no wire exceeds the limiting tension specified for any weather case.

5.3 PLAN AND PROFILE DRAWINGS

Plan and Profile (P & P) drawings comprise the most critical part of a transmission line design process and are the primary documents used by construction contractors in building the line. In general, all P & P drawings must, as a minimum, show the following:

a. Scale (generally drawn to V:H scale of 40:400)
b. Survey Station
c. Wire tensions of all wires for governing weather case
d. Sag of the lowest conductor at the maximum operating temperature (MOT) for the wire
e. Line angle at all running angle locations
f. Ground clearance at various line locations (usually at a selected level ground span, location of steep elevation change, highways and railroad tracks)
g. Wire clearance from other objects and line crossings
h. North direction arrow
i. Structure numbers, height and type listed above each structure
j. Span length between structures

5.4 STRUCTURE LOADING

Structure loading is discussed in Chapter 3 at length. As mentioned there, "deliverables" for steel structures (pole and lattice) and concrete poles consist of the loading schedules usually specified by the Loading Trees or Load and Design sheets (L&D). These sheets constitute an important communication between the design Engineer and the steel fabricator. Basic load cases are shown on these loading schedules; additional cases (example: uplift case at sub-zero temperature) can be added at the discretion of the design Engineer or as requested by the Owner.

5.5 STRUCTURE FRAMING DRAWINGS

Pole and structure framing drawings are discussed in Chapter 3 and Section 5.1.1.13, respectively. These framing drawings usually show important attachment points along the pole or structure along with drilling details. For steel and concrete poles, the fabricators usually generate drilling or welding detail drawings based on the pole framing outline. A typical RUS framing drawing of a 69 kV H-Frame is shown in Figure 3.31c. For lattice towers, member details show bolt holes and general connections to other members or gusset plates.

5.6 ASSEMBLY DRAWINGS

Assembly drawings constitute a subset of framing drawings and show details of various units that comprise the main assembly. For example, Fig. 3.33a shows the framing drawing of a 345 kV structure with bundled conductor support. The details of this conductor support assembly are shown in Figure 3.33d where all hardware items are listed. Similarly assembly details of a post insulator shown on Figure 3.33b are listed in Figure 3.33e.

Assembly units and drawings are discussed in Chapter 3 and Section 5.1.1.13, respectively. For project-specific non-standard assemblies, related drawings can be generated by modifying standard ones or obtaining them from the manufacturer.

Most utilities in the United States have databases of component assembly units with inventory numbers which are shown on the framing and assembly drawings. This assists in generating material lists required for a single structure or a complete line.

5.7 FOUNDATION DRAWINGS

General foundation design is discussed in Chapter 4. For standard RUS structures, foundation generally implies direct embedment of poles, with or without bearing pads and pole base shoes. In the case of steel structures, foundations cover direct embedment as well as other forms such as drilled shafts, isolated footings and grillages, depending on structural configuration. In all cases, these drawings show depth, size, rebar, base plates, anchor bolts (where required) and grounding details along with governing reactions and material data.

5.8 AS-BUILT RECORDS

These records refer to the final set of post-construction documents, mostly Plan and Profile and other construction documents, after all field activities are completed. The As-Built records are meant to highlight differences, if any, between what was designed and what was actually built. Plan and Profile drawings are marked up with the actual built information to update the actual design documents and PLS-CADD™ model. For example: structures are spotted a few feet away from their original intended survey location or the soil profile at a location necessitated a larger embedment (and therefore larger pole) or land use issues required minor adjustments in alignment etc. From a Project Management perspective, these records also defined as the Close Out stage of the project.

Chapter 6

Advanced topics

Over the past few decades, the advent of powerful digital computers, advances in numerical modeling of various climactic phenomena and newer materials have been facilitating accurate and rational analysis of transmission lines, structures and foundations. Programs such as PLS-CADD™ are enabling modeling and design of long, multi-circuit transmission lines with hundreds of structures (poles, frames and towers), miles of conductors and dozens of combinations of load cases. On the theoretical side, significant research is being conducted into the mathematical modeling of poles, cables, foundations, complex loading situations including seismic and high intensity wind events. Strong and durable composite materials were developed to provide a viable alternative to wood and steel along with special-purpose conductors to meet specific performance needs.

This chapter focuses on several advanced topics related to transmission line design not only from a design perspective but also from a research and development point of view. The state-of-the-art of current technology will be briefly discussed with specific reference to analysis methods, materials and response to extreme climatic events. Also discussed will be the process of Emergency Restoration of transmission lines after a catastrophic weather event and LRFD (Load and Resistant Factor Design) approaches to design of foundations.

6.1 ANALYSIS

Over the past decade, significant studies were conducted in several areas such as solid modeling, wind loading, broken wires and nonlinear behavior of both structures as well as wire systems.

Ashraf et al. (2005) analyzed steel transmission poles using a nonlinear finite element (NLFE) method which included geometric nonlinearities. Tapered thin-shell beam elements were used. This study compared results from a NLFE analysis to those from general purpose FE programs. The inference is that a thin shell element is the most accurate way to represent tapered steel poles, both multi-sided and circular.

Banik et al. (2010) evaluated code-mandated factored wind loads on lattice towers on the basis of wind speed, load and gust response factors and by using nonlinear static pushover and incremental dynamic analysis techniques. The major inference from the study is that for short duration high-intensity wind loads, consideration of material ductility could lead to an economic design while ensuring safety. Global capacities of towers were found to be 30% higher than the factored (calculated) design capacity indicating significant reserve strength in the structure that needs to be taken advantage of.

Der Kiureghian and Sackman (2005) studied the nonlinear behavior of inclined cables by formulating a procedure for determining the exact geometric stiffness of the cable with reference to the movement of the wire's supports, inclination and for cases involving slack tension.

Eidinger and Kempner (2012) presented reliability models of 230 kV, 345 kV and 500 kV lattice towers subject to extreme wind and ice loading. The models are used to help forecast failures. Two key findings of the study include the importance of tower-specific topographic factors (mountainous terrain, for example) and site-specific conditions (foundations).

Jayachandran et al. (2006) discussed the effects of snapped conductors on transmission structures and evaluated the subsequent dynamic response. A design proposal based on using dynamic load amplification factors (DLF) instead of residual static loads is proposed. It is recommended that a DLF of 1.40 be used to determine the dynamic effects due to broken conductors.

Lei and Chien (2005) presented a comprehensive modeling technique for seismic response of a series of lattice towers in a transmission line by considering both material and geometric nonlinearities. Tower members were modeled by 3-D beam-column elements instead of truss-type elements.

Stengel and Mehdianpour (2014) described development of a FE model of a single span overhead transmission line using cable elements with nonlinear characteristics and considering aerodynamic damping. A major inference drawn is that the first natural frequency of most long-span cables lies well below 1 Hz.

6.2 MATERIALS

Composite materials (i.e.) fiber-reinforced polymers are rapidly becoming a popular material in transmission and distribution lines. Special conductors with high tensile strength and superior sag performance are being developed for custom long-span applications such as river crossings with spans as high as 3000 ft (914 m) or more. Conductors coated with hydrophobic coatings are being specified in many locations to combat ice accumulations on HV lines. Optical ground wires can now carry more than 144 fibers, each fiber capable of transmitting thousands of channels of information and data, while possessing high tensile strengths for long-span applications.

6.2.1 Composites

Section 3.1.5 of Chapter 3 had earlier discussed composite transmission structures. Companies such as PUPI (2011), Shakespeare (2008) and RS Technologies (2012) are pioneering the use of fiber glass poles and cross arms, in lines ranging from 15 kV to 230 kV. Composite cross arms, both in tangent and deadend applications, now range in length up to 12 ft (3.7 m) and can be installed on the pole with minimum equipment. H-frames can be built up to a height of over

100 ft (30.5 m) and using modular poles, heights up to 155 ft. (47.2 m) can be obtained. The most recent advance in composite technology is the Iso-Truss system (2002) which contain structural forms fabricated with fiber-reinforced composite (FRC) – resin interwoven matrix of triangles and pyramids to distribute loads synergistically.

6.2.2 Special conductors

HTLS (High Temperature Low Sag) conductors are special HV conductors developed specifically for long span applications (see Section 3.4.7.2) where the design requirements include a wire capable of sustaining extremely large tensions but with superior sag-tension performance under all climatic conditions. They are also used on normal spans on existing lines to transmit more current (amperes). These conductors are rated to function even at extremely high operating temperatures of over 200°C (392°F) without losing mechanical strength. Research is underway to test the performance of various HTLS wires with compositions that include polymer matrix composites, metal matrix composites etc.

HTLS conductors such as ACCR™ (Aluminum Conductor Composite Reinforced) conductors offer a special-purpose wire for custom situations such as high ampacity demand, long span crossings, high corrosion areas and ability to withstand temperatures over 200°C (392°F).

6.3 EXTREME EVENTS

Transmission lines are subject to a wide variety of natural phenomena which include ice and snow storms, high intensity winds, earthquakes, tornadoes, downbursts and flooding. Though most design procedures contain provisions to resist climactic events (notably ice, wind and combined ice with wind), the uncertainty of nature means there is always a probability of severe events occurring during the life time of the line. This section focuses on four such situations: hurricanes, tornadoes, earthquakes and ice storms.

6.3.1 Hurricanes

A hurricane is a type of intense tropical weather system with a well-defined circulation and wind speeds exceeding 74 mph (119 kmph). For sustained speeds between 39 to 73 mph (63 to 118 kmph), the system is called a tropical storm. Hurricanes are categorized according to Saffir-Simpson scale from C1 (minimal damage) to C5 (catastrophic damage).

Dempsey (1996) discussed the damage potential of high intensity winds (HIWs) with specific focus on hurricanes and tornadoes. One major observation was that these winds put significantly larger loads on the structures and smaller loads on the supporting wires. Some utilities in the US, Australia, South Africa, Argentina and Canada have adopted specific HIW loads for towers in the 400 kV to 500 kV range.

Watson (2007) discussed the effects of oblique wind on transmission structures and suggested that oblique incidence of wind (including longitudinal wind) may pose a problem for H-frames in the 100 mph to 150 mph (161 kmph to 241 kmph) hurricane range.

6.3.2 Tornadoes

Tornadoes are short-lived, high intensity wind events characterized by extremely fast rotating winds ranging in speeds from 40 mph to 318 mph (64 kmph to 512 kmph), *inward and upward* winds and localized occurrence. Tornadoes are classified according to the Fujita (or Modified Fujita) scale from F0 (light damage) to F5 (incredible damage). The speeds associated with these scales are 40 to 72 mph (64 kmph to 116 kmph) for the F0 scale and 207 to 260 mph (333 kmph to 418 kmph) for the F4 scale. Speeds above 261 mph (420 kmph) are rated as F5. The recent tornado in Joplin, Missouri (2013) was measured at F5, created a 7-mile (11.3 km) long damage path, destroyed over 5000 buildings and caused over $1.3 billion in losses and over 150 deaths.

According to ASCE Manual 74 (2010), 86% of tornadoes in the U.S. are associated with Fujita-Pearson scales F0 to F2 (i.e.) wind speeds up to 157 mph (253 kmph), about 530 ft (162 m) wide and travel a few miles (maximum 10 miles or 16 km) before dissipating. Most design codes provide guidelines for extreme wind loads on wires and structures, concurrently. However, some utilities recommend and use a special tornado load on the structure alone (no wires) for a wind speed of 160 mph (258 kmph). Wire spans are considered to be significantly affected by tornado winds since the average width of the tornado path is more or less the same as that of an average wire span. For lattice towers, the effect of such a special load is the increase in shear bracing within the body of the tower.

Given the potential for damage to transmission lines under tornadoes, several recent studies focused on examining the performance of both transmission lines as well as structures for tornado-induced wind loads. The studies also considered the effects of downbursts and microbursts on transmission lines. As opposed to a tornado, a downburst is a strong downward current of air with straight line winds, directed *upward and outward* from a surface landing point. Downbursts create *vertical* wind shear. A microburst is a downburst on a very small scale.

Li (1999) proposed a stochastic model to accurately predict wind loads generated by thunderstorm downbursts for the design of transmission lines. The model is calibrated with existing meteorological records and includes consideration of the size effect of the downbursts.

Ahmad and Ansari (2009) derived mathematical expressions for tornado wind loads on a lattice tower and performed a time-history analysis of the tower for mode shapes and displacements. Chay et al. (2006) used an analytical/stochastic method of simulating downburst winds to explore quasi-static loading conditions that occur during downbursts. The results are interpreted with reference to several existing transmission tower design codes and with specific focus of structural adequacy. Ishac and White (1995) studied the effects of tornado loads on transmission lines to propose a design basis tornado loading for towers which also included additional shear bracing.

Matsui and Tamura (2009) discussed the development of a tornado flow simulator to examine the characteristics of flow fields in a tornado.

Ragan et al. (2014) studied the effects of tornado wind forces on H-frames and other pole structures. The focus of the study was to determine the force coefficients for smooth, cylindrical structures. A major observation was that the force coefficients are similar to those published in ASCE 7 (2010) for straight-line winds on smooth, round surfaces.

Savory et al. (2001) describe models for wind velocity time histories of transient tornado and microburst events and the resulting loading on a 168 ft (51.2 m)-tall double circuit transmission tower. The tower was analyzed for two HIW events including an F3 tornado and a microburst. Failure patterns clearly indicated the need and importance of shear bracing at lower levels of the tower.

Shehata et al. (2005) described a numerical model to predict the structural performance of a transmission tower subject to downbursts. Wind velocity data is scaled and transformed into forces applied to the tower. 3-dimensional elastic frame elements were used to model the tower members; 2-dimensional curved beam elements with geometric nonlinearity were employed to model the conductors and ground wires. The study asserted the importance of considering special HIW load cases for lattice towers in addition to code-mandated high wind cases.

6.3.3 Earthquakes

Earthquakes and subsequent ground motions usually do not govern the design of transmission lines; loads due to broken wires and wind/ice combinations are larger than earthquake loads (ASCE Manual 74). The one instance where failures occur is when soil liquefaction occurs and cause foundations to fail. The design shear capacity of foundations is generally adequate to handle any transverse motions. But substation structures must be designed to withstand seismic forces since deflections are more critical (ASCE Manual 113, 2008). Researchers, however, studied the effects of earthquakes on transmission lines and structures.

Long (1973) presented one of the earliest studies on seismic behavior of transmission structures. His main conclusion was that forces due to structure acceleration are small and forces on insulator supports are well within the structure capability when designed per NESC guidelines.

Dai and Chen (2008) studied the dynamic behavior of direct embedded concrete transmission poles with specific reference to influence of soil-structure interaction on pole vibration modes. Investigations included both numerical modeling and full scale tests. Soil effects were simulated by means of springs in the FE model. One important inference from the study is that the usual fixed-end assumption made for direct-embedded (cantilever) poles is not fully applicable when the surrounding soil is soft.

Ghoborah et al. (1996) investigated the effects of multiple support movement on the seismic response of transmission lines. Towers are modeled with space truss elements and conductors with nonlinear straight 2-node elements. The effects of wave propagation velocity and in-coherency of seismic waves are studied. Results indicate that all tower supports may not have the same uniform ground motion and larger internal forces and displacements can result from differential support movement. It is also inferred that wave propagation velocity has a significant effect on the lateral displacement of the tower. Mazda et al. (2010) also presented results similar results from an analytical study on transmission towers subject to differential support settlement. Vibration modes of 4-point and 3-point support towers were compared showing that as one support is weakened, it significantly affects the seismic performance of the structural system as a whole.

Tian et al. (2010) studied the behavior of transmission tower-line systems subjected to spatially-varying support excitations. Seismic input waves for vertical and horizontal ground motions were considered. Towers were modeled with beam elements while nonlinear cable elements were used for the wires. Results again showed that assuming uniform ground motion is unrealistic and does not provide the most critical case for response calculations.

Koser and Arora (1999 and 2002) formulated the design of a transmission towers and poles as discrete value optimization problems which included normal operating loads as well as seismic loads. Time-dependent design constraints are considered in the solution process. For the H-frames, structural nonlinearity is modeled but stresses were constrained to be elastic. Two cost functions – material and initial – were defined.

McClure and Lapointe (2003) discussed a macroscopic modeling approach to dynamic analysis of transmission lines with emphasis on propagation of unbalanced shock loads in a line due to wire rupture. The general purpose FE software program ADINA is employed to model the tower-cable-insulator system. As a means of validation, the model is applied to known tower failures in the Quebec region of Canada. Results indicate that large torsional moments are possible in the tower section with an increasing likelihood of plastic hinge formation and buckling.

Components of substation equipment, especially porcelain insulators, are generally considered weak links in the behavior of electrical systems during earthquakes. Nishenko et al. (2012) studied the structural adequacy of standard porcelain insulators under cyclic loading. Fujisaki (2009) discussed the seismic performance of electric transmission systems in the Bay Area region of San Francisco, California, during two specific earthquakes in 1983 and 1984. Observations from this study included damage to live tank breakers, disconnect switches, transformers, poles, rigid buses and connectors in switches.

6.3.4 Ice storms

The NESC extreme ice design weather condition with concurrent wind is generally defined in terms of 0 to 1¼ in (31.8 mm) of radial ice due to freezing rain which represents the standard 50-year return period for ice and is adequate for design. Some utilities use 1½ in (38 mm) radial ice thickness governed by local climatological experience. However, there were instances where larger radial ice thicknesses were recorded during failures. For example, the Midwest ice storm of January 2007 was associated with a radial ice of over 2½ in (64 mm) while 3 in to 4 in (76 mm to 102 mm) ice was recorded during the well-known Quebec snow storm of 1998. While it is impractical to design all lines and structures to such large ice accumulations, it still is an issue of much debate in North America. The local utility industry is approaching the issue from a line reliability perspective (i.e.) maintaining the safety of the line and ensuring reliable power transmission and exploring alternative ways of minimizing the danger of excess ice. Canadian utilities have employed a variety of ice melting methods to reduce the probability of line failures due to ice on wires or due to Aeolian vibration of iced wires.

6.4 EMERGENCY RESTORATION PLANS

Transmission lines and structures often suffer extensive damage and failures during extreme climatological events such as hurricanes, tornadoes and ice storms. In such situations, quick repair of the line and restoration of electric power supply to consumers is the responsibility of major utilities everywhere. In the USA, government agencies such as RUS/USDA, which finance the construction and maintenance of rural electrical transmission lines through small utility cooperatives, require them to submit a well-defined emergency restoration plan prior to approving funding.

An Emergency Restoration Plan (ERP) is a written, validated document that serves as a guideline for rapid restorative action in emergency situations. In case of a transmission line or grid failure under climactic and other catastrophic events, a good ERP guarantees quick installation of replacement structures, conductors and hardware and thereby quick restoration of power. Restoration plans usually consider standard round wood poles or steel poles as needed for a specific location. Line restoration involving high voltage lattice towers is more complex than that of wood or steel poles; the process becomes challenging if hilly or rolling terrain is encountered.

Brief descriptions of key components of an ERP are given below.

ERP Team – The composition and reporting structure (hierarchy) of the ERP Team must be clearly established to enable assignment of responsibilities. This team consists of personnel from engineering, operations and maintenance divisions of the utility and must undergo special training in emergency response procedures and communications.

Analysis of Hazards and Capabilities – The success of an ERP depends on basic understanding of the anticipated hazards in the area and the capabilities of the team responding to the emergency. This includes a complete database of previous events, challenges faced, equipment and personnel mobilized and the duration of the restoration process. The most important derivative of this analysis is the knowledge of the pattern and type of structure failures (i.e.) where and what kind of structure is statistically most likely to fail. This will help in maintaining stock of poles of a certain class which can be quickly used in any line restoration event.

Communications Plan – A clearly defined communications plan must be in place with important stakeholders identified along with channels and media of communication. All members of the ERP Team must participate in the communications plan. The plan should also include appropriate local, state and federal officials.

Restoration Kits – The primary goal of an ERP is rapid mobilization of assets and restoration of downed poles and lines. A restoration kit is an important tool of the process. A typical "restoration kit" contains the following:

(a) A list of damaged structures on the line and recommended replacements
(b) Framing and assembly drawings of all replacement structures
(c) Bill of Materials for the select group of structures

Restoration kits are normally customized for each line as a function of voltage and structural type. Often kits referring to voltages of 115, 138 and 161 kV can be combined into a single category for easier processing. The other technical issues that govern the kit structures are strength, clearances, embedment and insulators. Logistical parameters such as right of way, location, access, equipment mobilization, storage and transport are also critical importance.

Bledsoe et al. (2013) describe in detail the severe damage to a 500 kV line from a tornado and the steps comprising the emergency restoration of the structures and circuits. The reader is also referred to the advanced restoration techniques adopted by agencies such as Lindsey Manufacturing Company of Azusa, California, USA. Their emergency restoration system (ERS) employs a welded modular lightweight aluminum lattice structure which can be used as a basic module for various configurations: tangent, running angle and deadends, single or double circuit. These ERS's do not require any special foundations and can be erected in a relatively short time.

6.5 FOUNDATION STRENGTH

The existing NESC (2012) does not offer a clear definition of the strength factors used for foundations. Table 2.15c of Chapter 2 shows that all foundations have been assigned a Strength Factor of 1. There is no differentiation between structural component of a foundation and the soil component. Since geotechnical capacity is what typically controls the eventual strength of a foundation, whether it is a direct embedment or drilled shaft or an anchor, there is recognition that strength factors for foundations must be defined with specific reference to soil. As discussed in Section 4.5.2, FAD Tools adopted LRFD and incorporated strength factors for geotechnical capacity of foundations under moment, compression and uplift modes of failures. However, codes such as NESC are yet to provide guidance in this area.

References

A

Ahmad, S. and Ansari, M.E. (2009), Response of Transmission Towers Subjected to Tornado Loads, *7th Asia Pacific Conference on Wind Engineering*, Taipei, Taiwan.

American Concrete Institute (2014), ACI-318-2014, *Building Code Requirements for Reinforced Concrete*, Farmington Hills, Michigan, USA.

American Concrete Institute (2014), ACI-336-3R-14, *Design and Construction of Drilled Piers*, Farmington Hills, Michigan, USA.

Alcoa (2001), *SAG10™ Ver. 3.0, Users' Manual*, Alcoa Fujikura Ltd, Spartanburg, South Carolina, USA.

American Institute of Steel Construction (1995), *LRFD Manual*, AISC, 2nd Edition, Chicago, Illinois, USA.

American Institute of Steel Construction (1989), *Allowable Stress Design Manual*, AISC, 9th Edition, Chicago, Illinois, USA.

American Institute of Steel Construction (2013), *Steel Construction Manual*, AISC, 13th Edition, Chicago, Illinois, USA.

American Institute of Timber Construction (2007), Standard 109, *Standard for Preservative Treatment of Structural Glued Laminated Timber*, AITC, Centennial, Colorado, USA.

American Institute of Steel Construction (2006), *Anchor Rod and Base Plate Design*, AISC Steel Design Guide, 2nd Edition, Chicago, Illinois, USA.

American Society of Civil Engineers (2013), ASCE Special Report, *Joplin Missouri Tornado of May 22, 2011 – Structural Damage Survey and Case for Tornado-resilient Building Codes,* ASCE, Reston, Virginia, USA.

American Society of Civil Engineers (2010), ASCE Standard 7-10, *Minimum Design Loads for Buildings and Other Structures*, ASCE, Reston, Virginia, USA.

American Society of Civil Engineers (2011), ASCE Standard 48-11, *Design of Steel Transmission Pole Structures*, ASCE, Reston, Virginia, USA.

American Society of Civil Engineers (2010), ASCE Manual 74, *Guidelines for Electrical Transmission Line Structural Loading*, ASCE, Reston, Virginia, USA.

American Society of Civil Engineers (1997), ASCE Manual 91, *Design of Guyed Electrical Transmission Structures*, ASCE, Reston, Virginia, USA.

American Society of Civil Engineers (2015), ASCE Standard 10-15, *Design of Latticed Steel Transmission Structures*, ASCE, Reston, Virginia, USA.

American Society of Civil Engineers (2003), ASCE Manual 104, *Recommended Practice for Fiber-Reinforced Polymer Products for Overhead Utility Line Structures.*

American Society of Civil Engineers (2006), ASCE Manual 111, *Reliability-Based Design of Utility Pole Structures.*

American Society of Civil Engineers (2008), ASCE Manual 113, *Substation Structure Design Guide.*

American Society of Civil Engineers (2012), ASCE Manual 123, *Prestressed Concrete Transmission Pole Structures.*

AASHTO (2010), *LRFD Bridge Design Specifications*, American Association of State Highway and Transportation Officials, Washington, DC.

ANSI C29.11 (2012) *American National Standard Tests to Composite Suspension Insulators for Overhead Transmission Lines-Tests.*

ANSI O5.1 (2015), *American National Standard for Wood Poles – Specifications and Dimensions*, ANSI Standard 05-1-2008, New York, USA.

ANSI O5.2 (2012), *American National Standard for Structural Glued Laminated Timber for Utility Structures*, ANSI Standard 05-2-2006, New York, USA.

ANSI/TIA (2006), *Structural Standard for Antenna Supporting Structures, ANSI-TIA-222-G*, Telecommunications Industry Association, Arlington, Virginia, USA.

AS 3995 (1994), *Design of Steel Lattice Towers and Masts*, Council of Standards, Australia.

ASCE-PCI (1987), *Guide for the Design and Use of Concrete Poles*, JR 257, American Society of Civil Engineers, Reston, Virginia, USA.

ASCE-PCI (1997), *Guide for the Design of Prestressed Concrete Poles*, JR 412, ASCE – PCI (Prestressed Concrete Institute) Joint Committee, ASCE, Reston, Virginia, USA, 42 (6), 94–134.

AS/NZS 7000 (2010), *Australian-New Zealand Standard for Overhead Line Design – Detailed Procedures, AS/NZS 7000.*

Ashraf, M., Ahmad, H.M. & Siddiqui, Z.A. (2005), A Study of Power Transmission Poles, *Asian Journal of Civil Engineering,* 6 (6), 511–532.

B

Banik, S.S., Hong, H.P. & Kopp, G.A. (2010), Assessment of Capacity Curves for Transmission Line Towers under Wind Loading, *Wind and Structures*, 13 (1), 1–20.

Bhushan, K., Lee, L.J. & Grime, D.B. (1981), Lateral Load Tests on Drilled Piers in Sand, *Proceedings, ASCE Conference Session on Drilled Piers and Caissons*, St. Louis, Missouri, 98–113.

Bieniawiski, Z.T. (1973), Engineering Classifications of Jointed Rock Masses, *Transactions, South African Institute of Civil Engineers*, 15, 335–344.

Bieniawski, Z.T. (1976), Rock Mass Classification in Rock Engineering, *Exploration for Rock Engineering, Proceedings of the Symposium*, Bieniawski and Balkema (eds.), 1, 97–106.

Bledsoe, J.K., Yamatani, K. & Glover, D. et al. (2013), 500 kV Bradford-Sullivan Storm Restoration, *ASCE Electrical Transmission and Substation Structures Conference-2013*, Chicago, Illinois, USA.

Bragg, R.A., DiGioia, A.M., Jr. & Rojas-Gonzalez, L.F. (1988), *Direct Embedment Foundation Research*, Report EL-6309, Electric Power Research Institute (EPRI), Palo Alto, California, USA.

British Standard 8100 (1999), *Lattice Towers and Masts – Code of Practice for Strength Assessment of Members in Lattice Towers and Masts*, British Standards Institute, United Kingdom.

Broms, B.B. (1964), Lateral Resistance of Piles in Cohesive Soils, *ASCE Journal of Soil Mechanics and Foundations Division*, 90 (SM2), 26–37.

Broms, B.B. (1964), Lateral Resistance of Piles in Cohesionless Soils, *ASCE Journal of Soil Mechanics and Foundations Division*, 90 (SM3), 123–156.

Broms, B.B. (1965), Design of Laterally-Loaded Piles, *ASCE Journal of Soil Mechanics and Foundations Division*, 79–99.

BS EN 50341 (2013), *Overhead Electrical Lines exceeding AC 1 kV*. General Requirements. Common specifications, British Standards Institute.

Bulletin 62–4 (1976), *Electrostatic and Electromagnetic Effects of Overhead Transmission Lines*, Rural Utilities Service (formerly REA), United States Department of Agriculture (USDA).

Bulletin 1724E-150 (2014), *Unguyed Distribution Poles: Strength Requirements*, Rural Utilities Service, United States Department of Agriculture (USDA).

Bulletin 1724E-152 (2003), *The Mechanics of Overhead Distribution Line Conductors*, Rural Utilities Service, United States Department of Agriculture USDA.

Bulletin 1724E-153 (2001), *Electric Distribution Line Guys and Anchors*, Rural Utilities Service, United States Department of Agriculture (USDA).

Bulletin 1724E-200 (2015), *Design Manual for High Voltage Transmission Lines*, Rural Utilities Service, United States Department of Agriculture (USDA).

Bulletin 1724E-204 (2008), *Guide Specifications for Steel Single Poles and H-Frame Structures*, Rural Utilities Service, United States Department of Agriculture (USDA).

Bulletin 1724E-205 (1995), *Design Guide: Embedment Depths for Concrete and Steel Poles*, Rural Utilities Service, United States Department of Agriculture (USDA).

Bulletin 1724E-206 (2016), *Guide Specification for Prestressed Concrete Poles and Pole Structures*, Rural Utilities Service, United States Department of Agriculture (USDA).

Bulletin 1724E-214 (2009), *Guide Specifications for Standard Class Steel Transmission Poles*, Rural Utilities Service, United States Department of Agriculture (USDA).

Bulletin 1724E-216 (2016), *Guide Specification for Standard Class, Prestressed Concrete Transmission Poles*, Rural Utilities Service, United States Department of Agriculture (USDA).

Bulletin 1724E-224 (2007), *Electric Transmission Guide Specifications and Drawings for Steel Pole Construction – 34.5 to 230 kV*, Rural Utilities Service, United States Department of Agriculture (USDA).

Bulletin 1724E-226 (2007), *Electric Transmission Guide Specifications and Drawings for Concrete Pole Construction – 34.5 to 230 kV*, Rural Utilities Service, United States Department of Agriculture (USDA).

Bulletin 1724E-300 (2001), *Design Guide for Rural Substations*, Rural Utilities Service, United States Department of Agriculture (USDA).

Bulletin 1728F-700 (2011), *Specifications for Wood Poles, Stubs and Anchor Logs*, Rural Utilities Service, United States Department of Agriculture (USDA).

Bulletin 1728H-701 (2011), *Specifications for Wood CrossArms (Solid and Laminated), Transmission Timbers and Pole Keys*, Rural Utilities Service, United States Department of Agriculture (USDA).

Bulletin 1724F-702 (2012), *Specifications for Quality Control and Inspection of Timber Products*, Rural Utilities Service, United States Department of Agriculture (USDA).

Bulletin 1728F-803 (1998), *Specifications and Drawings for 24.9/14.4 kV Line Construction*, Rural Utilities Service, United States Department of Agriculture (USDA).

Bulletin 1724F-804 (2005), *Specifications and Drawings for 12.5/7.2 kV Line Construction*, Rural Utilities Service, United States Department of Agriculture (USDA).

Bulletin 1728F-810 (1998), *Electric Transmission Specifications and Drawings 34.5 kV Through 69 kV*, Rural Utilities Service, United States Department of Agriculture (USDA).

Bulletin 1728F-811 (1998), *Electric Transmission Specifications and Drawings 115 kV through 230 kV*, Rural Utilities Service, United States Department of Agriculture (USDA).

C

CAISSON (2011), *Analysis and Design of Reinforced Concrete Pier Foundations,* Power Line Systems, Inc., Madison, Wisconsin, USA.

CENELEC (2001), *Overhead Electric Lines Standard EN50341-1,* European Committee for Electro-technical Standardization, Brussels, Belgium.

CHANCE (1998), *Encyclopedia of Anchoring*, A.B. Chance/Hubbell Power Systems, Centralia, Missouri, USA.

Chay, M.T., Albermani, F. & Hawes, H. (2006), Wind Loads on Transmission Line Structures in Simulated Downbursts, *First World Congress on Asset Management*, Gold Coast, Australia.

CIGRE Technical Brochure No. 207 (2002), Thermal Behaviour of Overhead Conductors, CIGRE Study Committee B2.

CSA-S37 (2013), Canadian Standards for Antennas, Towers, and Antenna-Supporting Structures, CSA Group, Mississauga, Ontario, Canada.

CSA-C22.3 1-15 (2015), Canadian Standards for Overhead Systems, CSA Group, Mississauga, Ontario, Canada.

D

Dai, K.S. & Chen, S.E. (2008), Modal Characteristics of Direct Embedded Concrete Poles, *Proceedings of the 26th IMAC Conference & Exposition on Structural Dynamics*, Orlando, Florida, USA.

Das, B.M. & Seeley, G.R. (1982), Uplift Capacity of Pipe Piles in Saturated Clay, *Soils and Foundations, The Japanese Society of Soil Mechanics and Foundation Engineering*, 22 (1), 91–94.

Davidson, H.L. (1981), *Laterally Loaded Drilled Pier Research – Volume I and II*, Report El-2197, Electric Power Research Institute (EPRI), Palo Alto, CA.

Davisson, M.T. & Prakash, S. (1963), *A Review of Soil Pole Behavior,* Report 39, Highway Research Board, Washington DC, USA, 25–48.

Dempsey, D. (1996), *Winds Wreak Havoc on Lines,* Transmission and Distribution World, June, 32–42.

Der Kiureghian, A. & Sackman, J.L. (2005), Tangent Geometric Stiffness of Inclined Cables Under Self Weight, *ASCE Journal of Structural Engineering*, 131 (6), 941–945.

DiGioia, A.M. Jr. (1985), Design of Drilled Piers Subjected to High Overturning Moments, *Innovations in the Design of Electrical Transmission Structures, Proceedings, ASCE Conference*, Gene M. Wilhoite (eds.), Kansas City, Missouri, USA.

DiGioia, A.M. Jr., Rojas-Gonzalez, L.F. & Newman, F.B. (1989), Statistical Analysis of Drilled Shaft and Embedment Pole Models, *Foundation Engineering: Current Principles and Practices*, F.H. Kulhawy (eds.), New York, 1338–1352.

DiGioia, A.M. Jr. & Rojas-Gonzalez, L.F. (1994), Rock Socket Transmission Line Foundation Performance, *IEEE Transactions on Power Delivery*, 9 (3), 1570–1576.

Design Standard No.10 – Transmission Structures (1965), *United States Department of the Interior, Bureau of Reclamation*, Denver, Colorado.

E

ECCS (1985), Recommendations for Angles in Lattice Transmission Towers, Technical Committee 8 on Structural Stability.

ECCS (2013), *European Convention for Constructional Steelwork*, Ernst & Sohn, Berlin, Germany.

Eidinger, J.M. & Kempner, L. (2012), *Reliability of Transmission Towers under Extreme Wind and Ice Loading*, Paris Session, CIGRE.

EN 10025 (2004), European Structural Steel Standard, UK.

EN 1993 (2004), Eurocode 3: Design of steel structures – Part 1-1: General rules and rules for buildings, Belgium.

ENATS 43-125 (2005), *Design Guide and Technical Specification for Overhead Lines above 45 kV*, Engineering Directorate, Energy Networks Association, London, England, UK.

EPRI (1990), *Manual on Estimating Soil Properties for Foundation Design*, Report EL-6800, (Kulhawy, F.H. and Mayne, P.W.) EPRI, Palo Alto, California, USA.

ESAA C(b)1-2006 (2006), ESAA Guidelines for Design and Maintenance of Overhead Distribution and Transmission Lines, Energy Networks Association, Australia.

F

FAD Tools – Version 5.1, (2015), *Foundation Analysis and Design* Program, EPRI, Inc., Palo Alto, California, USA.

Fujisaki, E. (2009), *Seismic Performance of Electric Transmission Systems*, Seminar Presentation, Pacific Gas and Electric Company, San Francisco, California, USA.

G

Ghobarah, A., Aziz, T.S. & El-Attar, M. (1996), Response of Transmission Lines to Multiple Support Excitation, *Engineering Structures*, 18 (12), 936–946.

Gere, J.M & Carter, W.O. (1962), Critical Buckling Loads for Tapered Columns, *ASCE Journal of the Structural Division*, 88 (ST1), 1–12.

GO 6 (1969), *Rules for Overhead Electric Line Construction*, General Order No. 6, Public Utilities Commission of the State of Hawaii, USA.

GO 95 (2016), *Rules for Overhead Electric Line Construction*, General Order No. 95, Public Utilities Commission of California, Sacramento, California, USA.

H

Haldar, A., Prasad, Y.V.S.N. & Chari, T.R. (2000), Full Scale Tests on Directly-Embedded Steel Pole Foundations, *Canadian Geotechnical Journal*, 37 (2), 414–437.

Haldar, A., Chari, T.R. & Prasad, Y.V.S.N. (1997), *Experimental and Analytical Investigations of Directly Embedded Steel Pole Foundations*, CEA Project No. 384 T-971, Canadian Electricity Association, Montreal, Canada.

Hansen, J.B. (1961), *The Ultimate Capacity of Rigid Piles Against Trans- versal Forces*, Bulletin 12, The Danish Geotechnical Institute, Copenhagen, Denmark, 5–9.

Hansen, J.B. (1970). *A Revised and Extended Formula for Bearing Capacity*, Bulletin 28, The Danish Geotechnical Institute, Copenhagen, Denmark.

HeliCAP™ (2012), *Helical Capacity Design Software*, Users' Manual, Hubbell Power Systems, Centralia, Missouri, USA.

Hoyt, R.M. & Clemence, S.P. (1989), Uplift Capacity of Helical Anchors in Soil, *12th International Conference on Soil Mechanics and Foundation Engineering*, Rio de Janeiro, Brazil.

Hughes Brothers *Engineering Manual* (1943), Hughes Bros., Inc., Seward, Nebraska, USA.

Hughes Brothers *EHV Structures* (1953), Hughes Bros., Inc., Seward, Nebraska, USA.

Hughes Brothers *Product Catalog* (2012), T & D Materials, Hughes Bros., Inc., Seward, Nebraska, USA.

I

IEC-60826 (2010), *Design Criteria of Overhead Transmission Lines*, International Electro-technical Commission, Geneva, Switzerland.

IEC-61284 (1997), *Overhead Lines – Requirements and Tests for Fittings*, International Electro-technical Commission, Geneva, Switzerland.

IEEE-524 (2016), *Guide to the Installation of Overhead Transmission Line Conductors*.

IEEE-691 (2001), *Guide for Transmission Structure Foundation Design and Testing*, Institute of Electrical and Electronics Engineers, New York, USA.

IEEE-693 (2005), *Recommended Practices for Seismic Design of Substations*, Institute of Electrical and Electronics Engineers, New York, USA.

IEEE-738 (2013), *Standard for Calculating the Current-Temperature Relationship of Bare Overhead Conductors*, Institute of Electrical and Electronics Engineers, New York, USA.

IEEE-1724 (2011), *Guide for Preparation of a Transmission Line Design Criteria Document*, New York, USA.

Ireland, H.O. (1963), Discussion on Uplift Resistance of Transmission Tower Footing, *ASCE Journal of the Power Division*, 89 (1), 115–118.

IS-800 (2007), *General Construction in Steel – Code of Practice, Bureau of Indian Standards*.

IS-802 (1995), *Use of Structural Steel in Overhead Transmission Line Towers – Code of Practice*, Indian National Standards, Bureau of Indian Standards, New Delhi, India.

IS 4091 (2010), *Code of Practice for Design and Construction of Foundations for Transmission Line Towers and Poles*, Bureau of Indian Standards.

Ishac, M.F. & White, H.B. (1995), Effect of Tornado Loads on Transmission Lines, *IEEE Transactions on Power Delivery*, 10 (1), 445–451.

Iso-Truss Structures Inc. (2002), *Technical Overview of Iso-Truss Technology*, Center for Advanced Structural Composites, Brigham Young University, Provo, Utah, USA.

Ivey, D. L. (1968), Theory, *Resistance of a Drilled Shaft Footing to Overturning Loads*, Texas Transportation Institute, Research Report No. 105-1, Texas, USA.

J

Jayachandran, P., Hannigan, J.F., Browne, M.S. & Reynolds, B.M. (2006), *Dynamic Load Factors for Transmission Towers Due to Snapped Conductors,* (origins unknown).

K

Kandaris, P.M., DiGioia, A.M. Jr. & Heim, Z. (2012), Evaluation of Performance Criteria for Short Laterally Loaded Drilled Shafts, *GeoCongress 2012*, 165–174.

Keshavarzian, M. (2002), Self-Supported Wood Pole Fixity at the ANSI Ground Line, *Electrical Transmission in a New Age, Proceedings of the ASCE Conference*, Omaha, Nebraska, 268–281.

Koser, F.Y. & Arora, J.S. (1999), Optimal Design of H-Frame Transmission Poles for Earthquake Loading, ASCE Journal of Structural Engineering, 125 (11), 1299–1308.

Koser, F.Y. & Arora, J.S. (2002), Optimal Design of Latticed Tower Subjected to Earthquake Loading, *ASCE Journal of Structural Engineering*, 128 (2), 197–204.

Kulhawy, F.H. (1985), Uplift Behavior of Shallow Soil Anchors – An Overview, *Uplift Behavior of Anchor Foundations in Soil, Proceedings, ASCE Conference.*

Kulhawy, F.H., Trautmann, C.H., Beech, J.F., O'Rourke, T.D., McGuire, W., Wood, W.A. & Capano, C. (1983), Transmission Line Structure Foundations for Uplift-Compression Loading, Report EL-2870, EPRI, Palo Alto, California, 412 p.

L

Laminated Wood Systems (2012), *Engineered Laminated Wood Utility Structure Design*, LWS Seminar, Minneapolis, Minnesota, USA.

Lei, Y.-H. & Chien, Y.-L. (2005), Seismic Analysis of Transmission Towers Considering Both Geometric and Material Nonlinearities, *Tamkang Journal of Science and Engineering*, 8 (1), 29–42.

Li, C.Q. (1999), A Stochastic Model of Severe Thunderstorms for Transmission Line Design, *Journal of Probabilistic Engineering Mechanics,*15 (4), 359–364.

Long, L.W. (1973), Analysis of Seismic Effects on Transmission Structures, *Proceedings, IEEE-PES Summer Meeting*, Vancouver, British Columbia, Canada, 1–8.

LPILE (2015), *A Program for the Analysis of Piles and Drilled Shafts under Lateral Loads,* Users' Manual, Ensoft Inc., Austin, Texas, USA.

M

Matsui, M. & Tamura, Y. (2009), Influence of Incident Flow Conditions on Generation of Tornado-like Flow, *11th Americas Conference on Wind Engineering*, San Juan, Puerto Rico.

Mazda, T., Matsumoto, M., Oka, N. & Ishida, N. (2010), Evaluation of Seismic Behavior of Steel Transmission Towers with Different Boundary Conditions, *10th International Conference on Computational Structures Technology*, Stirlingshire, Scotland, UK.

McClure, G. & Lapointe, M. (2003), Modeling the Structural Dynamic Response of Overhead Transmission Lines, *Computers and Structures*, 81, 825–834.

Meyerhof, G.G. (1953), The Bearing Capacity of Foundations Under Eccentric and Included Loads, *Proceedings, 3rd International Conference on Soil Mechanics and Foundation Engineering*, Zurich, 1, 440–445.

Meyerhof, G.G. (1963), Some Recent Research on the Bearing Capacity of Foundations, *Canadian Geotechnical Journal*, 1 (1), 16–26.

Meyerhof, G.G., Mathur, S. K. & Valsangkar, A. J. (1981), Lateral Resistance and Deflection of Rigid Wall and Piles in Layered Soils, *Canadian Geotechnical Journal*, 18, 159–170.

Meyerhof, G.C. & Adams, J.I. (1968), The Ultimate Uplift Capacity of Foundations, *Canadian Geotechnical Journal*, 5 (4), 225–244.

MFAD 5.1 (2015), *Moment Foundation Analysis and Design*, Software, Electric Power Research Institute (EPRI), Palo Alto, California, USA.

N

NEMA-SG6 (2000), *Allowable Stress Design (ASD) Methods*, Part 36, National Electrical Manufacturers Association, Rosslyn, Virginia, USA.

NEMA-SG6 (2006), *Power Switching Equipment*, National Electrical Manufacturers Association, Rosslyn, Virginia, USA.

NESC (2012), *National Electrical Safety Code*, ANSI C2, Institute of Electrical and Electronics Engineers, New York, USA.

Nishenko, S., Mosalam, K., Takhirov, S. & Fujisaki, E. (2012), Seismic Performance of Insulators for Electric Substations, *NEES Quake Summit*, Boston, Massachusetts, USA.

P

Peck, R.B., Hanson, W.E. & Thornburn, T.H. (1974), *Foundation Engineering*, 2nd Edition, John Wiley and Sons, New York.

Petrasovits, G. & Award, A. (1972), Ultimate Lateral Resistance of a Rigid Pile in Cohesion-less Soil, *Proceedings, 5th European Conference on Soil Mechanics and Foundation Engineering*, Madrid, 3, 407–412.

PLS-POLE™ (2012), *Computer Program for the Analysis and Design of Transmission Poles*, Power Line Systems, Inc., Madison, Wisconsin, USA.

PLS-CADD™ (2012), *Computer-Aided Design and Drafting*, Power Line Systems, Inc., Madison, Wisconsin, USA.

PN-90 (1994), *Polish Standard PN-90 B-03200* Konstrukcje Stalowe Obliczenia Statyczne I Projektowanie (Steel Structures Design Rules), Warsaw, Poland.

Poulos, H.G. & Davis, E.H. (1980), *Pile Foundation Analysis and Design*, John Wiley & Sons, New York, NY, USA.

Prasad, Y.V.S.N. & Chari T.R. (1996), Rigid Pile with a Baseplate under Large Moments: Laboratory Model Evaluations, *Canadian Geotechnical Journal*, 33 (4), 1021–1026.

Prasad, Y.V.S.N. & Chari, T.R. (1999), Lateral Capacity of Rigid Piles in Cohesion-less Soils, *Soils and Foundations, Japanese Geotechnical Society*, 39 (2), 21–29.

Prasad, Y.V.S.N. (1997), Prediction of Lateral Capacity of Piles in Clays from Standard Penetration Tests, *Proceedings, 7th International Offshore and Polar Engineering Conference (ISOPE-97)*, Honolulu, Hawaii, 767–771.

Prasad, Y.V.S.N. & Haldar, A. (2001), Full Scale Tests on Grillage Anchor Foundations for High Voltage Transmission Lines, *Proceedings, 54th Canadian Geotechnical Conference*, Calgary, Alberta.

Pultruded Utility Products International (PUPI) (2011), *Fiber Glass Composite Cross Arms*, Geo-Tek Inc., Stewartville, Minnesota, USA.

R

Ragan, Q. S., Selvam, R.P. & Gorecki, P. (2014), Tornado-induced Wind Forces for Cylindrical, Transmission Line Structures, *Symposium on Transmission and Substation Design*, University of Texas, Arlington, USA.

Reese, L.C., & Matlcok, H. (1956), Non-Dimensional Solutions for Laterally Loaded Piles with Soil Modulus Assumed Proportional to Depth, *Proceedings, 8th Texas Conf. Soil Mechanics and Foundation Engineering*, Austin, Texas, USA.

Rose, A.T., DiGioia A.M. & Hirany A. (2001), Use of Rock Mass Ratings (RMR) to Estimate Rock Properties for Laterally Loaded, Rock Socketed Transmission Line Foundations, *Rock Mechanics in the National Interest*, Swets & Zeitlinger, Lisse, Netherlands.

RS Technologies (2012), *RS Pole Design Guide: General Methodologies and Procedures for Structure Design using RS Modular Composite Poles*, Calgary, Alberta, Canada.

S

SHAFT (2012), *A Program for the Study of Drilled Shafts under Axial Loads*, Users' Manual, Ensoft Inc., Austin, Texas, USA.

Shakespeare Composite Structures (2008), *Catalog of Composite Transmission and Distribution Products*, SCS, Newberry, South Carolina, USA.

Savory, E., Parke, G.A.R., Zeinoddini, M., Toy, N. & Disney, P. (2001), Modeling of Tornado and Microburst-Induced Wind Loading and Failure of a Lattice Transmission Tower, *Engineering Structures*, 23, 365–375.

Stas, C.V. & Kulhawy, F.H. (1984), *Critical Evaluation of Design methods for Foundations Under Axial Uplift and Compression Loading*, Report EL-3771, EPRI, Palo Alto, California, USA, 198 p.

Stengel, D. & Mehdianpour, M. (2014), Finite Element Modeling of Electrical Overhead Line Cables under Turbulent Wind, *Journal of Structures*, Hindawi Publishing Corp., 2014, 8 pp.

Shehata, A.Y., El-Damatty, A. & Savory, E. (2005), Finite Element Modeling of Transmission Lines Under Downburst Wind Loading, *Finite Elements in Analysis and Design*, 42 (1), 71–89.

Southwire (2014), *SAG10™ Ver. 4.03, Users' Manual*, Southwire Company, Carrollton, Georgia, USA.

Stresscrete (2009), *Utility Poles: Concrete Pole Catalog*, Stress Crete Group, Jefferson, Ohio, USA.

T

TC-07 (2013), Electricity Transmission Code, South Australia.

TERNA (2013) *130 Years of History for Electricity Transmission*, Directorate of External Relations and Communications, Rome, Italy.

Terzaghi, K. & Peck, R. B. (1967), *Soil Mechanics in Engineering Practice*, 2nd Edition, John Wiley and Sons, New York.

Terzaghi, K. (1943), *Theoretical Soil Mechanics*, John Wiley and Sons, New York.

Tian, L., Hongnan, L. & Guohuan, L. (2010), Seismic Response of Power Transmission Tower Line System Subjected to Spatially-Varying Ground Motions, *Journal of Mathematical Problems in Engineering*, Hindawi Publishing Corp., 2010, 20 pp.

TOWER™ (2012), *A Computer Program for the Analysis and Design of Steel Latticed Towers used in Transmission and Communication Facilities*, Power Line Systems, Inc., Madison, Wisconsin, USA.

TP.DL 12.01 (2003), Transmission Line Loading Code, Transpower, New Zealand.

TPNZ (2003), *Transmission Line Loading Code*, Transpower Standard TP-DL-12.01, Wellington, New Zealand.

Trautmann, C.H. & Kulhawy, F.H. (1988), Uplift Load Displacement Behavior of Spread Foundations, *ASCE Journal of Geotechnical Engineering*, 114 (20), 168–185.

V

Vesic, A.S. (1973), Analysis of Ultimate Loads of Shallow Foundations, *ASCE Journal of the Soil Mechanics and Foundations Division*, 99 (SM1), 45–73.

W

Watson, G. (2007), Oblique Wind on Structures: An Inconvenient Truth, *PLS Users Group Meeting*, San Diego, California, USA.

References – Additional reading

A

American Concrete Institute (2010), ACI-301, *Specifications for Structural Concrete*, Farmington Hills, Michigan, USA.

American Institute of Steel Construction (1953), *Historical Record – Dimensions and Properties: Iron and Steel Beams 1873 to 1952*, AISC, New York, USA.

American Institute of Timber Construction (2001), Standard 110, *Standard Appearance Grades for Structural Glued Laminated Timber*, AITC, Centennial, Colorado, USA.

American Institute of Timber Construction (2005), Standard 111, *Recommended Practice for Protection of Structural Glued Laminated Timber During Transit, Storage and Erection*, AITC, Centennial, Colorado, USA.

American Society of Civil Engineers (1987), Foundations for Transmission Line Towers, *Proceedings, Geotechnical Convention*, Atlantic City, New Jersey, USA.

ANSI/NEMA TT-1 (1983), *Tapered Tubular Steel Structures,* ANSI Standard, New York, USA.

B

Bowles, J. E. (1982), *Foundation Analysis and Design*, 3rd Edition, McGraw-Hill, Inc., New York, New York, USA.

Bulletin 1724E-203 (2016), *Guide for Upgrading Transmission Lines*, Rural Utilities Service, United States Department of Agriculture (USDA).

C

CAI (1976), *Foundation Design Guidelines for Transmission Line Structures*, Internal Report, Commonwealth Associates Inc., Jackson, Michigan, USA.

Code of Practice (2009), *Electricity Transmission and Distribution Asset Management*, Department of Water and Energy, New South Wales, Australia.

CRN ET-002 (2012), Engineering Standards – Transmission Lines, Country Regional Network, New South Wales, Australia.

D

Das, B.M. (1980), A Procedure for Estimation of Ultimate Uplift Capacity of Foundations in Clay, *Soils and Foundations, The Japanese Society of Sol Mechanics and Foundation Engineering,* 20 (1).

Das, Braja M. (1980), A Procedure for Estimation of Ultimate Uplift Capacity of Foundations in Clay, *Soils and Foundations, Japanese Society of Soil Mechanics and Foundation Engineering*, 20 (1), 77–82.

DiGioia, Jr, A. M., Hirany, A., Newman, F. B. & Rose, A. T. (1998), Rock-Socketed Drilled Shaft Design for Lateral Loads, *ESMO'98 Conference*, Orlando, Florida, USA.

E

ENATS 43-8 (2004), *Technical Specification for Overhead Line Clearances*, Engineering Directorate, Energy Networks Association, London, England, UK.

EPRI (1982), *Transmission Line Reference Book – 345 kV and Above*, Electric Power Research Institute, Palo Alto, California, USA.

EPRI (1997), *Fiber Optic Cables in Overhead Transmission Corridors*, Research Report No. TR-108959, EPRI, Haslet, Texas, USA.

EPRI (2005), AC *Transmission Line Reference Book – 200 kV and Above*, 3rd Edition, Electric Power Research Institute, Palo Alto, California, USA.

F

Farr, H.H. (1980), *Transmission Line Design Manual*, U.S. Department of the Interior, Denver, Colorado, USA.

G

Gaylord, E.H. (1973), *Design Criteria for Steel Transmission Poles,* Meeting Preprint 2021, Task Committee on Steel Transmission Poles, ASCE Structural Division, San Francisco, California, USA.

H

Haldar, A., Yenumula, P., Bursey, A. & Tucker, K. (2002), Residual Life Assessment of Existing Transmission Line with Particular Reference to Grillage Foundation Corrosion Problem, *Electrical Transmission in the New Age, ASCE Specialty Conference on Electrical Transmission Line and Substation Structures,* Omaha, Nebraska, USA.

Hansen, J.B. (1948), *The Stabilizing Effect of Piles in Clay*, CN-Post 3, Christiani and Nelson, Copenhagen, Denmark.

Horvath, R.G. & Kenney, T.C. (1979), Shaft Resistance of Rock-socketed Drilled Piers, *ASCE October Conference*, Preprint 3698, Atlanta, Georgia, USA.

Hughes Brothers *Product Catalog* (2012), T & D Materials, Hughes Bros., Inc., Seward, Nebraska, USA.

Hughes Brothers *EHV Structures* (1953), Hughes Bros., Inc., Seward, Nebraska, USA.

I

IEC-60652 (2002), *Loading Tests on Overhead Line Structures,* International Electro-technical Commission, Geneva, Switzerland.

IE-1956 (1956), *Indian Electricity Rules*, Ministry of Irrigation and Power, New Delhi, India.

IEEE-751 (1991), *Trial Use Design Guide for Wood Transmission Structures*, Institute of Electrical and Electronics Engineers, New York, USA.

IEEE-951 (1996), *Guide for Assembly and Erection of Metal Transmission Structures*, Institute of Electrical and Electronics Engineers, New York, USA.

IEEE-1283 (2002), *Guide for Determining the Effects of High Temperature Operation on Conductors, Connectors and Accessories*, New York, USA.

Ivey, D.L. (1968), Theory, *Resistance of a Drilled Shaft Footing to Overturning Loads*, Texas Transportation Institute, Research Report No. 105-1, Texas, USA.

J

Jacksonville Electric Authority (2003), *Overhead Transmission Construction Standards*, JEA, Jacksonville, Florida, USA.

K

Kalaga, S. (1992), Reliability Based Analysis and Design of Transmission Tower Structural Systems, *Journal of Structural Engineering, SERC*, 19 (2), 65–75.

Kalaga, S. (2010), *Design Guidelines for Transmission Line Structures: Vols. 1 and 2*, Personal Engineering Notes, Joplin, Missouri, USA.

Kalaga, S. & Adluri, S.M.R. (1999), End Restraints in Angle Columns, 6th *International Conference on Steel and Space Structures*, Singapore, 389–395.

Kalaga, S., Collins, R. & Mooberry, D. (2003), Reliability Assessment of Steel Transmission Poles, *Journal of Structural Engineering, SERC*, 30 (3), 197–200.

Kalaga, S., Collins, R. & Mooberry, D. (2004), Reliability Assessment of Transmission Structures, *Journal of Structural Engineering, SERC*, 30 (4), 225–237.

Kulhawy, F.H., Trautmann, C.H., Beech, J.F., O'Rourke, T.D., McGuire, W., Wood, W.A. & Capano, C. (1983), Transmission Line Structure Foundations for Uplift-Compression Loading, Report EL-2870, EPRI, Palo Alto, California, 412 p.

Kulhawy, F.H. & Jackson, C.S. (1989), Some Observations on Undrained Side Resistance of Drilled Shafts, *Proceedings, ASCE Foundation Engineering Conference: Current Principles and Practices*, 2, 1011–1025.

L

Li, C.Q. & Holmes, J.D. (1995), *Design of Transmission Line Structures in Australia: Present and Future*, CSIRO, Division of Building and Construction Engineering, South Clayton, Victoria, Australia.

Lings, R., Chartier, V. & Sarma, M. (2005), Overview of Transmission Lines Above 700 kV, *Proceedings, IEEE-PES Conference and Exposition in Africa*, Durban, South Africa, 33–44.

Lonestar Prestress (2009), *Prestressed Concrete Utility Poles Catalog*, Lonestar Prestress Manufacturing Inc., Houston, Texas, USA.

M

Mikhelson, I. (2004), *Structural Engineering Formulas,* Mc-Graw Hill, New York, USA.

Miller, M.F., Hosford, G.S. & Boozer, J.F. (1995), Fiberglass Distribution Poles: A Case Study, *IEEE Transactions on Power Delivery*, 10 (1), 497–503.

Mitch, M.P. & Clemence, S.P. (1985), The Uplift Capacity of Helix Anchors in Sand, *Proceedings, Special Session sponsored by ASCE Geotechnical Engineering Division*, Michigan, 26–47.

Mooney, J.S. & Adamczak, S.(1985), Uplift Capacity of Helix Anchors in Clay and Silt, *Proceedings, Special Session sponsored by ASCE Geotechnical Engineering Division*, Michigan.

N

Narasimha Rao, S. & Prasad, Y.V.S.N. (1993), Estimation of Uplift Capacity of Helical Anchors in Clay, *ASCE Journal of Geotechnical Engineering*, 119 (4), 352–357.

Narasimha Rao, S. & Prasad, Y.V.S.N. (1992), Behavior of Plate Anchors in Sloped Clayey Ground, *Soils and Foundations, Japanese Society of Soil Mechanics and Foundation Engineering*, 32 (4), 164–170.

Narasimha Rao, S. & Prasad, Y.V.S.N. (1992), Behavior of Plate Anchors Embedded in Two-layered Clayey Soil, *Soils and Foundations, Japanese Society of Soil Mechanics and Foundation Engineering*, 23 (2), 3–15.

Narasimha Rao, S., Prasad, Y.V.S.N. & Veeresh, C. (1994), Behavior of Embedded Screw Anchors in Soft Clays, *Geotechnique, Journal of Institution of Civil Engineers, London*, 43 (4), 605–614.

Narasimha Rao, S. & Prasad, Y.V.S.N. (1993), Uplift Behavior of Pile Anchors Subjected to Lateral Cyclic Loading, *ASCE Journal of Geotechnical Engineering*, 119 (4), 786–790.

Narasimha Rao, S. & Prasad, Y.V.S.N. (1993), Estimation of Uplift Capacity of Helical Anchors in Clay, *ASCE Journal of Geotechnical Engineering*, 119 (4), 352–357.

Narasimha Rao, S. & Prasad, Y.V.S.N. (1992), Discussion – Uplift Behavior of Screw Anchors in Sand 1: Dry Sand, *ASCE Journal of Geotechnical Engineering*, 118 (9), 1474–1478.

Narasimha Rao, S. & Prasad, Y.V.S.N. (1992), Behavior of Helical Anchors under Vertical Repetitive Loading, *International Journal of Sea-Floor Science and Engineering*, Taylor and Francis, 10, 203–228.

Narasimha Rao, S. & Prasad, Y.V.S.N. (1992), Behavior of Plate Anchors in Sloped Clayey Ground, *Soils and Foundations, Japanese Society of Soil Mechanics and Foundation Engineering*, 32 (4), 164–170.

Narasimha Rao, S., Prasad, Y.V.S.N. & Shetty, M.D. (1991), The Behavior of Model Screw Piles in Cohesive Soils, *Soils and Foundations, Japanese Society of Soil Mechanics and Foundation Engineering*, 31 (2), 35–50.

Narasimha Rao, S., Prasad, Y.V.S.N., Shetty, M.D. & Joshi, V.V. (1989), Uplift Capacity of Screw Pile Anchors, *Geotechnical Engineering, Journal of South East Asian Geotechnical Society*, 20, 139–159.

Narasimha Rao, S. & Prasad, Y.V.S.N. (1993), Experimental Studies on Plate Anchors in Layered Marine Soils, *3rd International Conference of Offshore and Polar Engineering*, Singapore.

Narasimha Rao, S. & Prasad, Y.V.S.N. (1992), Studies on Pile Anchors under Vertical Cyclic Loading in Soft Marine Clays, *Piletalk International-92*, Singapore, 1, 179–184.

Narasimha Rao, S., Rao K.M. & Prasad, Y.V.S.N. (1992), Behavior of Vertical Piles under Static Load and Cyclic Lateral Load in Marine Clays, *2nd International Conference of Offshore and Polar Engineering*, San Francisco, 475–482.

Narasimha Rao, S. & Prasad, Y.V.S.N. (1991), Behavior of Screw Piles under Compressive Loads, *Piletalk International-91*, Kuala Lumpur, Malaysia, 1, 107–113.

NDS (1980), *National Design Specification for Wood Structures,* National Forest Products Association, Washington D.C., USA.

P

Parker, F., Jr. & Reese, L.C. (1970), *Experimental and Analytical Studies of Behavior of Single Piles in Sand under Lateral and Axial Loading*, Research Report 117-1, Center for Highway Research, University of Texas at Austin, Texas, USA.

Peabody, A.B. & Wekezer, J.W. (1994), Buckling Strength of Wood Power Poles Using Finite Elements, *ASCE Journal of Structural Engineering*, 120 (6), 1893–1908.

PJM (2002), *Design and Application of Overhead Transmission Lines 69 kV and Above*, PJM Interconnection, LLC, Audubon, Pennsylvania, USA.

Prasad, Y.V.S.N. (1997), Standard Penetration Tests and Pile Behavior under Lateral Loads in Clays, *Indian Geotechnical Journal, Indian Geotechnical Society*, 27 (1), 12–21.

Prasad, Y.V.S.N. & Narasimha Rao, S. (1996), Lateral Capacity of Helical Piles in Clays, *ASCE Journal of Geotechnical Engineering,* 122 (11), 938–941.

Prasad, Y.V.S.N. (1996), Discussion – Drivability and Pullout Resistance of Helical Units in Saturated Sands, *Soils and Foundations, Japanese Journal of Soil Mechanics and Foundation Engineering,* 36 (2), 139–140.

Prasad, Y.V.S.N. (1996), Discussion – Cyclic Axial Loading of Drilled Shafts in Cohesive Soil, *ASCE Journal of Geotechnical Engineering*, 122 (3), 257–258.

Prasad, Y.V.S.N. & Narasimha Rao, S. (1995), Uplift Capacity of Helical Anchors under Vertical Cyclic Loading, *Indian Geotechnical Journal. Indian Geotechnical Society*, 25 (2), 12–21.

Prasad, Y.V.S.N. & Narasimha Rao, S. (1994), Pullout Behavior of Pile and Helical Pile Anchors Subjected to Lateral Cyclic Loading, *Canadian Geotechnical Journal,* 31 (1), 110–119.

Prasad, Y.V.S.N. & Narasimha Rao, S. (1994), Experimental Studies on Foundations of Compliant Structures: Volume I – Under Static Loading, *Ocean Engineering Journal, US Naval Academy*, PergamonPress, 21 (1), 1–13.

Prasad, Y.V.S.N. & Narasimha Rao, S. (1994), Experimental Studies on Foundations of Compliant Structures: Volume II – Under Cyclic Loading, *Ocean Engineering Journal, US Naval Academy*, Pergamon Press, 21 (1), 15–27.

Prasad, Y.V.S.N., Rao, K.M. & Narasimha Rao, S. (1991), Behavior of Screw Pile Anchors in Marine Environments, *Proceedings, 10th International Conference on Offshore Mechanics and Artic Engineering*, Stavanger, Norway.

Prasad, Y.V.S.N., Asim Haldar, Chari, T.R. & Ryan Phillips (1996), Centrifuge Modeling of Directly Embedded Steel Pole Foundations, *Geotechnical News, North American Geotechnical Community*, 14 (4), 34–36.

Phillips, D. (2014), Transmission Long Span Design, *MIPSYCON*, Minneapolis, Minnesota, USA.

Phillips, R. & Prasad, Y.V.S.N. (1995), *Direct Embedment of Steel Poles, Task 5-2: Centrifuge Model Tests,* Report No. 95-C29, C-Core Contract Report for Newfoundland and Labrador Hydro and Canadian Electricity Association.

R

RISA-3D (2013), RISA Technologies, Foothill Ranch, California, USA.

S

SAPS (1997), *A Computer Program for Nonlinear Structural Analysis of Power and Communication Systems,* Power Line Systems, Inc., Madison, Wisconsin, USA.

Sarkar, P., Haan, F., Gallus, W., Le, K. & Wurman, J. (2008), *Velocity Measurements in a Laboratory Tornado Simulator and Their Comparison with Numerical and Full-Scale Test Data,* Research Report, WST Laboratory, Iowa State University, Ames, Iowa, USA.

Sowers, G.B. & Sowers, G.F. (1970), *Introductory Soil Mechanics and Foundations*, 3rd Edition, Macmillan Publishing Company, London, UK.

Sowers, G.F. (1979), *Introductory Soil Mechanics and Foundations*, 4th Edition, MacMillan Publishing Company, New York.

T

Trans American Power Products (2005), *Catalog of Steel Transmission and Distribution Products*, TPP, Houston, Texas, USA.

U

Uniform Building Code (1997), International Conference of Building Officials.

U.S. Nuclear Regulatory Commission (2006), *Design Basis Tornado and Tornado Missiles for Nuclear Power Plants*, Regulatory Guide 1.76, Washington DC, USA.

V

Vesic, A.S. (1977), *Design of Pole Foundations,* NCHRP, Synthesis of Practice No. 42, Transportation Research Board, Washington, DC.

W

Wang, C.K. & Salmon, C.G. (1979), *Reinforced Concrete Design*, Harper and Row, New York, USA.

Appendix I

Analysis and design of a transmission line

This section illustrates the complete structural analysis and design of a High-Voltage Transmission Line using two different structural materials:

1. All Steel Structures (galvanized)
2. All Wood Structures (treated)

The model of the line serves to demonstrate several structural configurations, namely:

a. TAN Tangent (line angle less than 3°)
b. SA Small Angle (up to 15°)
c. LA Large Angle (45° and above)
d. DDE Full (Double) Deadends (line angle variable)
e. Substation Frames (not designed here; nominally shown only for illustration purposes. Substation structure design is outside the scope of this book).

All steel poles are either directly embedded (*a* and *b*) or supported on concrete piers (*c*, *d* and *e*). Configurations *b*, *c* and *d* (wood) are guyed poles with anchors.

The analysis and design process covers the following issues:

1. Structural performance of the entire system – strengths and usage
2. Conductor and OHGW performance – sags and clearances, galloping
3. Foundation Adequacy – direct embedment, piers, anchors

The line is modeled in PLS-CADD™, structures in PLS-Pole™ and foundations in CAISSON™. Load and Strength Factors conform to NESC Standards.

System data

Voltage 230 kV AC

Conductor 954 ACSR 54/7 Cardinal
Dia. = 1.196 in (30.4 mm), Area = 0.846 in^2 (545.8 mm^2),
Weight = 1.229 plf (17.94 N/m),
Rated Tensile Strength = 33,800 lbs (150.4 kN)

OHGW	3/8 in EHS Dia. = 0.36 in (9.14 mm), Area = 0.0792 in^2 (51.1 mm^2), Weight = 0.273 plf (3.98 N/m), Rated Tensile Strength = 15,400 lbs (68.5 kN)
Guy Wires	9/16 in EHS Dia. = 0.563 in (14.29 mm), Area = 0.248 in^2 (160.3 mm^2) Rated Tensile Strength = 35,000 lbs (155.75 kN) Strength Factor = 0.9 Usable Tensile Strength = 0.9 * 35.0 = 31.5 kips (140.2 kN)

Insulator data

Horizontal Post	Cantilever Rating 10,000 lbs (44.5 kN)
Strain/Suspension	Specified Rating 25,000 lbs (111.3 kN)

Soil data

30 ft (9.14 m):	Medium Dense Coarse Sand with Gravel, Blow Count N = 15 (Soil is approximately Type 5) Friction angle 30° Unit Weight = 100 pcf (15.7 kN/m^3) Ultimate Bearing Capacity = 10,000 psf. (478.9 kPa) Factor of Safety for Bearing = 3.0 Allowable Bearing Pressure = 10,000/3 = 3333.3 psf. (159.6 kPa)

Anchors

Helical Screw Anchors: 3-helix (8 in-10 in-12 in)
Maximum Holding Capacity in Type 5 Soil = 39,000 lbs (173.6 kN)
Strength Factor = 0.9
Usable Holding Capacity = 0.9 * 39.0 = 35.1 kips (156.2 kN)

Structural material

Steel Poles	ASTM A572-65	Yield Strength = 65 ksi (448 MPa)	12-sided polygons
Base Plates	ASTM A572-50	Yield Strength = 50 ksi (345 MPa)	
Substation Beams	ASTM A572-50	Yield Strength = 50 ksi (345 MPa)	
Wood Poles	Southern Yellow Pine (SYP) Maximum Flexural Strength = 8000 psi (55.2 MPa)		
Anchor Bolts	ASTM A615 Grade 75 2¼ in (57.2 mm) diameter		

Conductor tensions

Design Tension for 954 ACSR = 11,840 lbs (52.6 kN) at NESC Heavy Loading level)
Design Tension for 3/8 in EHS = 5,390 lbs (24 kN) at NESC Heavy Loading level)
Tensions in Slack Spans = 10% Rated Tensile Strength
Slack Tension for 954 ACSR = 3,380 lbs (15 kN)
Slack Tension for 3/8 in EHS = 1,540 lbs (6.9 kN)

Maximum tensions (as % of rated tensile strength)

Initial Unloaded = 35% (conductor), 25% (OHGW)
NESC Heavy = 50%
Extreme Wind = 70% (conductor), 80% (OHGW)
Extreme Ice = 70% (conductor), 80% (OHGW)
Extreme Ice with Concurrent Wind = 70% (conductor), 80% (OHGW)

Clearances

Maximum Operating Temperature (MOT) of Conductor = 212°F (100°C)
Required conductor ground clearance at MOT = 22.4 ft (6.83 m) without any buffer

System geometry

The layout of the transmission line selected for modeling is shown in Figure A1.1. The line runs from Substation 1 to Substation 2 and consists of two (2) double deadend structures DDE1 and DDE2, one large angle structure LA, one tangent structure TAN and one small angle structure SA. The two end spans are slack spans (i.e.) wires with reduced tension into the substations. The computed ruling span between DDE1 and DDE2 is about 518 ft (157.9 m). Right of Way is 150 ft (45.7 m). Tables A1.1 and A1.2

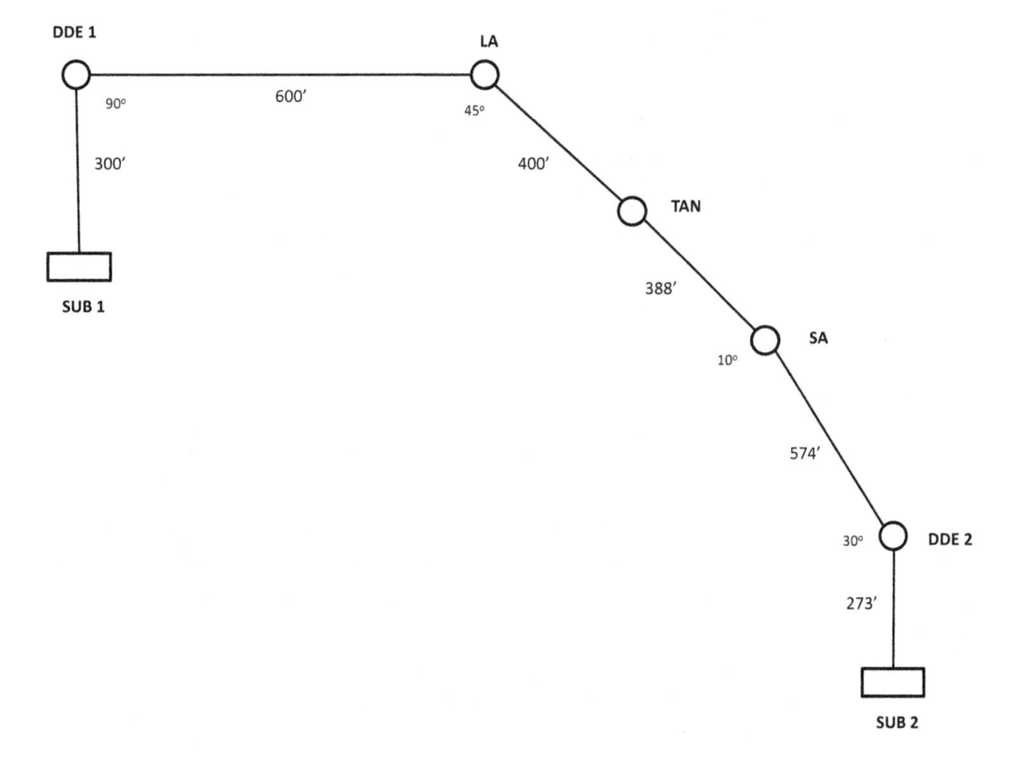

Figure A1.1 Layout of Design Transmission Line.

show details of the wood and steel structures considered for the line. The profile of the transmission line, as obtained from PLS-CADD™, is shown in Figure A1.2. The dotted line is the ground clearance line at 22.4 ft (6.83 m) height.

Table A1.1 Wood Structures Details.

No.	Name	Pole Height (ft) H_{AG}	D_E	Total	Pole Class	Material	d_{top} (in)	d_{butt} (in)	M_{GL} (kip-ft)	Number of Guys	Number of Anchors	Weight Pole (lbs)
1	SUB1*	55	N/A	55	N/A	65/50**	9	36	1285	–	–	8,390
2	DDE 1	83.5	11.5	95	H2	SYP	9.875	21.25	514	8	8	7,860
3	LA	70	10	80	H2	SYP	9.875	19.875	422	4	4	6,000
4	TAN	65.25	14.75	80	H3	SYP	10.5	21	454	N/A	N/A	6,740
5	SA	70	10	80	H1	SYP	9.25	18.5	361	4	4	5,370
6	DDE 2	88	12	100	H2	SYP	9.875	21.5	536	8	8	8,450
7	SUB2*	55	N/A	55	N/A	65/50**	9	36	1285	–	–	8,390

1 ft = 0.3048 m, 1 in = 25.4 mm, 1 lb = 4.45 N, 1 kip-ft = 1.356 kN-m.
H_{AG} = Height, Above Ground.
D_E = Embedment Depth.
d_{top} = Pole Diameter at Top
d_{butt} = Pole Diameter at Bottom.
M_{GL} = Ground Line Moment Capacity.
* = Steel Substation Deadend.
** = beam material is 50 ksi.

Table A1.2 Steel Structures Details.

No.	Name	Pole Height (ft) H_{AG}	D_E	Total	d_{top} (in)	d_{butt} (in)	Material (ksi)	Pole Thickness[1] (in)	M_{GL} (kip-ft)	Base Plate	Remarks	Structure Weight (lbs)
1	SUB 1**	55	N/A	55	9	36	65/50*	¼″	1285	Yes	Beam 8 × 8 × 3/8	8,390
2	DDE 1	83.5	N/A	83.5	9	48	65	¾″ (bottom) ½″ (top)	7350	Yes	Lap 3¾ ft	20,440
3	LA	70	N/A	70	9	48	65	½″ (bottom) 3/8″ (top)	4978	Yes	Lap 2½ ft	12,110
4	TAN	65.25	14.75	80	9	24	65	¼″	481	No	Lap 2½ ft	3,655
5	SA	70	17	87	9	36	65	¼″	1015	No	Lap 2½ ft	5,460
6	DDE 2	88	N/A	88	9	48	65	¾″ (bottom) ½″ (top)	7350	Yes	Lap 3¾ ft	20,740
7	SUB 2**	55	N/A	55	9	36	65/50*	¼″	1285	Yes	Beam 8x8x3/8	8,390

1 ft = 0.3048 m, 1 inch = 25.4 mm, 1 lb = 4.45 N, 1 kip-ft = 1.356 kN-m, 1 ksi = 6.89 MPa.
H_{AG} = Height, Above Ground.
D_E = Embedment Depth.
N/A implies pole is supported by a concrete drilled shaft.
* = beam material is 50 ksi. ** = shown here only for illustration purposes.
Lap is the overlap between the two pole shaft sections.
d_{top} = Pole Diameter at Top.
d_{butt} = Pole Diameter at Bottom.
M_{GL} = Ground Line Moment Capacity.
[1] 12-sided pole.

Geometrical configurations of all steel structures modeled are shown in Figures A1.5a to A1.5g and those of wood in Figures A1.6a to A1.6e. For illustrative purposes, guying for DDE wood poles is assumed to be identical in both directions; however, it can be often economical to use a single anchor on the slack tension side accepting all 4 guy wires. Bisector guying is used for small and large angle poles. For optimal guying performance, all guy wires are inclined to the pole at about 45°.

Analysis

The systems were analyzed for the following primary load cases:

a. NESC Heavy
b. Extreme Wind (wind speed of 90 mph)
c. Extreme Ice (1 in radial ice)
d. Extreme Ice with Concurrent Wind (1 in radial ice with 40 mph wind)

All analyses were performed using the nonlinear option of PLS-POLE™. A trial-and-error method with wire tensions was employed to determine optimum tensions which satisfy tension limits, clearances (sags) as well as galloping. A similar approach was used with wood and steel poles to ensure they meet the strength requirements (i.e.) keeping structural usage below 100%. For steel poles with base plates, the PLS-POLE™ program was used to determine base plate size and thickness compatible with the chosen ground line pole diameter. For guyed wood poles, the anchor locations were checked to see they fall within the available ROW. The non-linear option of PLS-POLE™ analysis process also includes built-in buckling checks for

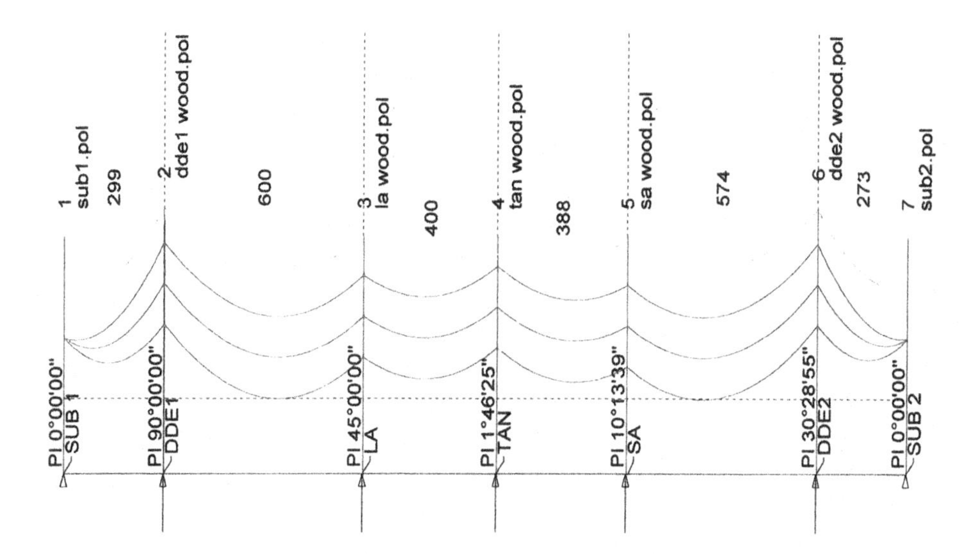

Figure A1.2 Profile of the Transmission Line.

Example T-Line
230 kV SC Transmission Line

Conductor: 954.0 Kcmil 54/ 7 Stranding ACSR "CARDINAL"

Area = 0.8462 Sq. in Diameter = 1.196 in Weight = 1.229 lb/ft RTS = 33800 lb
Data from Chart No. 1-838
English Units
Limits and Outputs in Average Tensions.

Span = 518.0 Feet Special Load Zone
Creep is NOT a Factor Rolled Rod

Design Points					Final			Initial		
Temp °F	Ice in	Wind psf	K lb/ft	Weight lb/ft	Sag Ft	Tension lb	RTS %	Sag Ft	Tension lb	RTS %
0.0	0.50	4.00	0.30	2.698	7.94	11412	33.8	7.65	11840	35.0*
15.0	1.00	4.00	0.00	4.101	10.23	13462	39.8	10.23	13462	39.8
32.0	1.00	0.00	0.00	3.960	10.58	12574	37.2	10.40	12792	37.8
32.0	0.50	0.00	0.00	2.284	8.35	9185	27.2	7.60	10086	29.8
60.0	0.00	20.70	0.00	2.401	9.53	8463	25.0	8.58	9399	27.8
-20.0	0.00	0.00	0.00	1.229	4.31	9573	28.3	3.90	10578	31.3
0.0	0.00	0.00	0.00	1.229	4.95	8326	24.6	4.27	9657	28.6
30.0	0.00	0.00	0.00	1.229	6.10	6756	20.0	4.96	8312	24.6
60.0	0.00	0.00	0.00	1.229	7.40	5573	16.5	5.83	7071	20.9
90.0	0.00	0.00	0.00	1.229	8.74	4719	14.0	6.87	6002	17.8
120.0	0.00	0.00	0.00	1.229	10.06	4105	12.1	8.03	5139	15.2
167.0	0.00	0.00	0.00	1.229	11.59	3563	10.5	9.91	4164	12.3
212.0	0.00	0.00	0.00	1.229	12.43	3325	9.8	11.66	3541	10.5

* Design Condition

Figure A1.3 Sag-Tension Data for Conductor.

Shield Wire Sags for Textbook Example
3/8 in EHS Wire

Conductor: 3/ 8x 7 Strand Steel EHS

Area = 0.0792 Sq. in Diameter = 0.360 in Weight = 0.273 lb/ft RTS = 15400 lb
Data from Chart No. 1-1293
English Units
Limits and Outputs in Average Tensions.

Span = 518.0 Feet Special Load Zone
Creep is NOT a Factor

Design Points					Final			Initial		
Temp °F	Ice in	Wind psf	K lb/ft	Weight lb/ft	Sag Ft	Tension lb	RTS %	Sag Ft	Tension lb	RTS %
0.0	0.50	4.00	0.30	1.226	7.80	5281	34.3	7.64	5390	35.0*
15.0	1.00	4.00	0.00	2.116	11.19	6356	41.3	11.19	6356	41.3
32.0	1.00	0.00	0.00	1.964	10.94	6033	39.2	10.88	6064	39.4
32.0	0.50	0.00	0.00	0.808	6.13	4420	28.7	5.89	4603	29.9
60.0	0.00	20.70	0.00	0.678	5.73	3971	25.8	5.44	4183	27.2
-20.0	0.00	0.00	0.00	0.273	2.07	4423	28.7	1.97	4652	30.2
0.0	0.00	0.00	0.00	0.273	2.19	4180	27.1	2.07	4426	28.7
30.0	0.00	0.00	0.00	0.273	2.40	3818	24.8	2.24	4087	26.5
60.0	0.00	0.00	0.00	0.273	2.64	3463	22.5	2.44	3750	24.3
90.0	0.00	0.00	0.00	0.273	2.94	3115	20.2	2.68	3416	22.2
120.0	0.00	0.00	0.00	0.273	3.30	2779	18.0	2.96	3089	20.1
167.0	0.00	0.00	0.00	0.273	4.01	2285	14.8	3.52	2599	16.9
212.0	0.00	0.00	0.00	0.273	4.90	1871	12.1	4.22	2170	14.1

* Design Condition

Figure A1.4 Sag-Tension Data for OHGW.

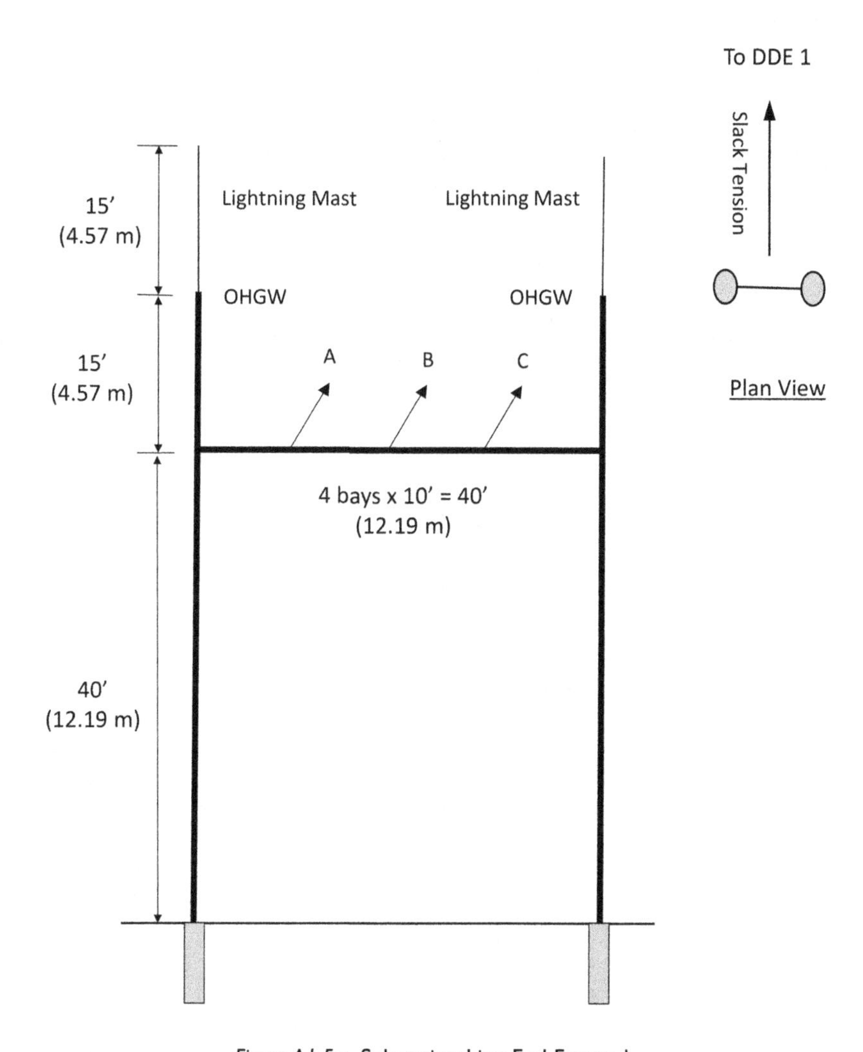

Figure A1.5a Substation Line End Frame 1.

all poles, especially guyed wood poles. Therefore, the buckling checks are not shown here.

For all poles, the bearing stress at the bottom of the poles was checked to determine the need for bearing pads to help distribute axial loads to the soil below.

Design

Tables A1.3 and A1.4 shows the results of the analyses for wood and steel options respectively. The usage values refer to the combined stress interaction check including axial and flexural stresses. Note that guyed wood poles are provided a concrete bearing pads to distribute the large resultant axial loads from the pole and guy tension

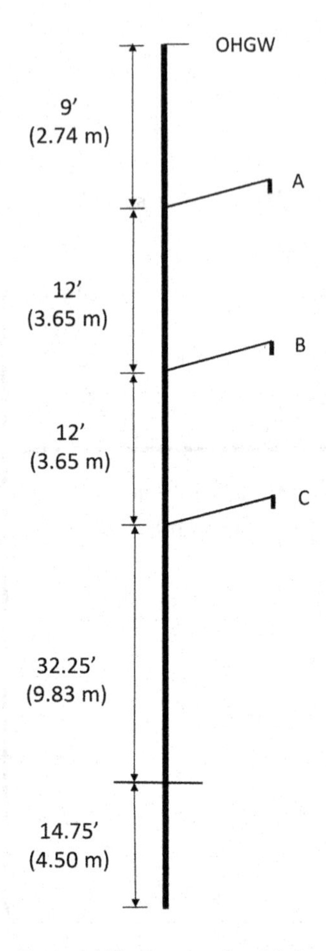

Figure A1.5b Tangent Steel Pole.

components. These pads were sized using the bearing strength of the soil. For the steel poles with pier foundations, only the length of the shaft needed is computed from CAISSON^{TM}.

For the steel poles where concrete piers were used, base plates were nominally sized for both diameter and thickness. (See Chapter 4 and 5 for guidelines for base plate and pier geometry). A trial-and-error process was used to arrive at an optimum plate thickness for the given loading.

The usage levels of all poles, guy wires, anchors, insulators and other items are within prescribed stress limits. Other checks (ground line deflections and rotations, skin friction resistance) were left to the user as an exercise.

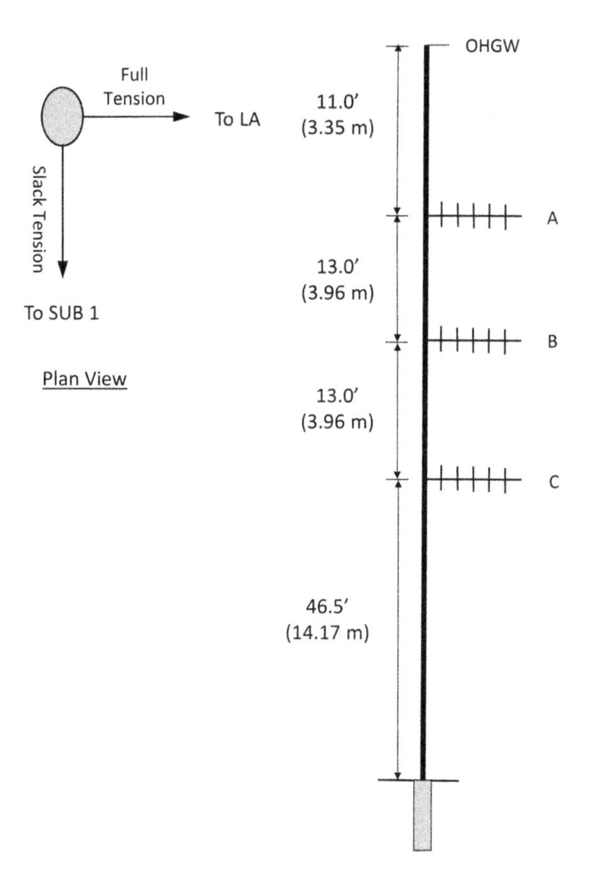

Figure A1.5c Double Deadend Steel Pole 1.

Discussion

The demand to resist large bending moments at line angles without any guying support resulted in heavy steel poles as can be seen in their dimensions. Note that by varying the pole diameters and thicknesses, a more optimal design can be possibly obtained (as an exercise to the user). Ground line pole diameters also affect the sizes of base plates and in turn the concrete pier diameter. In case of steel poles, a major portion of project cost is that of steel; so, even saving a few hundred pounds (kilos) off the material weight can lead to significant cost savings. For example, the total weight of all steel poles in this example is about 77,870 lbs (35,275 kgs) and any effort to reduce this further leads to a more economical design.

Wood poles, on the other hand, are lighter but involve large amounts of other material and associated labor. Guy wires and anchors add to costs and need extra

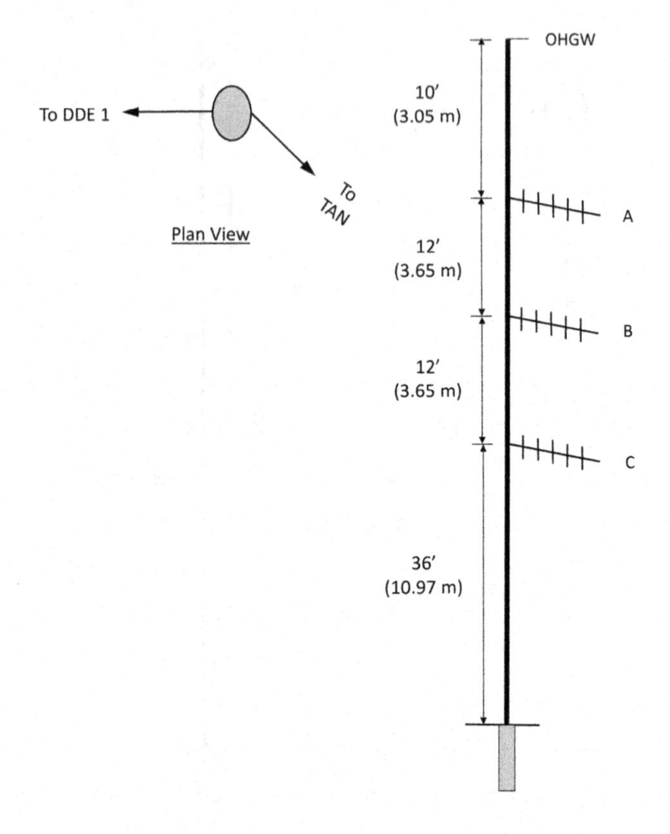

Figure A1.5d Large Angle Steel Pole.

installation labor. Since wood is a degradable material, maintenance is important. The classes of wood poles used in this example is assumed to be optimal; however, other combinations can also be tried (H1 or lower instead of H2 at large angles and deadends) if they can safely resist the applied loads.

Deflections are not evaluated qualitatively in this example. The user can re-work this example by assuming limitations placed on steel pole top deflections which may require stiffer poles (i.e.) thicker material and larger diameters.

Substation frames are usually checked for deflections which control design. The user may check the two frames of this example with reference to guidelines of ASCE Manual 113.

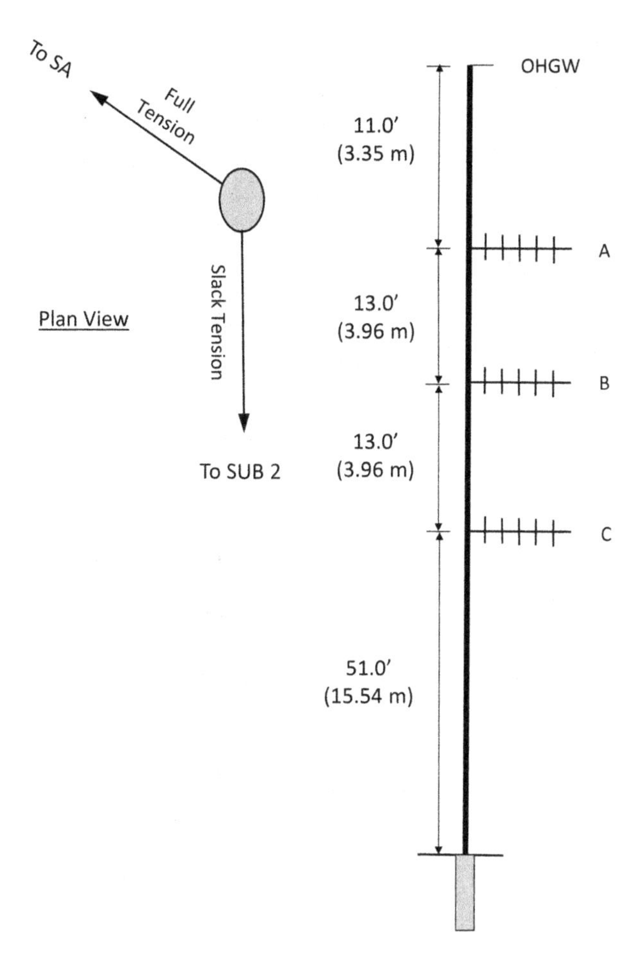

Figure A1.5e Double Deadend Steel Pole 2.

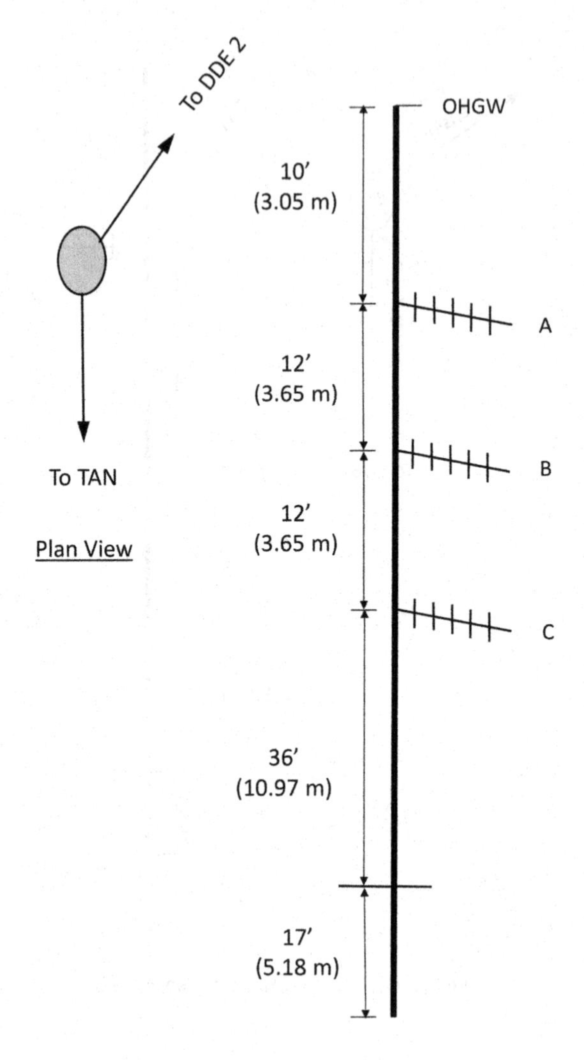

Figure A1.5f Small Angle Steel Pole.

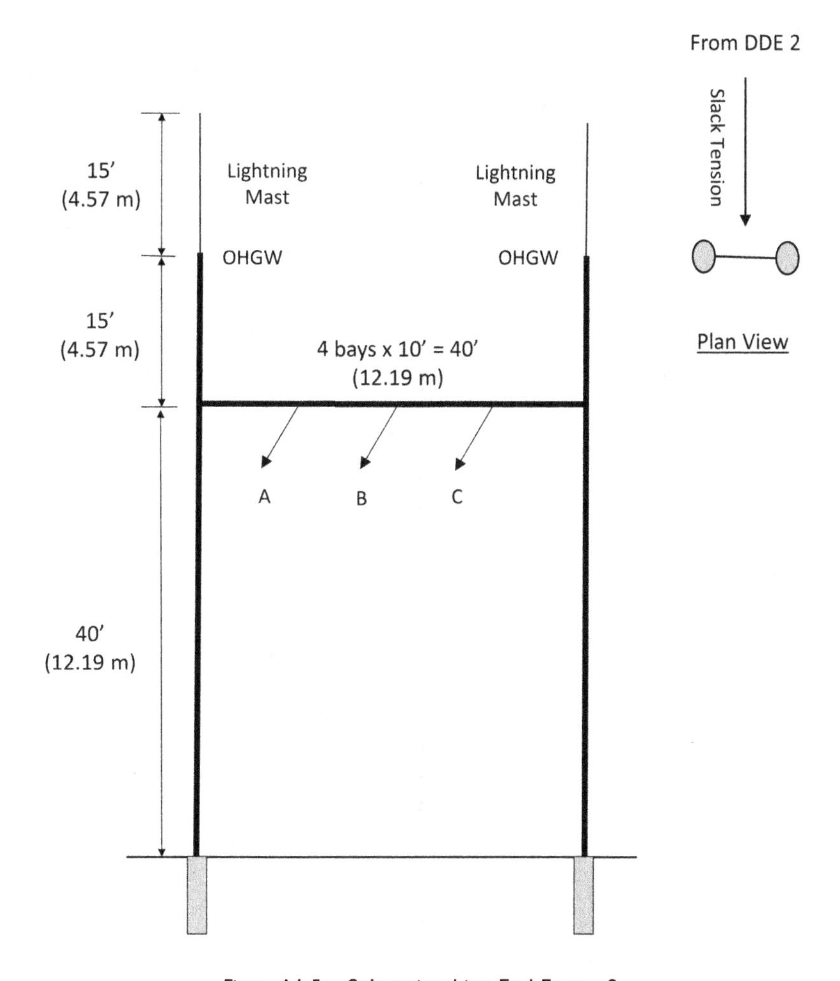

Figure A1.5g Substation Line End Frame 2.

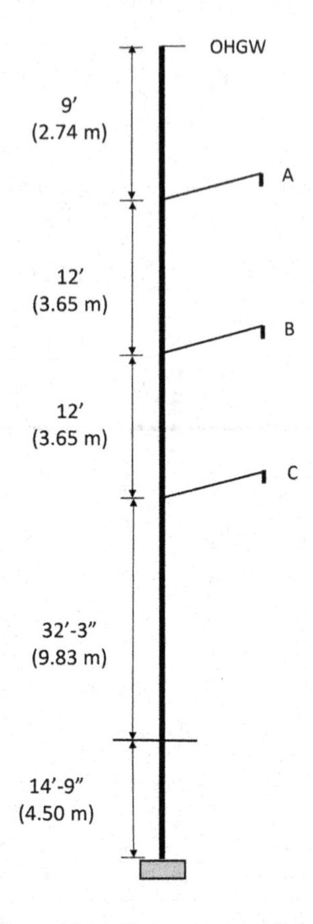

Figure A1.6a Tangent Wood Pole.

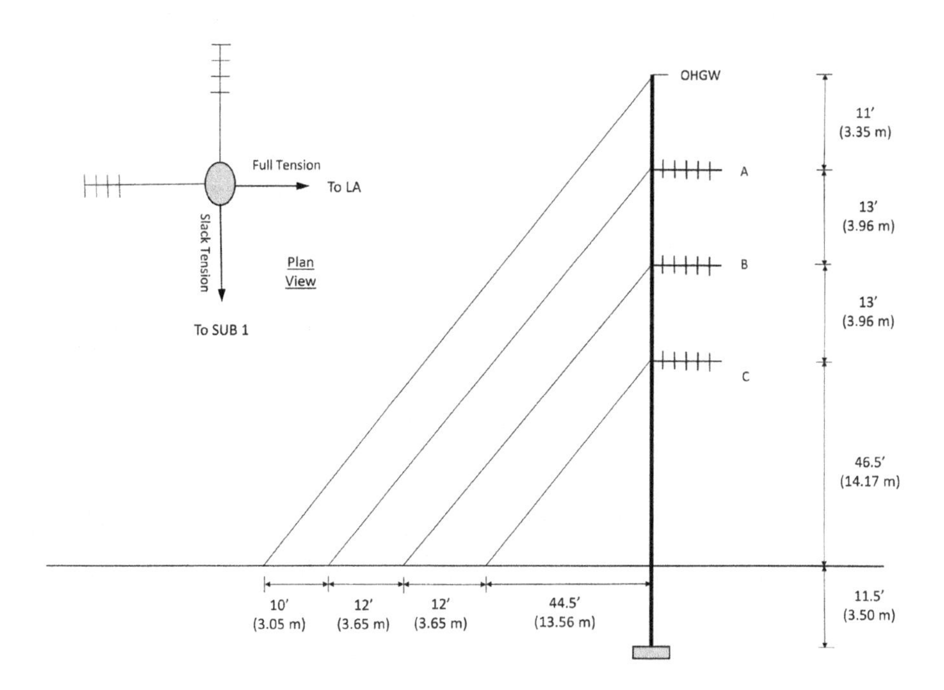

Figure A1.6b Double Deadend Guyed Wood Pole 1.

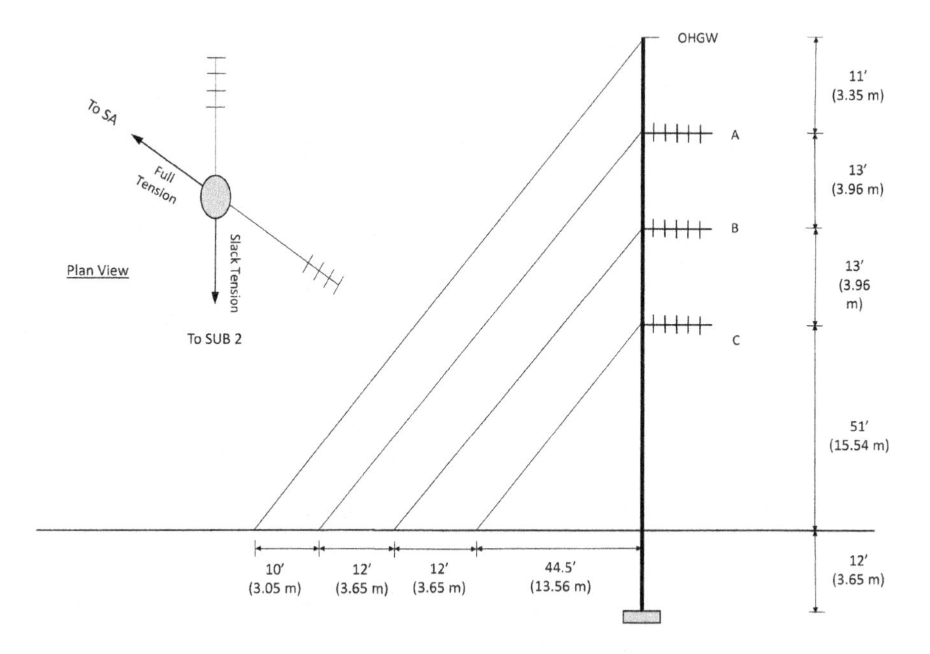

Figure A1.6c Double Deadend Guyed Wood Pole 2.

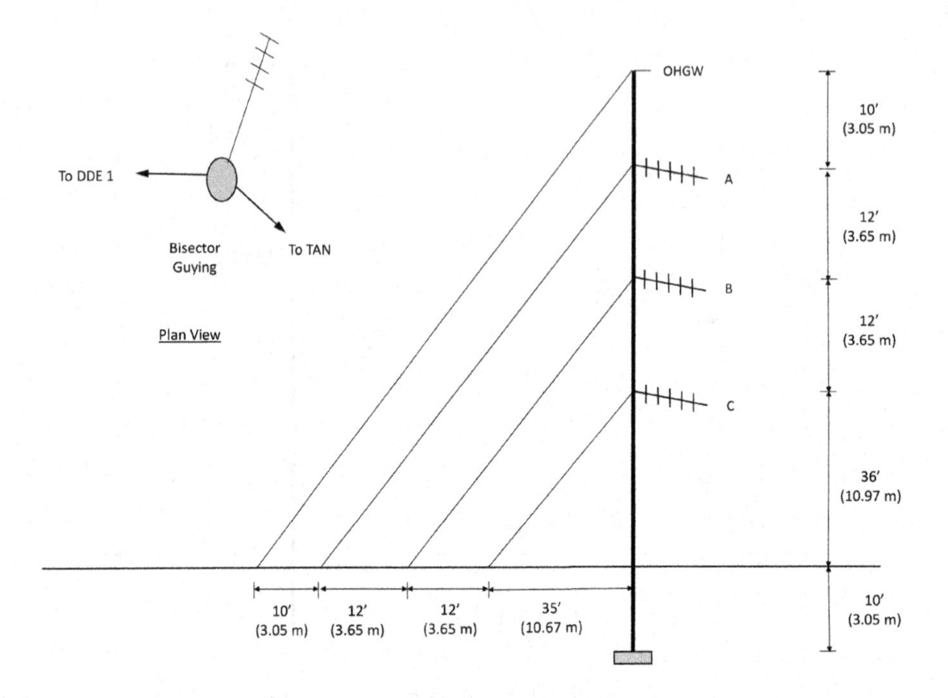

Figure A1.6d Large Angle Guyed Wood Pole.

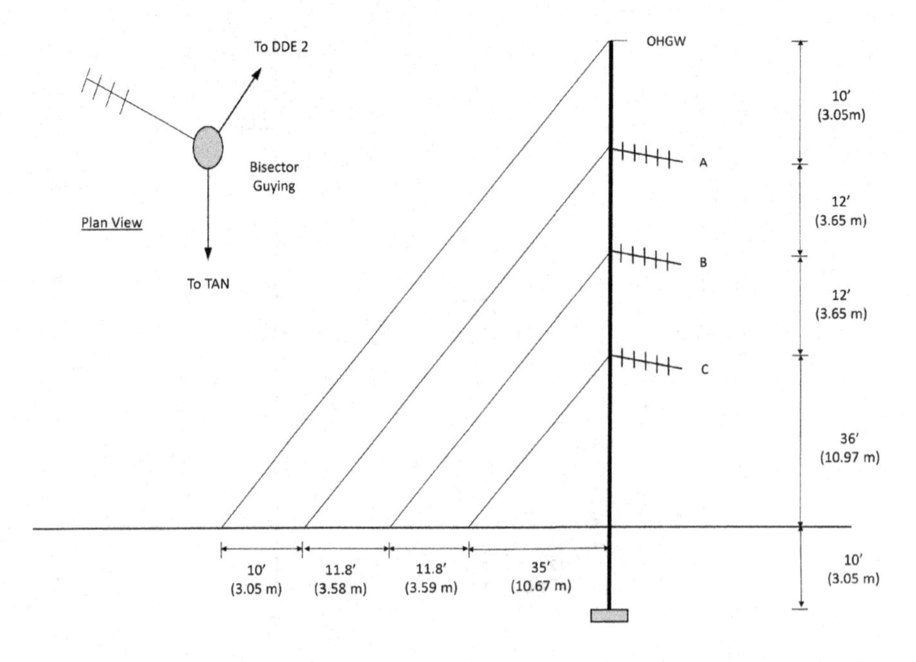

Figure A1.6e Small Angle Guyed Wood Pole.

Table A1.3 Results for Wood Structures.

No.	Name	Maximum Ground Line Reactions Moment (kip-ft)	Shear (kips)	Axial (kips)	Foundation	Overall Structure Usage*	Maximum Guy Usage	Maximum Anchor Usage**
1	DDE 1	39.7	<1	107	Direct Embedment 11.5 ft +3½ ft Dia. Pad	54.9%	96.5%	86.6%
2	LA	43.5	1.1	64.8	Direct Embedment 10 ft +3 ft Dia. Pad	41%	76.3%	68.4%
3	TAN	220.2	5.1	11.6	Direct Embedment 14.75 ft +2 ft Dia. Pad	75.5%	N/A	N/A
4	SA	12.4	<1	29.5	Direct Embedment 10 ft +2½ ft Dia. Pad	15.2%	26.9%	24.1%
5	DDE 2	186.5	1.8	100.3	Direct Embedment 12 ft +3½ ft Dia. Pad	54.6%	86.9%	78.0%

1 ft = 0.3048 m, 1 kip = 4.45 kN, 1 kip-ft = 1.356 kN-m.
*based on stress interaction (axial and bending).
** = force in guy wire/allowable anchor capacity.

Table A1.4 Results for Steel Structures.

No.	Name	Maximum Ground Line Reactions Moment (kip-ft)	Shear (kips)	Axial (kips)	Foundation*	Overall Structure Usage***	Base Plate (in)
1	SUB 1	631	16.6	6.5	— **	50.8%	— **
2	DDE 1	4631	74.7	38	Concrete Pier 6 ft Dia. x 24 ft	81.9%	3¾"
3	LA	2742	55.2	21.3	Concrete Pier 6 ft Dia. x 20.5 ft	57.7%	2⅜"
4	TAN	197	3.8	9.8	Direct Embedment 14.75 ft	42.3%	N/A
5	SA	835	16.9	9.2	Direct Embedment 17 ft	82.9%	N/A
6	DDE 2	4336	65.3	39.7	Concrete Pier 6 ft Dia. x 23 ft	77.1%	3¾"
7	SUB 2	469	11	6.5	— **	37.8%	— **

1 ft = 0.3048 m, 1 inch = 25.4 mm, 1 kip = 4.45 kN, 1 kip-ft = 1.356 kN-m.
*all concrete piers have 20 anchor bolts, 5 per quadrant; pier diameter rounded to the nearest foot.
**not computed; left as an exercise to student.
***based on stress interaction (axial, bending, shear etc.)

Appendix 2

Wood pole data

Table A2.1 Wood Pole Classes.

Class	*Lateral Load at Tip (lbs)***
H6	11,400
H5	10,000
H4	8,700
H3	7,500
H2	6,400
H1	5,400
1	4,500
2	3,700
3	3,000
4	2,400
5	1,900
6	1,500
7	1,200
9	740
10	370

*Per ANSI O5-1.
**Applied 2 ft. from the tip of the pole.
(Courtesy: RUS/USDA).
1 ft = 30.48 cm, 1 lb = 4.45 N.

Table A2.2 Typical Cross Arms Used in Wood H-Frames.

RUS Type	*Dimensions*	*Typical Weight (lbs)**
37	5¾″ × 7¾″ × 20 ft	286
41	5¾″ × 7¾″ × 22 ft	315
54	5¾″ × 7¾″ × 26 ft	372
60	5¾″ × 7¾″ × 27.5 ft	394
71	5¾″ × 7¾″ × 32 ft	458
82	5¾″ × 7¾″ × 40 ft	573

1 in = 25.4 mm, 1 ft = 0.3048 m, 1 lb = 4.45 N, 1 pcf = 0.157 kN/m^3.
*for Douglas-Fir, density taken as 50 pcf.
(Dimensions – Courtesy: RUS/USDA).

Table A2.3 Typical X-Braces Used on Wood H-Frames.

RUS Type	*Typical Line Voltage*	*Pole Spacing (ft)*	*Cross Section Dimensions*	*Brace Length L[1] (ft)*	*r (in)*	*kL/2r**	*Nominal Compressive Capacity (lbs)***
TM-110A	69 kV to 230 kV	10½	3-3/8″ × 4-3/8″	13.43	0.974	82.8	38,300
		12½	3-3/8″ × 4-3/8″	16.26	0.974	100.1	26,100
		14	3-3/8″ × 4-3/8″	18.38	0.974	113.2	21,000
TM-110B		15½	3-3/8″ × 5-3/8″	20.51	0.974	126.3	20,200
TM-110C		19½	3-5/8″ × 7-1/2″	26.16	1.046	150.1	21,600

1 in = 25.4 mm, 1 ft = 30.48 cm, 1 lb = 4.45 N, 1 psi = 6.895 kPa.
*Mid-length clamp reduces effective length to ½ L.
**Using $E = 1.8 \times 10^6$ psi.
[1] Brace length L includes allowance of 12 in (30.48 cm) for pole diameter.
(Partial Courtesy: RUS/USDA).

Table A2.4 Wood Pole Dimensions.*

Class		*4*	*3*	*2*	*1*	*H1*	*H2*	*H3*	*H4*	*H5*	*H6*
Top Dia. (in)		*6.7*	*7.3*	*8.0*	*8.6*	*9.2*	*9.9*	*10.5*	*11.1*	*11.8*	*12.4*
Length L (ft)	Min. D_e (ft)	Minimum Diameter (in) at 6 ft from Pole Bottom									
20	4	8.0	8.6	9.2	9.9						
25	4.5	8.8	9.4	10.0	10.7						
30	5	9.4	10.2	10.8	11.6						
35	5.5	10.0	10.8	11.6	12.4	13.2	13.9				
40	6	10.7	11.5	12.3	13.1	13.9	14.6	15.4	16.2		
45	6.5	11.1	11.9	12.9	13.7	14.5	15.4	16.2	17.0	17.8	18.6
50	7	11.6	12.4	13.4	14.3	15.1	16.1	16.9	17.7	18.6	19.4
55	7.5	12.1	12.9	13.9	14.8	15.8	16.6	17.5	18.5	19.3	20.2
60	8	12.4	13.4	14.3	15.3	16.2	17.2	18.1	18.9	19.9	20.9
65	8.5	12.9	13.9	14.8	15.8	16.7	17.7	18.6	19.6	20.6	21.5
70	9	13.2	14.3	15.3	16.2	17.2	18.1	19.3	20.2	21.2	22.0
75	9.5		14.6	15.6	16.7	17.7	18.8	19.7	20.7	21.7	22.6
80	10		15.0	16.1	17.2	18.1	19.1	20.2	21.2	22.1	23.1
85	10.5		15.3	16.4	17.5	18.6	19.6	20.7	21.7	22.8	23.7
90	11		15.6	16.9	17.8	18.9	20.1	21.2	22.1	23.2	24.2
95	11.5			17.2	18.1	19.4	20.5	21.5	22.6	23.7	24.7
100	12			17.5	18.6	19.7	20.9	21.9	23.1	24.2	25.2
105	12.5			17.8	18.9	20.1	21.3	22.4	23.6	24.5	25.6
110	13			18.1	19.3	20.5	21.7	22.8	23.9	25.0	26.1
115	13.5			18.5	19.6	20.9	22.0	23.1	24.4	25.5	26.6
120	14			18.8	19.9	21.2	2.3	23.6	24.7	25.8	27.1
125	14.5			18.9	20.2	21.5	22.6	23.9	25.0	26.3	27.4
Class		*4*	*3*	*2*	*1*	*H1*	*H2*	*H3*	*H4*	*H5*	*H6*

(Values rounded off to the nearest decimal).
1 in = 25.4 mm, 1 ft = 30.48 cm, D_e = Embedment Depth = 10% L + 2′.
* Douglas Fir & Southern Yellow Pine.

Appendix 3

Steel pole data

Table A3.1a Structural Steels Used in Transmission Structures (U.S.).

Material	*Yield Stress (ksi)*	*Ultimate Stress (ksi)*
A-7	33	55
A-36	36	58
A-500 (42) Grade B	42	58
A-500 (46) Grade B	46	58
A-572 (50)	50	65
A-572 (60)	60	75
A-572 (65)	65	80
A-588 (50)	50	70
A-871 (65)	65	80

1 ksi = 6.895 MPa.

Table A3.1b Structural Steels Used in Other Countries.

Material	*Yield Stress (MPa or kg/mm^2)*	*Yield Stress (ksi)*	*Country*
S-235	235 MPa	34.1	UK, Europe For thicknesses less than 40 mm (1.57 inches) (Available in Type N, M, W, Q etc.)
S-275	275 MPa	40.0	
S-355	355 MPa	51.5	
S-450	440 MPa	63.8	
Grade 300	300 MPa	43.5	Canada (Available in Type G, W, WT, R, A etc.)
Grade 350	350 MPa	50.8	
Grade 400	400 MPa	58.0	
Mild Steel (6 mm to 40 mm thick)	23 to 26 kg/mm^2	32.7 to 37.0	India (Available in mild and high-strength types)
High Tensile Steel (6 mm to 63 mm thick)	30 to 36 kg/mm^2	42.6 to 51.1	
AS-250	250 MPa	36.0	Australia (Grades shown are equivalent to American steel grades A36, A572-50 and A572-65)
AS-350	350 MPa	50.0	
AS-450	450 MPa	65.0	

Table A3.2 Commonly Used Bolts on Lattice Towers.

Bolt Type	*Diameter (in)*	*Typical Ultimate Shear Strength – Through Threads (kips)**	*Typical Ultimate Shear Strength – Through Shaft (kips)**
A394 Type 'O'	1/2	6.95	9.00
	5/8	11.15	14.10
	3/4	16.65	20.25
	7/8	23.15	27.60
	1	30.4	36.05
A307 Grade A	5/8	8.28	11.04
	3/4	11.92	15.90
	7/8	16.23	21.65

1 in = 25.4 mm, 1 kip = 4.45 kN.
*In Single Shear.

Table A3.3 High Strength Bolts Used on Steel Transmission Structures.

Bolt Type	*Diameter (in)*	*Nominal Ultimate Tensile Strength (kips)**	*Typical Ultimate Shear Strength – Through Shaft (kips)***
A325	1/2	17.67	11.70
	5/8	27.63	18.40
	3/4	39.73	26.50
	7/8	54.11	36.10
	1	70.67	47.00
A490	1/2	22.20	14.70
	5/8	34.67	23.10
	3/4	49.87	33.20
	7/8	68.00	45.10
	1	88.80	58.90

1 in = 25.4 mm, 1 kip = 4.45 kN, 1 ksi = 6.895 MPa.
*Nominal Tensile Stress: A325-90 ksi, A490-113 ksi.
**In Single Shear.

Table A3.4 Steel Shapes and Geometric Properties.

Shape	*Cross Section Area, A*	*Moment of Inertia, I*	*Radius of Gyration, r*	*Angle* α	η	*Maximum Q/It*	*Maximum C/J*	*Approx. Flat Width of Face, w**
	3.14 D t	0.393 D^3 t	0.354 D	N/A	0.500	0.637/D t	0.637 (D + t)/D^3t	N/A
	4.00 D t	0.666 D^3 t	0.408 D	45°	0.707	0.563/D t	0.5 (D + t)/D ^{3}t	D–9 t
6	3.46 D t	0.481 D^3 t	0.373 D	30°	0.577	0.606/D t	0.577 (D + t)/D^3 t	0.577 (D–9t)
8	3.32 D t	0.438 D^3 t	0.364 D	22.5°	0.541	0.618/D t	0.603 (D + t)/D^3 t	0.414 (D–9t)
12	3.22 D t	0.411 D^3 t	0.358 D	15°	0.518	0.631/D t	0.622 (D + t)/D^3 t	0.268 (D–9t)
16	3.19 D t	0.403 D^3 t	0.356 D	11¼°	0.510	0.634/D t	0.628 (D + t)/D^3 t	0.199 (D–9t)

(With permission from ASCE).

D = mean diameter, D_o - t; D_o = outside diameter, across flats (for polygonal sections); t = thickness.

α = angle between x-axis and corner of the polygon, C_x and C_y = distance from y and x axes to point.

J = polar moment of inertia; Q/It = value for determining maximum flexural shear stress.

C/J = value for determining maximum torsional shear stress.

$C_x = \eta\ (D + t)\ \cos(\alpha)\ C_y = \eta\ (D + t)\ \sin(\alpha)$.

*Bending Radius assumed as 4t.

Table A3.5 Typical Standard RUS Steel Pole Classes.

Steel Class Designation	*Minimum Ultimate Moment Capacity 5 ft from Pole Top (kip-ft)*	*Lateral Load at Tip (lbs)**
S-12.0	96	12,000
S-11.0	88	11,000
S-10.0	80	10,000
S-09.0	72	9,000
S-08.0	64	8,000
S-07.4	57	7,410
S-06.5	50	6,500
S-05.7	44	5,655
S-04.9	38	4,875
S-04.2	32	4,160
S-03.5	27	3,510
S-02.9	23	2,925
S-02.4	19	2,405
S-02.0	15	1,950

*Applied **2 ft** from the tip of the pole.*
(Courtesy: RUS/USDA).
1 ft = 30.48 cm, 1 lb = 4.45 N, 1 kip-ft = 1.356 kN-m.

Table A3.6 Typical Standard Class Steel Poles – Class 1.

Total Pole Length (ft)	*Base Diameter at Bearing Plate (in)*	*Ground Line Moment Capacity (ft-kips)*	*Estimated Pole Weight (lbs)*
40	13.34	93	959
45	13.92	106	1101
50	14.50	119	1248
55	14.59	132	1427
60	15.17	145	1580
65	15.74	158	1763
70	16.32	171	1929
75	16.90	184	2102
80	17.48	197	2280
85	18.06	210	2463
90	18.64	223	2653
95	19.23	236	2848
100	19.31	248	3106
105	19.89	261	3310
110	20.48	274	3520
115	21.06	287	3734
120	21.64	299	3955

(Courtesy: Trinity-Meyer Utility Structures, LLC. 2015).
GL is defined by embedment of 10% of Pole Length plus 2 ft
1 inch = 25.4 mm, 1 ft = 30.48 cm, 1 lb. = 4.45 N, 1 ksi = 6.895 MPa, 1 kip-ft = 1.356 kN-m.
Pole Taper = 0.116104 in/ft
Pole Top Diameter = 8.69 in
Thickness = 3/16 in
Steel Yield Strength = 65 ksi.

Table A3.7 Typical Standard Class Steel Poles – Class H1.

Total Pole Length (ft)	*Base Diameter at Bearing Plate (in)*	*Ground Line Moment Capacity (ft-kips)*	*Estimated Pole Weight (lbs)*
40	13.99	112	990
45	14.64	127	1139
50	15.31	143	1293
55	15.47	159	1485
60	16.14	174	1648
65	16.80	190	1843
70	17.46	205	2021
75	18.12	221	2205
80	18.79	236	2396
85	19.44	252	2595
90	20.11	267	2800
95	20.77	283	3012
100	20.93	298	3299
105	21.60	313	3520
110	22.25	329	3748
115	22.92	344	3980
120	23.58	359	4222

(Courtesy: Trinity-Meyer Utility Structures, LLC. 2015).
GL is defined by embedment of 10% of Pole Length plus 2 ft
1 inch = 25.4 mm, 1 ft = 30.48 cm, 1 lb. = 4.45 N, 1 ksi = 6.895 MPa, 1 kip-ft = 1.356 kN-m.
Pole Taper = 0.132332 in/ft
Pole Top Diameter = 8.69 in
Thickness = 3/16 in
Steel Yield Strength = 65 ksi.

Table A3.8 Typical Standard Class Steel Poles – Class H2.

Total Pole Length (ft)	*Base Diameter at Bearing Plate (in)*	*Ground Line Moment Capacity (ft-kips)*	*Estimated Pole Weight (lbs)*
40	14.80	132	1027
45	15.56	151	1185
50	16.32	170	1351
55	16.59	188	1557
60	17.36	207	1732
65	18.12	225	1947
70	18.88	244	2139
75	19.65	262	2340
80	20.41	280	2549
85	21.17	299	2764
90	21.94	317	2988
95	22.70	335	3220
100	22.96	354	3543
105	23.72	372	3786
110	24.49	390	4037
115	25.25	408	4296
120	263.01	426	4560

(Courtesy: Trinity-Meyer Utility Structures, LLC. 2015).
GL is defined by embedment of 10% of Pole Length plus 2 ft
1 inch = 25.4 mm, 1 ft = 30.48 cm, 1 lb. = 4.45 N, 1 ksi = 6.895 MPa, 1 kip-ft = 1.356 kN-m.
Pole Taper = 0.152616 in/ft
Pole Top Diameter = 8.69 in
Thickness = 3/16 in
Steel Yield Strength = 65 ksi.

Table A3.9 Typical Standard Class Steel Poles – Class H3.

Total Pole Length (ft)	*Base Diameter at Bearing Plate (in)*	*Ground Line Moment Capacity (ft-kips)*	*Estimated Pole Weight (lbs)*
40	15.76	155	1072
45	16.65	177	1242
50	17.53	199	1421
55	17.92	222	1642
60	18.81	242	1833
65	19.69	264	2072
70	20.57	286	2283
75	21.45	307	2502
80	22.34	329	2730
85	23.22	350	2968
90	24.11	372	3213
95	24.99	393	3469
100	25.37	415	3840
105	26.26	436	4108
110	27.14	458	4387
115	28.03	479	4675
120	28.91	501	4969

(Courtesy: Trinity-Meyer Utility Structures, LLC. 2015).
GL is defined by embedment of 10% of Pole Length plus 2 ft
1 inch = 25.4 mm, 1 ft = 30.48 cm, 1 lb. = 4.45 N, 1 ksi = 6.895 MPa, 1 kip-ft = 1.356 kN-m.
Pole Taper = 0.176764 in/ft
Pole Top Diameter = 8.6 in
Thickness = 3/16 in
Steel Yield Strength = 65 ksi.

Table A3.10 Typical Standard Class Steel Poles – Class H4.

Total Pole Length (ft)	*Base Diameter at Bearing Plate (in)*	*Ground Line Moment Capacity (ft-kips)*	*Estimated Pole Weight (lbs)*
40	17.20	180	1180
45	18.15	206	1364
50	19.09	231	1559
55	19.54	256	1809
60	20.48	281	2019
65	21.42	307	2278
70	22.37	332	2507
75	23.31	357	2745
80	24.25	382	2994
85	25.20	407	3252
90	26.14	432	3520
95	27.08	457	3797
100	27.53	482	4211
105	28.48	507	4503
110	29.42	532	4819
115	30.36	557	5131
120	31.31	582	5453

(Courtesy: Trinity-Meyer Utility Structures, LLC. 2015).
GL is defined by embedment of 10% of Pole Length plus 2 ft
1 inch = 25.4 mm, 1 ft = 30.48 cm, 1 lb. = 4.45 N, 1 ksi = 6.895 MPa, 1 kip-ft = 1.356 kN-m.
Pole Taper = 0.188645 in/ft
Pole Top Diameter = 9.66 in
Thickness = 3/16 in
Steel Yield Strength = 65 ksi.

Table A3.11 Typical Standard Class Steel Poles – Class H5.

Total Pole Length (ft)	*Base Diameter at Bearing Plate (in)*	*Ground Line Moment Capacity (ft-kips)*	*Estimated Pole Weight (lbs)*
40	18.31	207	1231
45	19.40	236	1430
50	20.48	265	1640
55	21.07	294	1910
60	22.15	323	2136
65	23.23	353	2423
70	24.31	382	2672
75	25.39	410	2932
80	26.48	439	3202
85	27.56	468	3485
90	28.64	497	3779
95	29.72	526	4097
100	30.30	555	4574
105	31.38	584	4897
110	32.46	613	5884
115	33.55	642	6340
120	34.63	671	6812

(Courtesy: Trinity-Meyer Utility Structures, LLC. 2015).
GL is defined by embedment of 10% of Pole Length plus 2 ft
1 inch = 25.4 mm, 1 ft = 30.48 cm, 1 lb. = 4.45 N, 1 ksi = 6.895 MPa, 1 kip-ft = 1.356 kN-m.
Pole Taper = 0.216367 in/ft
Pole Top Diameter = 9.66 in
Thickness = 3/16 in
Steel Yield Strength = 65 ksi.

Table A3.12 Typical Standard Class Steel Poles – Class H6.

Total Pole Length (ft)	*Base Diameter at Bearing Plate (in)*	*Ground Line Moment Capacity (ft-kips)*	*Estimated Pole Weight (lbs)*
40	17.08	236	1539
45	17.98	270	1781
50	18.88	303	2035
55	19.16	336	2352
60	20.07	369	2623
65	20.97	402	2964
70	21.88	435	3260
75	22.78	468	3568
80	23.68	501	3888
85	24.58	533	4221
90	25.49	566	4566
95	26.39	599	4924
100	26.67	632	5445
105	27.58	664	5819
110	28.48	697	6207
115	29.38	730	6620
120	30.28	762	7032

(Courtesy: Trinity-Meyer Utility Structures, LLC. 2015).
GL is defined by embedment of 10% of Pole Length plus 2 ft
1 inch = 25.4 mm, 1 ft = 30.48 cm, 1 lb. = 4.45 N, 1 ksi = 6.895 MPa, 1 kip-ft = 1.356 kN-m.
Pole Taper = 0.180628 in/ft
Pole Top Diameter = 9.85 in
Thickness = 0.25 in
Steel Yield Strength = 65 ksi.

Table A3.13 Anchor Bolt Types Used for Drilled Shaft Foundations.

Bolt Type	*Yield Strength (ksi)*	*Tensile Strength (ksi)*
A36	36	58
A307	36	60
A325	92	120
A490	120	150
A615 Grade 75	75	100
F1554 Grade 36	36	58
F1554 Grade 55	55	75
F1554 Grade 105	105	125

1 ksi = 6.89 MPa.

Appendix 4

Concrete pole data

Table A4.1 Standard Spun Concrete Pole Classes of RUS/USDA.

Concrete Class Designation	*Minimum Ultimate Moment Capacity at 5 feet from Pole Tip (kip-ft)*	*Lateral Load at Tip (lbs)**
C-12.0	96	12,000
C-11.0	88	11,000
C-10.0	80	10,000
C-09.0	72	9,000
C-08.0	64	8,000
C-07.4	57	7,410
C-06.5	50	6,500
C-05.7	44	5,655
C-04.9	38	4,875
C-04.2	32	4,160
C-03.5	27	3,510
C-02.9	23	2,925
C-02.4	19	2,405
C-02.0	15	1,950

(Courtesy: RUS/USDA).
1 kip-ft = 1.356 kN-m, 1 lb = 4.45 N, 1 ft = 30.48 cm.
*Applied 2 ft from the tip of the pole.

Table A4.2 Characteristics of Prestressing Steel.

Nominal Strand Diameter (in)	*Nominal Steel Area (in^2)*	*Steel Grade (ksi)*	*Minimum Breaking Load (lbs)*	*Minimum Load at 1% Extension (lbs)*
3/8	0.080	250	20,000	17,000
7/16	0.108	250	27,000	23,000
1/2	0.144	250	36,000	30,600
1/2	0.153	270	41,300	35,100

(with permission from ASCE).
1 in = 25.4 mm, 1 lb = 4.45 N, 1 ksi = 6.89 MPa.

Appendix 5

Pole deflection limitations

A5.1 CONCRETE POLE DEFLECTIONS (RUS BULLETIN 1724E-216)

Although significant horizontal pole deflection limitations are considered to be beyond the scope of the standard class concrete pole specification, some allowances can be made for these effects. They should be considered during the analysis of the actual loading conditions applied to the concrete pole. Typically, this type of analysis should be accomplished by nonlinear structural analysis techniques. Since the electrical clearances must be assured in the operation of transmission lines, deflections must remain within an acceptable range.

RUS Bulletin 216 limits the *allowable pole deflection to 15% of the pole height* above the point of fixity when the tip load specified in Bulletin Section 4.1.2 is applied under a horizontal testing procedure *under short term loading conditions*. Long term loading will cause continued deflection due to the plastic deformation of the concrete.

The owner should recognize that the actual pole deflection for an application will be less than the specified deflection limit of 15% of the pole height. With the standard class pole, all of the loading is applied near the pole tip. In a typical transmission line application, the actual horizontal loading will be some distance from the pole tip. As such, the actual deflection at the conductor under short term ultimate loading conditions can be expected to be less than 10% of the height above ground.

The specification also limits the *allowable pole deflection to 5% of the pole height* above the point of fixity when 40% of the tip load specified in Bulletin Section 4.1.2 is applied under a horizontal testing procedure under long term loading conditions. This 40% loading approximates *the un-factored NESC loading conditions* as is discussed in the commentary on cracking strength.

The NESC requires that electrical clearances be maintained under a wind loading of 6 psf. It is expected that the deflection of a standard class pole under this 6 psf loading condition will be less than 3% of the height above ground.

For situations where the owner wishes to know the deflection for a standard class pole, they should use a suitable structural analysis computer program in which the actual design loading conditions and concrete pole properties are input into the program. Another option would be to ask the pole manufacturer to provide the analysis. If the project has special deflection limitations, it is recommended that RUS Bulletin

206 be utilized instead of this specification. In doing so, there will be little doubt as to what the actual pole deflections will be under all loading conditions.

A5.2 STEEL POLE DEFLECTIONS (RUS BULLETIN 1724E-214)

Although significant horizontal pole deflection limitations are considered to be beyond the scope of the standard class steel pole specification, some allowances can be made for these effects. They should be considered during the analysis of the actual loading conditions applied to the pole. Typically, this type of analysis should be accomplished by nonlinear structural analysis techniques. Since the electrical clearances must be assured in the operation of transmission lines, deflections must remain within an acceptable range.

This specification limits the *allowable pole deflection to 15% of the pole height* above the point of fixity when the tip load specified in Bulletin Table 1, Section 5.1 is applied under a horizontal testing procedure *under short term loading conditions.* The testing procedure is also discussed in ASCE Steel Pole Manual 48-11.

The owner should recognize that the actual pole deflection for an application will be less than the specified deflection limit of 15% of the pole height. With the standard class pole, all of the loading is applied near the pole tip. In a typical transmission line application, the actual horizontal loading will be some distance from the pole tip. As such, the actual deflection at the conductor under short-term ultimate loading conditions can be expected to be *less than 10% of the height above ground.*

The NESC requires that electrical clearances be maintained under a wind loading of 6 psf. It is expected that the deflection of a standard class pole under this 6 psf loading condition will be less than 3% of the height above ground.

For situations where the owner wishes to know the deflection for a standard class pole, the owner should use a suitable structural analysis computer program, in which the actual design loading conditions and steel pole properties are input. Or, the owner should ask the pole manufacturer to provide the analysis.

Note: Please see Chapter 2 for more information on deflection criteria.

Appendix 6

Conductor data

Table A6.1 Commonly Used Transmission Conductors.

#	*Label*	*Size/Conductor Type/ Stranding*	*Diameter (in)*	*Weight (lb/ft)*	*Ultimate Tensile Capacity (lbs)*
1	Penguin	266.8 ACSR 6/1	0.563	0.2911	8,350
2	Linnet	336.4 ACSR 26/7	0.721	0.4630	14,100
3	Pelican	477 ACSR 18/1	0.814	0.5180	11,800
4	Hawk	477 ACSR 26/7	0.858	0.6570	19,500
5	Dove	556.5 ACSR 26/7	0.927	0.7660	22,600
6	Orchid	636 AAC 37	0.918	0.5970	11,400
7	Drake	795 ACSR 26/7	1.108	1.0940	31,500
8	Tern	795 ACSR 45/7	1.063	0.8960	22,100
9	Rail	954 – 45/7 ACSR	1.165	1.0750	25,900
10	Cardinal	954 ACSR 54/7	1.196	1.2290	33,800
11	Bittern	1272 ACSR 45/7	1.345	1.4340	34,100
12	Falcon	1590 ACSR 54/19	1.545	2.0440	54,500
13	Kiwi	2167 ACSR 72/7	1.735	2.3005	49,800
14	Bluebird	2156 ACSR 84/19	1.762	2.5110	60,300
15	Drake	795 ACSS 26/7	1.108	1.0934	25,900 (Std)
16	Cardinal	954 ACSS 54/7	1.196	1.2270	26,000 (Std)

Note: 1 in = 25.4 mm, 1 lb/ft = 14.6 N/m, 1 lb = 4.45 N, 1 ft = 0.3048 m.

Appendix 7

Shield wire data

Table A7.1 Common Overhead Ground Wires.

#	*Label*	*Material*	*Diameter (in)*	*Area (in^2)*	*Weight (lb/ft)*	*Ultimate Tensile Capacity (lbs)*
1	3 # 6 AW	Alumoweld	0.35	0.077	0.178	10,280
2	7 # 8 AW	Alumoweld	0.385	0.091	0.262	15,930
3	7 # 9 AW	Alumoweld	0.343	0.072	0.208	12,630
4	3/8 HS	Steel	0.36	0.079	0.273	10,800
5	3/8 EHS	Steel	0.36	0.079	0.273	15,400
6	7/16 HS	Steel	0.435	0.149	0.399	14,500
7	7/16 EHS	Steel	0.435	0.149	0.399	20,800

Note: 1 inch = 25.4 mm, 1 lb/ft = 14.6 N/m, 1 lb = 4.45 N, 1 ft. = 0.3048 m.

Appendix 8

Optical ground wire data

Table A8.1 Typical Optical Ground Wires.

#	*Label*	*Maker*	*No. of Fibers*	*Diameter (in)*	*Weight (lb/ft)*	*Ultimate Tensile Capacity (lbs)*
1	DNO-9551	AFL	12	0.403	0.238	12,474
2	DNO-4629	AFL	24	0.587	0.446	23,050
3	49AY84ACS	BRUGG	24	0.587	0.450	24,075
4	DNO-6071	AFL	36	0.555	0.383	18,960
5	59AY101ACS	BRUGG	48	0.646	0.517	24,621
6	AC-129/724	AFL	48	0.724	0.706	35,034
7	AC-86/646	AFL	48	0.646	0.509	25,098
8	AC-102/691	AFL	72	0.691	0.582	29,555
9	SC-193/102	AFL	96	0.917	0.845	36,143
10	AC-125/726	AFL	144	0.726	0.692	34,882
11	SC-207/110	AFL	144	0.953	0.910	39,652

Note: 1 inch = 25.4 mm, 1 lb/ft = 14.6 N/m, 1 lb. = 4.45 N, 1 ft. = 0.3048 m.

Appendix 9

Guy wire data

Table A9.1 Database of Common Guy Wires

#	*Label*	*Diameter (in)*	*Area (in^2)*	*Unit Weight (lb/ft)*	*Ultimate Tensile Capacity (lbs)*
1	3/16 EHS	0.1875	0.028	0.08	4,000
2	3/8 HS	0.375	0.091	0.273	10,800
3	3/8 EHS	0.375	0.094	0.273	15,400
4	7/16 HS	0.435	0.132	0.39	14,500
5	7/16 EHS	0.435	0.137	0.39	20,800
6	9/16 HS	0.5625	0.258	0.652	24,500
7	9/16 EHS	0.5625	0.258	0.652	35,000
6	½ EHS	0.50	0.177	0.517	26,900
7	19 #8 AWG	0.642	0.257	0.714	43,200
8	20M AW	0.444	0.1204	0.347	20,000

Note: 1 inch = 25.4 mm, 1 lb/ft = 14.6 N/m, 1 lb = 4.45 N
HS = High Strength EHS = Extra High Strength

Appendix 10

Guy anchor data

Table A10.1 Common Transmission Guy Anchors*,**.

#	*RUS Label*	*Description*	*Ultimate Holding Capacity (lbs)*	*Minimum Soil Class for Ultimate Capacity*
Log Anchors				
1	TA-2L	8″ × 5′ Log 3/4″ × 8′ Rod	16,000	5
2	TA-4L	8″ × 8′ Log 3/4″ × 8′ Rod	32,000	5
Helical Screw Anchors				
3	TA-2H	Double Helix	Variable***	See Manufacturer Specifications
4	TA-3H	Triple Helix	Variable***	See Manufacturer Specifications
5	TA-4H	Quadruple Helix	Variable***	See Manufacturer Specifications
Plate Anchors				
6	TA-2P	Plate Anchor 3/4″ × 8′ Rod	16,000	5
7	TA-3P	Plate Anchor 1″ × 8′ Rod	24,000	4

*Reference: RUS Bulletins 1724F-810,811.
**Rock anchors are used in rocky soils. Holding power depends on type of rock, installation procedures and grout used.
***Based on size and location of helices per manufacturer specifications.
Note: 1 in = 25.4 mm, 1 in^2 = 6.452 cm^2, 1 ft = 30.48 cm, 1 lb = 4.45 N.

Table A10.2 Common Distribution Guy Anchors*.

#	*RUS Label*	*Description*	*Ultimate Holding Capacity (lbs)*	*Minimum Soil Class for Ultimate Capacity*
1	F1.10	Expanding Anchor Min. Area 120 in^2	10,000	5
2	F1.12	Expanding Anchor Min. Area 135 in^2	12,000	5
3	F3.8	Plate Anchor 5/8″ Min. Area 100 in^2	8,000	5
4	F3.10	Plate Anchor 3/4″ Min. Area 120 in^2	10,000	5
5	F3.12	Plate Anchor 3/4″ Min. Area 135 in^2	12,000	5
6	F4.1	Expanding Type Service Anchor	2,500	Sandy Soil
7	F4.2	Screw Type Service Anchor	2,500	Sandy Soil

*Reference: RUS Bulletin 1724E-153 and 1728F-803 and 804.
Note: 1 in = 25.4 mm, 1 in^2 = 6.452 cm^2, 1 ft = 30.48 cm, 1 lb = 4.45 N.

Appendix 11

Insulator data

Table A11.1a Typical Polymer Insulators.

#	*Type*	*Voltage (kV)*	*Maker*	*No. of Sheds*	*L (in)*	*Weight (lbs.)*	*Specified Mechanical Load (lbs.)*
1	Suspension	69	OB	16	35	5	25,000
2		115	OB	20	41	6	25,000
3		138	OB	28	53	7	25,000
4		161	OB	36	65	9	25,000
5		230	OB	44	77	10.4	25,000
6		345	NGK	86	112	29	25,000
7		500	NGK	117	151	38	30,000
8		765	NGK	153	194	49	30,000
9	Strain	69	OB	16	35	5	30,000
10		115	OB	20	41	6	30,000
11		138	OB	28	53	7	30,000
12		161	OB	36	65	9	30,000
13		230	OB	44	77	10.4	30,000
14		345	NGK	86	110	29	30,000
15		500	NGK	117	153	49	50,000
16		765	NGK	153	196	63	50,000

Note: 1 inch = 25.4 mm, 1 lb. = 4.45 N OB = Ohio Brass, NGK = NGK Locke.
Specified Mechanical Load (SML) – is the ultimate tension that the insulator can withstand. For design, recommended working load or Rated Mechanical Load (RML) is used. RML is usually 40% to 50% of SML, depending on type of material and loading.

Table A11.1b Approximate Weights and Lengths of Suspension Insulators.

*Number of Insulators**	*Length of String** (includes Suspension Hardware) (ft.)*	*Weight of String (includes Suspension Hardware) (lbs.)*	*Maximum Voltage for the Number of Insulators (kV)*
3	2.00	45	34.5 kV, 46 kV
4	2.50	58	69 kV
5	3.00	71	
6	3.50	84	
7	3.92	96	115 kV
8	4.42	109	
9	4.92	122	
10	5.33	135	161 kV
11	5.83	147	
12	6.33	160	230 kV
13	6.83	173	
14	7.25	186	
15	7.75	198	
16	8.25	211	

Note: 1 in. = 25.4 mm, 1 ft. = 0.3048 m, 1 lb. = 4.45 N.
(Source: RUS/USDA).
*5-3/4 in. x 10 in. Standard Suspension Bells with a Ball Hook.
**Exact length and weight will vary depending on the conductor suspension hardware used.

Table A11.2 Typical Polymer Horizontal Post and Braced Horizontal Post Insulators.

Type	*Nominal Voltage (kV)*	*Maker*	*Approximate Cantilever Length (in)*	*Approximate Brace Length (in.)*	*Rated Cantilever Load or Design Load (lbs.)*
Horizontal Post	69	OB	34	–	2,500
	115	OB	44	–	2,135
	138	OB	55	–	1,650
	161	OB	65	–	1,350
	230	OB	75	–	1,140
	345	NGK	115	–	1,790
	500	NGK	128	–	1,595
Braced Horizontal Post**	69	MP	35	42	10,200*
	115	NGK	58	70	V 10,000, L 1,530 T 7,500, C 7,500
	138	NGK	64	78	V 10,000, L 1,375 T 7,500, C 7,500
	161	NGK	74	92	V 10,000, L 1,165 T 7,500, C 7,500
	230	NGK	96	122	V 8,450, L 880 T 7,500, C 7,500
	345	MP	134	210	V 19,500, L 600 T 12,000 C 10,000

Note: 1 inch = 25.4 mm, 1 lb. = 4.45 N V = Vertical, L = Longitudinal, T = Tension, C = Compression, OB = Ohio Brass, NGK = NGK Locke, MP = MacLean Power.
Rated Cantilever Load (RCL) – is the maximum continuous vertical load that the insulator can withstand. RCL is usually 40% to 50% of the ultimate cantilever load, depending on type of material and loading.
*Maximum Design Cantilever Load; check interaction curves for effect of longitudinal loads.
**For BHP insulators, the load rating is also dependent on the effects of longitudinal wire load, if any. Most manufacturers give interaction diagrams for braced horizontal posts as a function of longitudinal load.

Appendix 12

Soil classification

Table A12.1 Simplified Soil Classification.

Soil Class	*Description*	*Geological Classification*	*Typical Blow Count 'N'*
0	Sound hard rock, un-weathered	Granite, Basalt, Massive Limestone	n/a
1	Very dense and/or cemented sands; coarse gravel and cobbles	Caliche (Nitrate-bearing gravel/rock)	60 to 100+
2	Dense fine sand; very hard silts and clays (may be preloaded)	Basal till, boulder clay, caliche, weathered laminated rock	45 to 60
3	Dense sands and gravel; hard silts and clays	Glacial till, weathered shales, schist, gneiss, siltstone	35 to 50
4	Medium dense sand and gravel, very stiff to hard silts and clays	Glacial till, hardpan, marls	24 to 40
5	Medium dense coarse sand and sandy gravels; stiff to very stiff silts and clays	Saprolites, residual soils	14 to 25
6	Loose to medium dense fine to coarse sand; to stiff clays and silts	Dense hydraulic fill, compacted fill, residual soils	7 to 14
7	Loose fine sands;Alluvium; loess; medium to stiff and varied clays; fill	Flood Plains soils, lake clays, adobe, gumbo, fill	4 to 8
8	Peat, organic silts, inundated silts, fly ash, very loose sands, very soft to soft clays	Miscellaneous fill, swamp marsh	0 to 5

(Courtesy: Hubbell Power Systems).

Table A12.2 ASTM Unified Soil Classification System.

Major Divisions			*Group Symbols*	*Typical Descriptions*
Coarse-Grained Soils – more than 50% retained on # 200 sieve*	Gravels – 50% or more of coarse fraction retained on # 4 sieve	Clean Gravels	GW	Well-graded gravels and gravel-sand mixtures. Little or no fines
			GP	Poorly-graded gravels and gravel-sand mixtures. Little or no fines
		Gravels with Fines	GM	Silty gravels. Gravel-sand-silt mixtures
			GC	Clayey gravels. Gravel-sand-clay mixtures
	Sands – 50% or more of coarse fraction passes # 4 sieve	Clean Sands	SW	Well-graded sands and gravelly sands. Little or no fines
			SP	Poorly-graded sands and gravelly sands. Little or no fines
		Sands with Fines	SM	Silty Sands. Sand-silt mixtures
			SC	Clayey sands. Sand-clay mixtures
Fine-Grained Soils – 50% or more passes # 200 sieve*	Silts and Clays – Liquid Limit less than 50		ML	Inorganic silts, very fine sands, rock flour, silty or clayey fine sands
			CL	Inorganic clays of low to medium plasticity, gravelly clays, sandy clays, silty clays, lean clays
			OL	Organic Silts and organic silty clays of low plasticity
	Silts and Clays – Liquid Limit 50 or more		MH	Inorganic silts, micaceous or diatomaceous fine sands or silts, elastic silts
			CH	Inorganic clays of high plasticity, fat clays
			OH	Organic clays of medium to high plasticity
Highly Organic Soils			Pt	Peat, muck and other highly organic soils

*Based on the material passing the 3″ (76 mm) sieve.
(Courtesy: Hubbell Power Systems).

Appendix 13

ASTM and other standards

SELECTED STANDARDS

American Society for Testing and Materials (*ASTM*) Specifications

A6	Standard Specification for General Requirements for Rolled Steel Bars, Plates, Shapes and Sheet Piling
A7	Standard Specification for Steel for Bridges and Buildings
A36	Standard Specification for Carbon Structural Steel
A123	Standard Specification for Zinc (Hot-Dip Galvanized) Coatings on Iron and Steel Products
A143	Standard Practice for Safeguarding Against Embrittlement of Hot-Dip Galvanized Structural Steel Products and Procedure for Detecting Embrittlement
A153	Standard Specification for Zinc Coatings (Hot-Dip) on Iron and Steel Hardware
A242	Standard Specification for High Strength Low Alloy Structural Steel
A283	Standard Specification for Low & Intermediate Tensile Strength Carbon-Steel Plates of Structural Quality
A307	Standard Specification for Carbon Steel Bolts and studs ranging from ¼″ to 4″ diameter
A325	Standard Specification for Structural Bolts, Steel, Heat Treated, 120/105 ksi Minimum Tensile Strength
A354	Standard Specification for Quenched and Tempered Alloy Steel Bolts, Studs, and Other Externally Threaded Fasteners
A370	Standard Testing Methods and Definitions for Mechanical Testing of Steel Products
A385	Standard Practice for Providing High-Quality Zinc Coatings (Hot-Dip)
A394	Standard Specification for Steel Transmission Tower Bolts Zinc-Coated and Bare
A416	Standard Specification for Steel Strand, Uncoated Seven-Wire for Prestressed Concrete
A449	Standard Specification for Hex Cap Screws, Bolts and Studs, Steel, Heat Treated, 120/105/90 ksi Minimum Tensile Strength, General Use
A475	Standard Specification for Zinc-Coated Steel Wire Strand
A490	Standard Specification for Structural Bolts, Alloy Steel, Heat Treated, 150 ksi Minimum Tensile Strength

A529 Standard Specification for High Strength Carbon Manganese Steel of Structural Quality
A563 Standard Specification for Carbon and Alloy Steel Nuts
A572 Standard Specification for High-Strength Low-Alloy Columbium-Vanadium Structural Steel
A588 Standard Specification for High-Strength Low-Alloy Structural Steel with 50 ksi [345 MPa] Min. Yield Point with Atmospheric Corrosion Resistance
A615 Standard Specification for Deformed and Plain Carbon Steel Bars for Concrete Reinforcement
A633 Standard Specification for Normalized High-Strength Low Alloy Structural Steel Plates
A673 Standard Specification for Sampling Procedure for Impact Testing of Structural Steel
A871 Standard Specification for High-Strength Low-Alloy Structural Steel Plate with Atmospheric Corrosion Resistance
B695 Standard Specification for Coatings of Zinc Mechanically Deposited on Iron and Steel
C31 Standard Practice for Making and Curing Concrete Test Specimens in the Field
C33 Standard Specification for Concrete Aggregates
C39 Standard Test Method for Compressive Strength of Cylindrical Concrete Specimens
C94 Standard Specification for Ready-Mixed Concrete
C143 Standard Test Method for Slump of Hydraulic Cement Concrete
C150 Standard Specification for Portland Cement
C172 Standard Practice for Sampling Freshly Mixed Concrete
C231 Standard Test Method for Air Content of Freshly Mixed Concrete by the Pressure Method
C260 Standard Specification for Air-Entraining Admixtures for Concrete
C494 Standard Specification for Chemical Admixtures for Concrete
C595 Standard Specification for Blended Hydraulic Cements
C618 Standard Specification for Coal Fly Ash and Raw or Calcined Natural Pozzolan for Use in Concrete
C935 Specification for General Requirements for Prestressed Concrete Poles Statically Cast
C1089 Standard Specification for Spun Cast Prestressed Concrete Poles
D422 Standard Test Method for Particle-Size Analysis of Soils
D1586 Standard Test Method for Penetration Test (SPT) and Split-Barrel Sampling of Soils
D1587 Standard Practice for Thin-Walled Tube Sampling of Fine-grained Soils for Geotechnical Purposes
D1883 Standard Test Method for California Bearing Ratio of Laboratory Compacted Soils
D2113 Standard Practice for Rock Core Drilling and Sampling of Rock for Site Exploration
D2166 Standard Test Method for Unconfined Compressive Strength of Cohesive Soil
D2216 Standard Test Methods for Laboratory Determination of Water (Moisture) Content of Soil and Rock by Mass
D2435 Standard Test Methods for One-Dimensional Consolidation Properties of Soils Using Incremental Loading

D2487	Standard Practice for Classification of Soils for Engineering Purposes (Unified Soil Classification System)
D2488	Standard Practice for Description and Identification of Soils (Visual-Manual Procedure)
D2573	Standard Test Method for Field Vane Shear Test in Saturated Fine-Grained Soils
D2850	Standard Test Method for Unconsolidated-Un-drained Tri-axial Compression Test on Cohesive Soils
D3080	Standard Test Method for Direct Shear Test of Soils Under Consolidated Drained Conditions
D4318	Standard Test Methods for Liquid Limit, Plastic Limit, and Plasticity Index of Soils
D4546	Standard Test Methods for One-Dimensional Swell or Collapse of Soils
D7012	Standard Test Methods for Compressive Strength and Elastic Moduli of Intact Rock Core Specimens under Varying States of Stress and Temperatures
E165	Standard Practice for Liquid Penetrant Examination
E709	Standard Guide for Magnetic Particle Testing
F436	Standard Specifications for Hardened Steel Washers
F1554	Standard Specification for Anchor Bolts, Steel, 36, 55 and 105 ksi Yield Strength

American National Standards Institute (ANSI)

ANSI O5.1	Specifications and Dimensions for Wood Poles
NEMA TT 1-1983	Tapered Tubular Steel Structures

American Wood Preservers Association (AWPA)

AWPA C1-91	Standards for Preservative Treatment by Pressure Processes

American Welding Society (AWS)

D1.0	Specifications for Design Stress and Joint Design for Welding
D1.1	Structural Welding Code-Steel

Steel Structures Painting Council (SSPC)

SSPC-PA 1	Shop, Field, and Maintenance Painting of Steel
SSPC-SPCOM	Surface Preparation Specifications: Surface Preparation, Commentary for Steel and Concrete Substrates
SSSPC-CS-23.00	(I) Interim Specification for the Application of Thermal Spray Coatings (Metallizing) of Aluminum, Zinc, and Their Alloys and Composites for the Corrosion Protection of Steel
SSPC-PS Guide 8.00	Guide to Top-coating Zinc-Rich Primers
SSPC-PS Guide 12.00	Guide to Zinc-Rich Coating Systems
IFI	Industrial Fasteners Institute Fastener Standards

Appendix 14

Composite pole data

Table A14.1 Wood and Composite Pole Classes.

	*Lateral Load (lbs)**	
ANSI Wood Pole Class	*ANSI Wood Pole*	*Grade B Composite Pole***
H6	11,400	7,410
H5	10,000	6,500
H4	8,700	5,655
H3	7,500	4,875
H2	6,400	4,160
H1	5,400	3,510
1	4,500	2,925
2	3,700	2,405
3	3,000	1,950
4	2,400	1,560
5	1,900	1,235

(Courtesy: RS Technologies).
Applied **2 ft from the tip of the pole.*
***Factored by the SF ratio of 1.0 / 0.65.*
1 ft = 30.48 cm, 1 lb = 4.45 N.

Table A14.2 Typical Material Properties of Composite Cross Arms.

Beam Series[1,2]	Bending Stress – Tangent* (ksi)	Bending Stress – Deadend* (ksi)	Modulus of Elasticity E* (ksi)	Shear Modulus G* (ksi)	Tensile Stress* (ksi)	Bearing Stress* (ksi)	Wall Thickness** (in)
A	37.3	49.6	4900	281	86.1	52.2	0.18, 0.20
B	50.0	64.9	3600	219	64.9	52.7	0.24, 0.27
C	67.0	74.5	5200	233	92.3	54.1	0.24, 0.27
D	94.0	104.0	5500	252	104.0	56.7	0.32, 0.35

[1] Cross Section dimensions of all beams: 3.625 in x 4.625 in.
[2] Poisson's Ratio = 0.32.
*Average values determined from a series of tests.
**The two thickness values represent the thickness along long and short sides.
1 in = 25.4 mm, 1 ksi = 6.89 MPa.

Appendix 15

Design checks by other codes

This section gives the design check formulae for steel transmission tower angles in Europe, Canada, Australia, UK and India.

Note: Rupture is used here to define the various ways in which a connection can fail under member tension (excluding net section failure) or the occurrence of inadequate end, edge and bolt hole spacing. Rupture failures can be longitudinal wedge tearing at the end of the member, cracks between holes and edge of the member, block shear, insufficient bearing strength of bolts etc.

1 ECCS – European Convention for Constructional Steel Work

Compression Capacity based on L/r

The effect of cross section slenderness is accounted for by calculating a stress F_{cr} as:

$F_{cr} = F_y$ if $(b/t) = (b/t)_{lim1}$

$F_{cr} = \{2 - (b/t)/(b/t)_{lim1}\}\ F_y$ if $(b/t)_{lim1} < b/t < (b/t)_{lim2}$

$F_{cr} = \pi^2 E/\left((5.1)\left(\frac{b}{t}\right)\right)^2$ if $b/t > (b/t)_{lim2}$

where

$(b/t)_{lim1} = 0.567(E/F_y)^{1/2}$

$(b/t)_{lim2} = 1.333(b/t)_{lim1}$

b/t = longest leg width to thickness ratio (will become a/t if angle is connected by short leg of width 'a')

F_y = steel yield stress

The three member slenderness ratios (L/r for each of the three angle axes) are first normalized as:

Normalized Slenderness Ratios, $\Lambda = (L/r)/\{\pi(E/F_{cr})^{1/2}\}$

The three Λ values are adjusted to new values Λ_{eff} to account for selected member end conditions. The angle is considered 'long' if any Λ (about a given axis) is larger than 1.414. In this case, the capacity relative to a particular axis is based on the Restraint

Code. If Λ about a given axis is less than or equal to 1.414, the member is considered as 'short' and the capacity relative to that particular axis is based on Eccentricity Code.

If Λ about a given axis is less than or equal to 1.414, the member is considered as 'short' and the effective slenderness Λ_{eff} is calculated as follows:

When Eccentricity Code C1 = 1 — No Eccentricity

$\Lambda_{eff} = \Lambda$ All Axes

When Eccentricity Code C1 = 2 — Eccentricity only at One End

$$\Lambda_{eff} = 0.25 + 0.8232\Lambda \quad Z-\text{axis}$$
$$\Lambda_{eff} = 0.50 + 0.6464\Lambda \quad X \text{ or } Y \text{ axes}$$

When Eccentricity Code C1 = 3 — Eccentricity at Both Ends

$$\Lambda_{eff} = 0.50 + 0.6464\Lambda \quad Z-\text{axis}$$
$$\Lambda_{eff} = 0.707 + 0.6464\Lambda \quad X \text{ or } Y \text{ axes}$$

If Λ about a given axis is larger than 1.414, the member is considered as 'long' and the effective slenderness Λ_{eff} is calculated as follows:

When Restraint Code C2 = 4 — Continuous at Both Ends

$$\Lambda_{eff} = 0.50 + 0.6464\Lambda \quad Z-\text{axis}$$
$$\Lambda_{eff} = \Lambda \quad X \text{ or } Y \text{ axes}$$

When Restraint Code C2 = 5 — Continuous at One End/One Bolt at the Other

$\Lambda_{eff} = \Lambda$ All axes

When Restraint Code C2 = 6 — Continuous at One End/Two or More Bolts at the Other

$$\Lambda_{eff} = 0.50 + 0.6464\Lambda \quad \text{All axes}$$

When Restraint Code C2 = 7 — One Bolt at Each End

$$\Lambda_{eff} = \Lambda \quad Z-\text{axis}$$
$$\Lambda_{eff} = 0.40 + 0.8635\Lambda \quad X \text{ or } Y \text{ axes}$$

When Restraint Code C2 = 8 — Two or More Bolts at Each End

$$\Lambda_{eff} = 0.50 + 0.6464\Lambda \quad Z-\text{axis}$$
$$\Lambda_{eff} = 0.707 + 0.6464\Lambda \quad X \text{ or } Y \text{ axes}$$

Maximum Effective Slenderness

$(\Lambda_{eff})max = 3$ for all members

Design Compressive Stress

If $\Lambda_{eff} \leq 0.20 \quad F_a = F_{cr}$

If $\Lambda_{eff} > 0.20 \quad F_a = F_{cr} / \left\{\varphi + sqrt\left[\varphi^2 - \Lambda_{eff}^2\right]\right\}$

where:

$$\varphi = \frac{\left\{1 + 0.125\ \left(\Lambda_{eff} - 0.2\right) + \ \Lambda_{eff}^{2}\right\}}{2}$$

Tension Capacity based on Net Section

A member tension capacity based on tearing of the member across its net area A_{net} is calculated as:

$N_{CAP} = (F_t)\ (A_{net})$
F_t = design tensile stress as defined
A_{net} is often defined as
$A_{net} = A_g - (d)(t)(n_h)$

where:

A_g = gross area of the angle section
d = bolt hole diameter
n_h = number of holes
t = thickness of angle

If $n_h \geq 0$, use the above equations with:

$F_t = F_y$	
$A_{eff} = A_g$	if Both legs are connected
$A_{eff} = A_g - (a/2)(t)(n_a)$	if Long leg is connected
$A_{eff} = A_g - (b/2)(t)(n_a)$	if Short leg is connected ‘
n_a = number of angles	

Connection Bearing Capacity

$B_{CAP} = (F_b)(d)(t)(n_b)(n_{bear})$

where:

F_b = design bearing stress = 2 F_y

$n_{bear} = n_a$	if number of angles $n_a = 1$ or 2
$= {}^{n_a}/2$	if number of angles $n_a = 4$

The above equations assume that minimum edge, end and spacing distances for bolt holes are satisfied.

Connection Rupture Capacity

Use equations given under Section 2 – CENELEC without any partial safety factor.

Connection Shear Capacity

$S_{CAP} = S_{1bolt}\ (n_s)(n_b)$

where:

S_{1bolt} = Shear strength of a single bolt
n_s = number of shear planes
n_b = number of bolts

2 CENELEC – EN 50341-1

Compression Capacity based on L/r

The effect of cross section slenderness is accounted for by calculating a stress F_{cr} as:

$$F_{cr} = \beta_a F_y$$

where β_a the reduction factor is calculated as:

$\beta_a = \frac{A_{eff}}{A}$
$A_{eff} = A - (1 - \rho)\, btn_a\}$ if CL = Long Only or Short Only
$A_{eff} = A - 2\,(1 - \rho)\, btn_a\}$ if CL = Both

where:

$\rho = 1$ if $\mu_r \leq 0.91$
$\rho = 2 - (\mu_r/0.91)$ if $0.91 < \mu_r < 1.213$
$\rho = 0.98/\mu_r^2$ if $\mu_r > 1.213$

$$\mu_r = \frac{\left(\frac{b}{t}\right)}{(b/t)_{lim1}}$$

$(b/t)_{lim1} = 0.623(E/F_y)^{1/2}$

$(b/t)_{lim1}$ = longest leg width to thickness ratio

F_y = steel yield stress

The three member slenderness ratios (L/r for each of the three angle axes) are normalized as:

Normalized overall slenderness $\Lambda = (L/r)/\{\pi\ (E/F_{cr})^{1/2}\}$

The three Λ values are adjusted to new values Λ_{eff} to account for selected member end conditions. The angle is considered 'long' if any Λ (about a given axis) is larger than

1.414. In this case, the capacity relative to a particular axis is based on the Restraint Code. If Λ about a given axis is less than or equal to 1.414, the member is considered as 'short' and the capacity relative to that particular axis is based on Eccentricity Code.

If Λ about a given axis is less than or equal to 1.414, the member is considered as 'short' and the effective slenderness Λ_{eff} is calculated as follows:

When Eccentricity Code C1 = 1 — No Eccentricity

$$\Lambda_{eff} = EXP\,[-1.98 + 1.747\Lambda] \quad \text{All axes if } \Lambda < 1.035$$
$$\Lambda_{eff} = -0.287 + 1.091\Lambda \quad \text{All axes } \Lambda > 1.035$$

When Eccentricity Code C1 = 2 — Eccentricity only at One End

$$\Lambda_{eff} = 0.02 + 0.88\Lambda \quad Z - \text{axis}$$
$$\Lambda_{eff} = 0.30 + 0.68\Lambda \quad X \text{ or } Y \text{ axes}$$

When Eccentricity Code C1 = 3 — Eccentricity at Both Ends

$$\Lambda_{eff} = 0.30 + 0.68\Lambda \quad Z - \text{axis}$$
$$\Lambda_{eff} = 0.52 + 0.68\Lambda \quad X \text{ or } Y \text{ axes}$$

If Λ about a given axis is larger than 1.414, the member is considered as 'long' and the effective slenderness Λ_{eff} is calculated as follows:

When Restraint Code C2 = 4 — Continuous at Both Ends

$$\Lambda_{eff} = -0.287 + 1.091\Lambda \quad \text{All axes}$$

When Restraint Code C2 = 5 — Continuous at One End/One Bolt at the Other

$$\Lambda_{eff} = -0.287 + 1.091\Lambda \quad \text{All axes}$$

When Restraint Code C2 = 6 — Continuous at One End/Two or More Bolts at the Other

$$\Lambda_{eff} = 0.30 + 0.68\Lambda \quad \text{All axes}$$

When Restraint Code C2 = 7 — One Bolt at Each End

$$\Lambda_{eff} = -0.287 + 1.091\Lambda \quad Z - \text{axis}$$
$$\Lambda_{eff} = 0.16 + 0.94\Lambda \quad X \text{ or } Y \text{ axes}$$

When Restraint Code C2 = 8 — Two or More Bolts at Each End

$$\Lambda_{eff} = 0.30 + 0.68\Lambda \quad Z - \text{axis}$$
$$\Lambda_{eff} = 0.52 + 0.68\Lambda \quad X \text{ or } Y \text{ axes}$$

Maximum Effective Slenderness

$$(\lambda_{eff})_{max} = 120 \text{ for Leg members}$$
$$(\lambda_{eff})_{max} = 200 \text{ for Other members}$$
$$(\lambda_{eff})_{max} = 240 \text{ for Redundant members}$$

Design Compressive Stress

If $\Lambda_{eff} \leq 0.20 \quad F_a = F_{cr}$

If $\Lambda_{eff} > 0.20 \quad F_a = F_{cr} / \left\{\varphi + sqrt\left[\varphi^2 - \Lambda_{eff}^2\right]\right\}$

where:

$$\varphi = \frac{\left\{1 + 0.34\ (\Lambda_{eff} - 0.2) + \Lambda_{eff}^2\right\}}{2}$$

CENELEC also provides an additional conservative equation if Λ_{eff} is always equal to Λ. For this, the above equations for eccentricity and restraint are ignored and the following substitution is made for the design compressive stress:

$$\varphi = \frac{\left\{1 + 0.49\ (\Lambda_{eff} - 0.2) + \Lambda_{eff}^2\right\}}{2}$$

Tension Capacity based on Net Section

$N_{CAP} = (F_t)\ (A_{net})$

If $n_b = 0$, use the above equation with $F_t = F_u$ and A_{net} calculated as follows:

If connected by Long leg only:

$n_b = 1 \quad A_{net} = (b - d)(t)(n_a)$
$n_b > 1 \quad g = 0 \quad A_{net} = (b + ½a - d)(t)(n_a)$
$g > 0 \quad A_{net} = Min\ [A_{net1}, A_{net2}]$
$A_{net1} = (b + ½a - d)(t)(n_a)$
$A_{net2} = \{b + ½a - 2d + s^2/16/(g - f)\}(t)(n_a)$

n_b = number of bolts
s = bolt spacing or pitch
g = gauge length or transverse spacing of bolt holes
f = edge distance

If connected by Short leg only:

Use same equations as above for Long leg only except that 'b' is replaced by 'a' and 'a' is replaced by 'b'.

If connected by Both Legs (assumes bolts are in adjacent parallel lines):

$g = 0 \quad A_{net} = (0.9)\ Min\ [A_{net1},\ A_{net2}]$
$A_{net1} = A - (h)(t)(n_a)$
$A_{net2} = A - \{2h - s^2/16/(a + b - 2f - t)\}(t)(n_a)$

$g > 0$ $\quad A_{net} = (0.9)\ Min\ [A_{net1},\ A_{net2},\ A_{net3}]$

$A_{net1} = A - (2h)(t)(n_a)$

$A_{net2} = A - \{3h - s^2/16/(g - f)\}(t)(n_a)$

$A_{net3} = A - \{4h - 2s^2/16/(g - f)\}(t)(n_a)$

If $n_b = 0$, use the basic equation $N_{CAP} = (F_t)\ (A_{net})$ with F_t and A_{eff} calculated as follows:

$F_t = 0.9\ F_u$	if Both legs are connected
$F_t = F_u$	if Long or Short only are connected
$n_b = 1$	
$A_{eff} = A$	if Both legs are connected
$A_{eff} = A - (a - t)(t)(n_a)$	if Long leg is connected
$A_{eff} = A - (b - t)(t)(n_a)$	if Short leg is connected
n_a = number of angles	
$n_b > 1$	
$A_{eff} = A$	if Both legs are connected
$A_{eff} = A - (a/2)(t)(n_a)$	if Long leg is connected
$A_{eff} = A - (b/2)(t)(n_a)$	if Short leg is connected

Connection Bearing Capacity

$B_{CAP} = (F_b)(d)(t)(n_b)(n_{bear})$

where:

F_b = design bearing stress $= 1.5\ F_u$

$n_{bear} = n_a$ if number of angles $n_a = 1$ or 2

$= n_a/2$ if number of angles $n_a = 4$

The above equations assume that minimum edge, end and spacing distances for bolt holes are satisfied.

Connection Rupture Capacity

$R_{CAP} = (F_b)(d)(t)(n_b)(n_{bear})$

where:

F_b = design bearing stress $= \alpha\ F_u$

$n_{bear} = n_a$ if number of angles $n_a = 1$ or 2

$= {}^{n_a}/_2$ if number of angles $n_a = 4$

α = adjustment factor used to account for the checking of edge, end and bolt hole spacing.

$\alpha = \text{Minimum}\ [\alpha_1, \alpha_2, \alpha_3, \alpha_4]$

$\alpha_1 = 1.2\ (e/h)$

$\alpha_2 = 1.85\ (e/h - 0.5)$

$\alpha_3 = 2.3\ (f/h - 0.5)$
$\alpha_4 = \alpha_1$
$= 0.96\ (s/h - 0.5)$ if NOT connected by Both legs and $g = 0,\ n_b > 1$
if NOT connected by Both legs and $g > 0,\ n_b > 2$
if connected by Both legs and $g = 0,\ n_b > 3$
if connected by Both legs and $g > 0,\ n_b > 4$

Connection Shear Capacity

$S_{CAP} = S_{1bolt}\ (n_s)(n_b)$

where:

S_{1bolt} = Shear strength of a single bolt
n_s = number of shear planes
n_b = number of bolts

3 Canada CSA - S37

Compression Capacity based on L/r

The design equations are similar to those shown per ASCE 10-15 except that the limits on width-to-thickness ratios are defined as:

$$\left(\frac{w}{t}\right)_{lim1} = 200/\sqrt{F_y}$$

$$\left(\frac{w}{t}\right)_{lim2} = 380/\sqrt{F_y}$$

where F_y is in MPa units.

In all cases the *w/t* ratio shall not exceed 25.

Tension Capacity based on Net Section

$N_{CAP} = (F_t)\ (A_{net})$
$A_{net} = A_g - (d)(t)(n_b)$

If $n_b > 0$, use the above equations with:

$F_t = 0.765\ F_u$
$A_{eff} = A$ if connected by Both Legs
$A_{net} = 0.70\ [A - (d)(t)(n_b)]$ if connected by Long or Short

The net section capacity shall not exceed $N_{cap-max} = 0.9\ F_y\ A_g$

Connection Bearing Capacity

$B_{CAP} = (F_b)(d)(t)(n_b)(n_{bear})$

where:

F_b = design bearing stress = 1.92 F_u
$n_{bear} = n_a$ if number of angles $n_a = 1$ or 2
$= {}^{n_a}/2$ if number of angles $n_a = 4$

The above equations assume that minimum edge, end and spacing distances for bolt holes are satisfied.

Connection Rupture Capacity

R_{CAP} = Minimum [0.80 R_{DIS}, 0.75 R_{BSH}]
$R_{DIS} = F_u\ [dist_1 + dist_2\ (nb - 1)](t)(n_{bear})$

where:

dist_1 = Minimum [1.2 (e − ½h + ¼d), 2.4 d]
dist_2 = Minimum [1.2 (s − h + ¼d), 2.4 d]

R_{BSH} = Connection Block Shear Capacity, checked only for members connected by one leg (Long only or Short only) and calculated as follows:

If $(F_u)(A_t) \geq 0.60\ (F_u)(A_v)$:
$R_{BSH} = (F_y)(0.6)(A_{gv}) + (F_u)(A_t)$

If $0.6\ (F_u)(A_v) = 0.60\ (F_u)(A_t)$:
$R_{BSH} = (F_u)(0.6)(A_v) + (F_y)(A_{gt})$

In all cases, R_{BSH} should not exceed $(F_u)(0.6)(A_v) + (F_u)(A_t)$.
The other parameters are defined as follows:

A_v = Minimum net area in shear along a line of transmitted force
$= (e - ½h)(t)(n_a)$ for $n_b = 1$
$= \{(e - ½h)+(s - h)(n_b - 1)\}(t)(n_a)$ for $n_b > 1$
A_{gv} = Gross area in shear along a line of transmitted force
$= (e)(t)(n_a)$ for $n_b = 1$
$= \{e + (s)(n_b - 1)\}(t)(n_a)$ for $n_b > 1$
A_t = Minimum net area in tension from the hole to the toe of the angle perpendicular to the line of force
$= (f - ½h)(t)(n_a)$
A_{gt} = Gross area in tension from the centerline of the bolt to the toe of the angle perpendicular to the line of force
$= (f)(t)(n_a)$

Connection Shear Capacity

$S_{CAP} = 0.80\ V\ (n_s)(n_b)$ with adjustment factor of 0.80

where:

V = Shear strength of a single bolt
n_s = number of shear planes
n_b = number of bolts

4 Australia – AS 3995

Compression Capacity based on L/r

The equations are identical to those of ASCE 10-15 with one difference: the allowable compressive stress F_a (Equation 3.17a) is reduced by a factor of 0.90.

Tension Capacity based on Net Section

$N_{CAP} = (F_t)\ (A_{net})$
$A_{net} = A_g - (d)(t)(n_b)$

$NCAP$ is the smallest of $[(0.9)(A_g)(F_y),\ (0.9)(A_{net})(0.85)(K_t)(F_u)]$ where K_t is a constant depending on the Leg Connect Code:

K_t = 0.75 if connected by Short leg only or None
= 0.85 if connected by Long leg only
= 1.00 if connected by Both Legs

Connection Bearing Capacity

$B_{CAP} = (F_b)(d)(t)(n_b)(n_{bear})$

where:

F_b = design bearing stress = $2.025\ F_y$
$n_{bear} = n_a$ if number of angles $n_a = 1$ or 2
$= n_a/2$ if number of angles $n_a = 4$

The above equations assume that minimum edge, end and spacing distances for bolt holes are satisfied.

Connection Rupture Capacity

No specific checks available. Use Equation (3.23) of Chapter 3.

Connection Shear Capacity

$S_{CAP} = 0.90\ V\ (n_s)(n_b)$ with adjustment factor of 0.90

where:

V = Shear strength of a single bolt
n_s = number of shear planes
n_b = number of bolts

5 UK – BS 8100

Compression Capacity based on L/r

The equations for the British designs are identical to those of ECCS shown above with some minor differences. The effect of cross section slenderness is accounted for by compressive stress F_{cr} as:

$F_{cr} = F_y$ if $(b/t) = (b/t)_{lim1}$
$F_{cr} = \{2 - (b/t)/(b/t)_{lim1}\}F_y$ if $(b/t)_{lim1} < b/t < (b/t)_{lim2}$
$F_{cr} = \pi^2 E/(5.1 * b/t)^2$ if $b/t > (b/t)_{lim2}$

where

$(b/t)_{lim1} = 0.567\ (E/F_y)^{1/2}$
$(b/t)_{lim2} = 1.333(b/t)_{lim1}$
b/t = longest leg width to thickness ratio (will become a/t if angle is connected by short leg of width 'a')
F_y = Yield Stress

Normalized slenderness ratio

The three member slenderness ratios (L/r for each of the three angle axes) are normalized as:

$$\Lambda = (L/r)/\{\pi\ (E/F_y)^{1/2}\}$$

The three Λ values are adjusted to new values Λ_{eff} to account for the structural configuration of the member (leg, bracing, end conditions etc.) These adjustments are implemented by choosing a numerical value for the Restraint Code. It must be noted that the effects of these Restraint Codes in BS-8100 on member capacities are completely different from Restraint Codes used by other standards.

Effective slenderness is defined as $\Lambda_{eff} = kL$ where the parameter 'k' depends on the input Restraint Code.

When Restraint Code C2 = 10 or blank

k = 1 All Axes

When Restraint Code C2 = 11

$k = 0.80 + 0.10\ \Lambda$ Z – axis, but $0.90 \le k \le 1.0$
$k = 1.0$ X or Y axes (symmetrical bracing)

When Restraint Code C2 = 12

$k = 0.80 + 0.10\ \Lambda$ Z – axis, but $0.90 \le k \le 1.0$
$k = 0.80 + 0.10\ \Lambda$ X or Y axes (staggered bracing)

When Restraint Code C2 = 13

$k = 1.20\ (0.80 + 0.10\ \Lambda)$ All Axes but $1.08 \le k \le 1.2$

(This is a check with asymmetrical bracing)

When Restraint Code C2 = 14

$k = 0.75 + 0.35/\Lambda$ All Axes

When Restraint Code C2 = 15

$k = 0.70 + 0.40/\Lambda$ All Axes

When Restraint Code C2 = 16

$k = 0.70 + 0.58\ /\Lambda$ All Axes

When Restraint Code C2 = 17

$k = 0.80$ All Axes

When Restraint Code C2 = 18

$k = 0.85$ All Axes

When Restraint Code C2 = 19

$k = 0.90$ All Axes

When Restraint Code C2 = 20

$k = 0.95$ All Axes

The Effective Slenderness Λ_{eff} shall not exceed the limits shown below:

$(\Lambda_{eff})_{max} = 120\ \{\sqrt{E/F_y}\}$ for Leg members

$$(\Lambda_{eff})_{max} = 180 \left\{\sqrt{E/F_y}\right\} \qquad \text{for Other members}$$

$$(\Lambda_{eff})_{max} = 350 \left\{\sqrt{E/F_y}\right\} \qquad \text{for Tension – Only members}$$

The design compressive stress is calculated as:

If $\Lambda_{eff} \leq 0.20 \quad F_a = J\ F_{cr}$

If $\Lambda_{eff} > 0.20 \quad F_a = J\ F_{cr} / \left\{\varphi + sqrt\left[\varphi^2 - \Lambda_{eff}^2\right]\right\}$

where:

$$\varphi = \frac{\left\{1+0.34\ (\Lambda_{eff}-0.2)+\ \Lambda_{eff}^2\right\}}{2}$$

J = adjustment factor based on Eccentricity Codes shown below:

When Eccentricity Code C1 = 21 or blank J = 1
When Eccentricity Code C1 = 22 J = 0.9
When Eccentricity Code C1 = 23 J = 0.8

Tension Capacity based on Net Section

Same as in Section 1 – ECCS

Connection Bearing Capacity

Same as in Section 2 – CENELEC, except for the following definitions:
α = Minimum $[\alpha_1, \alpha_2, \alpha_3, \alpha_4]$
$\alpha_1 = 1.33\ (e/h)$
$\alpha_2 = 2.00\ (e/h - 0.5)$
$\alpha_3 = 2.67\ (f/h - 0.5)$
$\alpha_4 = \alpha_1$
$= (s/h - 0.5)$ if NOT connected by Both legs and $g = 0,\ n_b > 1$
if NOT connected by Both legs and $g > 0,\ n_b > 2$
if connected by Both legs and $g = 0,\ n_b > 3$
if connected by Both legs and $g > 0,\ n_b > 4$

The equations assume that minimum edge, end and spacing distances for bolt holes are satisfied.

Connection Shear Capacity

$S_{CAP} = S_{1bolt}\ (n_s)(n_b)$

where:

S_{1bolt} = Shear strength of a single bolt
n_s = number of shear planes
n_b = number of bolts

6 India – IS 802

Compression Capacity based on L/r

The equations for India are identical to those of ASCE 10-15 with one difference: the maximum slenderness ratio for Legs is 120 instead of 150.

Tension Capacity based on Net Section

If n_b = 0, use the basic equation $N_{CAP} = (F_t)\ (A_{net})$ with F_t and A_{eff} calculated as follows:

$F_t = F_y$	if Both legs are connected
$F_t = F_y$	if Long or Short leg only are connected
$A_{eff} = A$	if Both legs are connected
$A_{eff} = A - [b/(3a+b)](b)(t)(n_a)$	if Short leg is connected
$A_{eff} = A - [b/(3b+a)](a)(t)(n_a)$	if Long leg is connected
n_a = number of angles	

If $n_b = 0$, use the basic equation $N_{CAP} = (F_t)\ (A_{net})$ with $F_t = F_y$ and:
Long or Short Legs only are connected -

$A_{net} = A1 + (A2)(k)$
$k = 3A1/(3A1 + A2)$

$A1$ is the effective sectional area of the connected leg calculated as follows:

If Long leg only is connected -

$n_b = 1\ A1 = (b - (0.5t) - h)\ (t)(n_a)$
$n_b > 1\ g = 0\ A1 = (b - (0.5t) - h)\ (t)(n_a)$
$g > 0\quad A1$ *is Minimum of* $[A11, A12]$
$A11 = (b - (0.5t) - h)\ (t)(n_a)$
$A12 = [b - (0.5t) - 2h + s^2/16/(g\text{-}f)](t)(n_a)]$

If Short leg only is connected -
Use the same equations for Long leg only is connected, except that ‘b’ is replaced by ‘a’.
$A2$ is the gross sectional area of the unconnected leg calculated as follows:

$A2 = (a - (0.5t) - h)\ (t)(n_a)$	if Long leg only is connected
$A2 = (b - (0.5t) - h)\ (t)(n_a)$	if Short leg only is connected

If Both legs are connected -

$g = 0$ $A_{net} = Minimum\ [A_{net1}, A_{net2}]$
$A_{net1} = A - (h)(t)(n_a)$
$A_{net2} = A - [2h - s^2/16/(a + b - 2f - t)](t)(n_a)$
$g > 0$ $A_{net} = Minimum\ [A_{net1}, A_{net2}, A_{net3}]$
$A_{net1} = A - (2h)(t)(n_a)$
$A_{net2} = A - [3h - s^2/16/(g - f)](t)(n_a)$
$A_{net3} = A - [4h - 2s^2/16/(g - f)](t)(n_a)$

Connection Bearing Capacity

$B_{CAP} = (F_b)(d)(t)(n_b)(n_{bear})$

where:

F_b = design bearing stress $= 1.5F_u$
$n_{bear} = n_a$ if number of angles $n_a = 1$ or 2
$= {}^{n_a}/_2$ if number of angles $n_a = 4$

The above equations assume that minimum edge, end and spacing distances for bolt holes are satisfied.

Connection Rupture Capacity

Use Equation (3.23) of Chapter 3.

Connection Shear Capacity

$S_{CAP} = S_{1bolt}\ (n_s)(n_b)$

where:

S_{1bolt} = Shear strength of a single bolt
n_s = number of shear planes
n_b = number of bolts

Answers to problems

2.1 79.5 ft (24.2 m)
2.2 40 ft (12.2 m); Swing is negligible for HP insulators facilitating shorter ROW.
2.3 1079. 68 ft (329.1 m)
2.4 60 ft (18.3 m); $D_e = 7.8$ ft (2.38 m)
2.5 1512 ft (460.9 m); even a small increase in horizontal separation leads to a large increase in allowable horizontal span.
2.6 1367.5 ft (416.8 m)
2.7 800 ft (243.8 m)
3.1 (a) Class S-4.9 (4,875 lbs)
(b) Class S-5.7 (5,625 lbs)
3.2 9.26 ft (2.82 m)
3.3 NO, Not Adequate
3.4 3/16 (0.1875) in. (4.76 mm), Satisfies local buckling criteria
3.5 55-Class H2
3.6 (a) 1,834 lbs (8.16 kN)
(b) 3,844 lbs (17.1 kN)
(c) Guy Force = 5,436 lbs (24.2 kN)
3.7 80 ft (24.4 m) pole (70 ft above ground, 10 ft embedment)
3.8 180.7 kips (47.6 kN)
3.9 (a) 4.31 in^2 (2782 mm^2)
(b) 139.7 kips (621.8 kN)
(c) Yes, both end, edge distances are adequate
3.10 34.44 kips (153.1 kN)
3.11 118.3 kips (526 kN)
3.12 172.9 kip-ft (234.4 kN-m)
3.13 26 bolts
3.14 (a) 700.4 ft (213.5 m)
(b) 637.4 ft (194.1 m)
(c) 380.4 ft (115.9 m)
A one step increase in pole class led to a 20.7% increase in allowable spans.
4.1 4881.4 kip-ft (6618.8 kN-m); 0.98
4.2 27.16 ft, rounded to 28 ft (8.53 m)
4.3 3527.1 kip-ft (4782.51 kN-m)
4.4 (a) 12 ft (3.66m)
(b) 12 ft (3.66m)

4.5 385.2 kip-ft (522.31 kN-m)
4.6 514.2 kip-ft (697.22 kN-m)
4.7 23.5 ft (7.16m); 21 ft (6.40 m)
4.8 28 ft (8.53m); 21 ft (6.40 m)
4.9 600 kip-ft (813.56 kN-m)
4.10 96115 lb/ft^2 (4602 kN/m^2); 4709.64 kips (20950.36 kN)
4.11 48134.4 lb/ft^2 (2304.69 kN/m^2); 2358.6 kips (10491.99 kN)
4.12 22950 lb/ft^2 (1098.85 kN/m^2); 1124.55 kips (5002.45 kN)
4.13 128800.5 lb/ft^2 (6167 kN/m^2); 5152 kips (22918.15 kN)
4.14 5253.29 lb/ft^2 (251.53 kN/m^2); 340.64 kips (1515.30 kN)
4.15 29.17 kips (129.76 kN)
4.16 313.38 kips (1394.04 kN)

Index

Printed in the United States
by Baker & Taylor Publisher Services